MOBILE TERMINAL RECEIVER DESIGN

MOBILE TERMINAL RECEIVER DESIGN

LTE AND LTE-ADVANCED

Sajal Kumar Das
ERICSSON, Bangalore, India

WILEY

This edition first published 2017

Registered Office
John Wiley & Sons Singapore Pte. Ltd., 1 Fusionopolis Walk, #07-01 Solaris South Tower, Singapore 138628.

For details of our global editorial offices, for customer services and for information about how to apply for permission to reuse the copyright material in this book please see our website at www.wiley.com.

Library of Congress Cataloging-in-Publication Data

Names: Das, Sajal Kumar, author.
Title: Mobile terminal receiver design : LTE and LTE-advanced / Sajal Kumar Das.
Description: Singapore ; Hoboken, NJ : John Wiley & Sons, 2016. | Includes bibliographical references and index.
Identifiers: LCCN 2016026712 (print) | LCCN 2016035373 (ebook) | ISBN 9781119107309 (cloth) | ISBN 9781119107439 (pdf) | ISBN 9781119107446 (epub)
Subjects: LCSH: Cell phones–Design and construction. | Mobile communication systems–Technological innovations. | Long-Term Evolution (Telecommunications)
Classification: LCC TK6564.4.C45 D37 2016 (print) | LCC TK6564.4.C45 (ebook) | DDC 621.3845/6–dc23
LC record available at https://lccn.loc.gov/2016026712

Set in 10/13pt Times by SPi Global, Pondicherry, India

10 9 8 7 6 5 4 3 2 1

Contents

Preface

Mobile systems have evolved over several generations, from 1G to 4G and beyond, due to an ongoing demand for higher data rates, for better quality, for more complex applications, and for seamless intersystems handover and lower latency. As a result, the mobile phone has changed from a simple telephone to a complex smartphone. Today, a smartphone encapsulates computing capabilities with cellular network access functionality in a single integrated system, with high-quality graphics, a portable size, support for complex user applications, and multimode connectivity features. The demand for supporting various complex applications, new applications, smaller form factors, lower power consumption, and multi-RAT support, has meant that the challenges in mobile phone design have been manifold. In particular, new challenges have arisen in the design of innovative mobile handset solutions, which can offer smaller sizes, low power consumption, low cost, and tremendous flexibility, while supporting more advanced features and providing an improved data rate and higher performance.

This book has been written to address these challenges. Its aim has been to equip mobile phone system designers and students with an all-in-one guide, starting from basic concepts and progressing to advanced system design, and introducing readers to various innovative solutions. It walks readers through 2G, 3G, and 4G mobile-phone system architectures and their basic building blocks, the different air-interface standards, operating principles, hardware anatomy, software and protocols, internal modules, components, and circuits for legacy and next-generation smartphones, including various research areas in 4G and 5G systems.

Mobile Terminal Receiver Design explains basic working principles, system architecture, and specification details of legacy and next-generation mobile systems. It covers in detail RF transmitter and receiver blocks, digital baseband processing blocks, receiver and transmitter signal processing, protocol stacks, AGC, AFC, ATC, the power supply, clocking, connectivity modules, and application modules with different design solutions for exploring various tradeoffs. It also explains the internal blocks, hardware and software components, and anatomy of legacy and LTE/LTE-Advanced smartphones, from principle to practice to product. Multi-RAT design requirements are also discussed, together with key design attributes such as low power consumption, slim form factors, seamless I-RAT handover, sensitivity, and selectivity.

This book is based on my experiences as a design engineer in the field of wireless and mobile communications and modelled from an academic course developed for electronics communication engineering students, and from a useful design handbook for practicing engineers and technicians. It is intended to help software, hardware and RF design engineers, researchers, product managers, as well as industry veterans in the areas of mobile phone system and chipset design to understand the evolution of radio access technologies and emergent trends, and also to help them make innovative and competitive next-generation mobile devices.

I express my sincere thanks to my colleagues, friends and family members for their valuable suggestions. Any constructive criticisms and suggestions for improving the book will be gratefully received and should be sent to sajal_das@yahoo.com.

Dr Sajal Kumar Das

Abbreviations

3GPP – Third-Generation Partnership Project
ACK – acknowledgment (in ARQ protocols)
ADC – analog to digital converter
AM – amplitude modulation
AMPS – advanced mobile phone service
AMR – adaptive multirate (speech codec)
APN – access point name
ARQ – automatic repeat request
AWGN – additive white Gaussian noise
BCCH – broadcast control channel
BCH – broadcast channel
BER – bit error rate
CDMA – code division multiple access
CFO – carrier frequency offset
CMOS – complementary metal oxide semiconductor
CN – core network
CPC – continuous packet connectivity
CQI – channel quality indicator
CRC – cyclic redundancy check
CS – circuit switched
DCCH – dedicated control channel
DECT – digital European cordless telephone
DFE – digital front end
DigRF – digital RF interface standard
DL – downlink
DL-SCH – downlink shared channel
DPCCH – dedicated physical control channel

DRX – discontinuous reception
DS-CDMA – direct sequence code division multiple access
DSP – digital signal processor
DTCH – dedicated traffic channel
DTX – discontinuous transmission
DwPTS – downlink pilot time slot
EDGE – enhanced data rates for GSM evolution
eNB – E-UTRAN Node B
EPC – evolved packet core
EPS – evolved packet system
ETACS – extended total access communication system
EUTRA – evolved universal terrestrial radio access
E-UTRAN – evolved UTRAN
FCC – Federal Communication Commission
FDD – frequency division duplex
FDMA – frequency division multiple access
FEC – forward error correction
FER – frame error rate
FFT – fast Fourier transform
FTP – file transfer protocol
GaAS – gallium arsenide
GERAN – GSM EDGE radio access network
GP – guard period
GPRS – general packet radio services
GSM – Global System for Mobile Communications
GSM-EFR – GSM enhanced full rate
HARQ – hybrid ARQ
HSDPA – high-speed downlink packet access
HSPA – high-speed packet access
HSUPA – high-speed uplink packet access
ICI – intercarrier interference
IFFT – inverse FFT
IMT – International Mobile Telecommunication
IP – Internet protocol
LTE – Long-Term Evolution
MAC – medium access control
MBMS – multimedia broadcast and multicast service
MCH – multicast channel
MCS – modulation and coding scheme
MIMO – multiple input multiple output
MTCH – MBMS traffic channel
NACK – negative acknowledgment (in ARQ protocols)

NAS – nonaccess stratum
OFDM – orthogonal frequency division multiplexing
OFDMA – orthogonal frequency division multiple access
PAPR – peak-to-average power ratio
PBCH – physical broadcast channel
PCCH – paging control channel
PCFICH – physical control format indicator channel
PCS – personal communication standard
PDC – personal digital cellular
PDCP – packet data convergence protocol
PDCCH – physical downlink control channel
PDC-EFR – PDC-enhanced full rate
PDCP – packet-data convergence protocol
PDSCH – physical downlink share channel
PDN – packet data network
P-GW – packet data network gateway
PHICH – physical hybrid ARQ indicator channel
PMCH – physical multicast channel
PMI – precoding matrix indicator
POCSAG – Post Office Code Standard Advisory Group
PRB – physical resource block
PSHO – packet switched handover
P-SS – primary synchronization signal
PUSCH – physical uplink shared channel
QAM – quadrature amplitude modulation
QoS – quality of service
QPSK – quadrature phase-shift keying
RB – radio bearer
RB – resource block
RF – radio frequency
RF-BB – radio frequency and baseband module
RL – radio link
RLC – radio link control
ROHC – robust header compression
RRC – radio resource control
RS – reference signal
RTT – radio transmission technology
RV – redundancy version
SAE – system architecture evolution
SC-FDMA – single carrier frequency division multiple access
SCTP/IP – stream control transmission protocol / IP
SMS – short message service

SR – scheduling request
SRB – signal radio bearer
S-SS – secondary synchronization signal
TDD – time division duplex
TDMA – time division multiple access
TDMA-EFR – TDMA enhanced full rate
TE – transverse electric
TEM – transverse electromagnetic
TM – transverse magnetic
TR – technical release
TTI – transmission time interval
UE – user equipment
UL – uplink
UL-SCH – uplink shared channel
UMTS – universal mobile telecommunications system
UTRA – universal terrestrial radio access
UTRAN – universal terrestrial radio access network
VoIP – voice over IP
WDT – watchdog timer

1

Introduction to Mobile Terminals

1.1 Introduction to Mobile Terminals

A mobile communication device is a small, portable electronic device, with wireless communication capabilities, which is easy to carry around. There are several types of mobile communication devices, like cell phones or mobile phones, WLAN devices, and GPS navigation devices, but it is the mobile phone that has adopted the term "mobile device," and gradually its purpose has shifted from a verbal communication tool to a multimedia tool.

A mobile phone, which is also known as mobile terminal (MT), cellular phone, cell phone, hand phone, or simply a phone, is a device that can send and receive telephone calls over a radio link while being connected to a cellular base station operated by a cellular network operator. It provides user mobility around a wide geographic area. A feature phone is a low-end mobile phone with limited capabilities and it provides mainly voice calling, text messaging, multimedia, and Internet functionality. In addition to telephone calls, modern multifunctional mobile phones with more computing capabilities, which support a wide variety of other applications and services like SMS, MMS, e-mails, Internet, Web browsing, news, gaming, playing music, movies, calendar management, contact, video, photography, short-range connectivity, location-specific information, WLAN connectivity, and GPS connectivity, are considered as smartphones. Smartphones offer all these services in single device, so they are becoming increasingly important as work tools for users who rely on these services. Today, they have become universal replacements for personal digital assistant (PDA) devices. Typically, a smartphone incorporates handheld computer functionalities along with the communication capabilities of a cell phone by providing support

Mobile Terminal Receiver Design: LTE and LTE-Advanced, First Edition. Sajal Kumar Das.

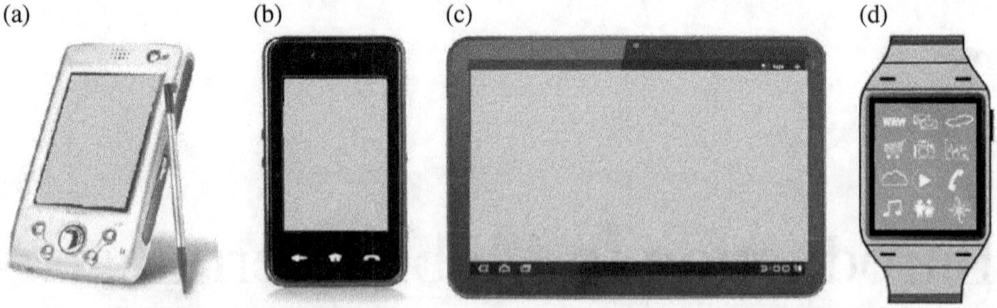

Figure 1.1 (a) PDA, (b) smartphone (c) tablet (d) wearable device

of multimodal, multi-RAT connectivity and user customized applications. Personal digital assistants/enterprise digital assistants, tablet computers, ultramobile PCs, and a lot of wearable devices also provide mobile communication capabilities by integrating communication modems in them. Various types of these devices are shown in Figure 1.1.

1.1.1 Building Blocks of a Smartphone

A system-level block diagram of a smartphone is shown in Figure 1.2. Smartphones are equipped with various functional blocks as given below:

- Mobile terminal modem unit. This unit (cellular systems modem) interfaces with the cellular base stations, and sends/receives user information (voice, data) generated by the application unit. So it interacts with a base station using different cellular air interface standards like GSM, WCDMA, LTE etc. to send/receive information to distantly located called party or server. It also interacts locally with its application units, like speech, video, and data transfer applications for getting/providing the user application data. This is discussed in Chapters 2, 3 and 4. This consists of two main submodules: Radio Frequency (RF) unit and Baseband (BB) unit.
  - RF unit. The RF analog front-end unit's transmitter circuit helps to upconvert the low-frequency baseband signal to a high-frequency amplified RF signal for transmission, and the receiver circuit helps to downconvert the analog amplified received high-frequency signal to a low-frequency baseband signal. The RF unit is discussed in detail in Chapter 6.
  - Baseband unit. The baseband unit helps for digital bit detection, system protocol processing for proper and reliable communication with the network. These are discussed in detail in Chapter 4 and 5.
  - SIM. A subscriber identification module (SIM) is an integrated circuit that securely stores the international mobile subscriber identity (IMSI) and the related key used to identify and authenticate subscribers on mobile telephony devices. A SIM circuit is embedded into a removable plastic card, called "SIM card." This is discussed in detail in Chapter 5.

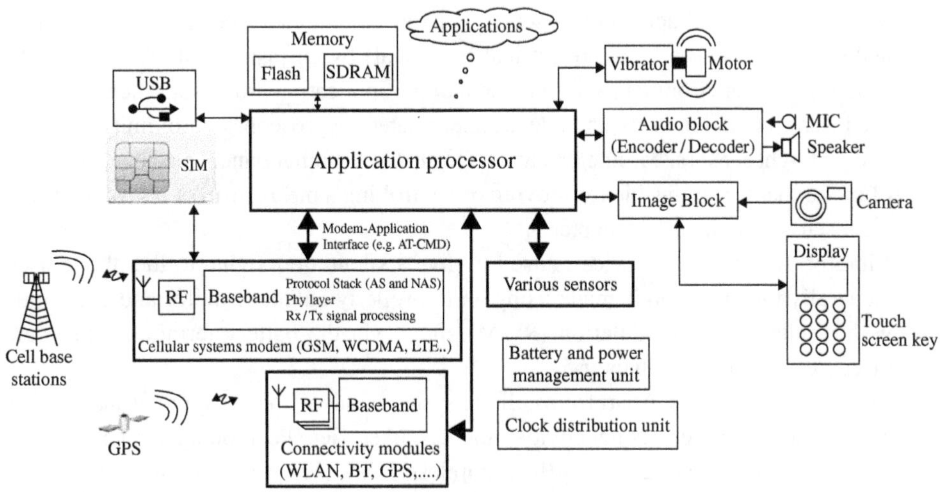

Figure 1.2 System-level block diagram of a typical smartphone

- Application unit. This unit is in charge of running various applications. It interacts with the modem and connectivity modules to send/receive information from remote devices, and uses that data to drive various applications. It provides the functions that users want to execute on the smart phone and these may include speech, audio playback, fax transmission/reception, Internet, e-mail, Web browsing, image reproduction, streaming video, games, and so forth. This unit also handles the interface functions such as keyboard, display, and speech recognition, and it interfaces and manages other connectivity modules such as GPS and WLAN. Depending on the smartphone device complexity, there could one or several application processors in a mobile phone. The architecture design and selection details are provided in Chapter 5 and 7. The application processor consists of components like the processor core and device interfaces, which communicate with other peripheral devices attached to the application processor like the LCD screen, camera, keypad, universal serial bus (USB), and multimedia card (MMC) via interfaces. These are discussed in detail in Chapter 5.
  - Peripheral devices. There are several peripheral devices placed in the smart phone for different purposes. Like, for data transfer with other devices or PC, an USB device is placed in the phone. Similarly, UART, I2S etc. are used for intermodule or interdevice communication. The other devices are like, SD/MMC, LCD display, keyboard, microphone, and speaker are also used in a mobile phone. These are discussed in detail in Chapter 5.
  - Multimedia modules. It performs multimedia related functions like, speech encoding/decoding, audio encoding/decoding, video encoding/decoding by employing various multimedia standards (MP3, JPEG, MPEG, and so forth). As multimedia-related functions are time consuming, so these are generally implemented in dedicated hardware block. Also, smartphone contain graphics processing unit (GPU) for rapid processing of multimedia functions. These are discussed in detail in Chapters 5 and 7.

  - Various sensors and actuators. A sensor is a device that measures a physical quantity and converts it into a signal (electrical or optical) by an instrument. They sense the changes in the environment and send them to the application processor. The commonly used sensors in handsets include accelerometers, gyroscopes, proximity sensors, ambient light sensors, barometers, and so forth. On the other hand, an actuator is a type of motor that is responsible for moving or controlling a mechanism or a system. These are discussed in detail in Chapter 5.
  - Vibrator. A vibra alert device is used to give a silent alert signal to the phone user. Generally the vibration is made using an improperly balanced motor and controlled with a pulse width modulation (PWM) signal via the battery terminal. These are discussed in detail in Chapter 5.
  - Connectivity modules. Apart from cellular system modem, the smart phone also houses several other wireless connectivity modules like, Geo Positioning System (GPS), Bluetooth (BT), FM radio, ZigBee, Wireless LAN (WLAN), and so forth. These individual submodules have RF and digital baseband processing unit and interact with the other devices, peripherals like, headset or server through radio interface. These are discussed in detail in Chapter 5.
- Power management module. This unit is responsible for distributing the regulated battery power among various modules, conversion of the battery voltage (generally 3.6 V) according to the different voltage level needed by different modules, which means up or down conversion to various voltages (such as 4.8 V, 2.8 V, 1.8 V and 1.6 V) using, for example, a DC-DC converter, a battery power consumption control device, sleep-related functionalities management, battery-charging control. The battery-charging component is responsible for charging the battery of the smartphone. These are discussed in detail in Chapter 8.
- Clock distribution module. This distributes a clock signal to the mobile phone. The clock signal is required in every digital blocks in the system and also it is required in RF unit for scheduling transmission and reception at a specific time. These are discussed in detail in Chapter 5.
- Memory. Various types of memory are used in the mobile phone for storing code and data. Generally, Flash memory, EPROM, and DRAM memory are used in a mobile phone. These are discussed in detail in Chapter 5.

Apart from all these hardware blocks, firmware and software components reside in the memory and are executed by processors to configure, control, and process different hardware modules, applications, and protocols. These are discussed in Chapter 7.

1.2 History of the Mobile Phone

Prior to 1973, mobile telephony was limited to phones installed in cars, trains and other vehicles, mainly due to the larger size and weight of the equipment. On April 3, 1973, Martin Cooper, a senior engineer at Motorola, made the first mobile telephone call from handheld subscriber equipment, which was around 23 cm long, 13 cm deep and 4.45 cm wide and weighed 1.1 kg and offered a talk time of just 30 min with 10 h of recharge time. Since then, mobile phones have evolved dramatically, with enriched features like audio, and video players, video cameras, handheld gaming devices and support for Internet access, augmented reality, commercial services and a whole host of other applications. They also reduced in size, weight, and cost.

In 1992 Motorola introduced the first digital palm-size mobile telephone named Motorola 3200. In 1992, Nokia developed Nokia 1011, which was first mass-produced GSM phone. In 1992 IBM introduced Simon, a personal communicator with PDA and phone combination, which had a monochrome touchscreen and a stylus. In 1996, Nokia introduced the communicator 9000 series as a smart phone with outward facing dial pad, navigation keys, and monochromatic display. Nokia 7120 supported WAP browser. One year later, Ericsson released the GS 88 smart phone with a touchscreen inside and a stylus. Samsung Uproar cell phone was introduced with MP3 music capabilities. Nokia 8310 was having several premium features such as FM radio, infrared, and a fully functional calendar. Ericsson T39 was a tiny Bluetooth-capable handset. In 1999, NTT DoCoMo pioneered the first mobile Internet service in Japan on existing 2G technologies, which was soon replaced by the first 3G handsets in October 2001. In 2002, the first phones with built-in cameras became publicly available in the Nokia 7650 and the Sanyo SPC-5300. In 2004, Motorola introduced Razor V3, which is a very lightweight sleek phone. In January 2007, Apple launched its first iPhone, combining three products into one handheld device: a mobile phone, an iPod, and a wireless communication device, which had an autorotate and a multitouch sensor. This device helped Apple to capture a significant market share. In 2008, Nokia released a GPS-enabled smartphone with sleek, compact design.

The mobile phone continues to evolve. In 2008 LTE standardization was released and today the most recent phone comes with fourth-generation (4G) technology. This allows users to download music, watch videos, make video calls and join video conferences at much faster speeds. Today, this magical portable technology box has become an essential part of interpersonal communication and its significance is further increasing over time.

1.3 Growth of the Mobile Phone Market

The first mobile subscriptions took place in the early 1980s. During that period the total number of mobile phones in the market were around 0.023 million. Since then aided by affordability of cheap mobile phones and support of newer features fueled the mobile phone growth year after year. Figure 1.3(a) shows the growth of mobile subscribers since 1980 (according to ITU published figures). In 2014, the number of worldwide mobile users reached more than 5.6 billion (whereas world human population was 7.1 billion).

Low-end mobile phones are often referred to as feature phones. They are limited in their capabilities and primarily designed for basic telephony services. Handsets with more advanced computing ability, hosting a lot of other features apart from voice communication,

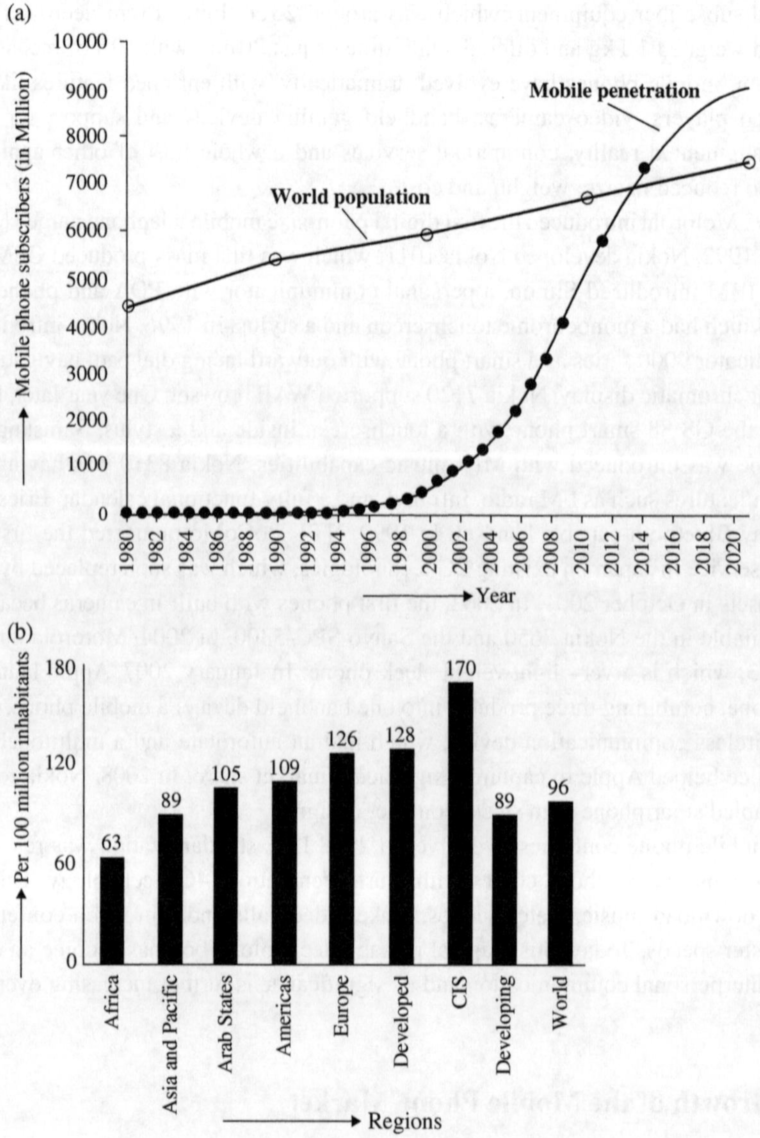

Note: CIS – Commonwealth of Independent States

Figure 1.3 (a) Growth of mobile subscribers over years. (b) Mobile cellular subscriptions by regions in 2014

Table 1.1 Smartphone usage data

Smartphone users	Around 80% of the world population now has a mobile phone and the number of mobile phones used is more than 5 billion. The number of smartphone users in the United States is 92 million. Ninety percent of the users use their smartphones throughout the day.
Owners by age group and gender	Age group: 13–17: 7%; 18–24: 18%; 25–34: 27%; 35–44: 22%; 45–54: 14%; 55–64: 7%; 66+: 3%. Gender of users: 47% women; 53% men
Primary usage	92% SMS; 84% Internet browsing; 70% e-mail; 65% games; 60% social networking; 50% music and videos
Community type	Urban: 65% Rural: 35%

are known as smartphones. Recently, smartphone penetration has increased significantly due to greater use of the Internet and complex applications. Global smartphone users surpassed the 1 billion mark in 2012 and in 2014 touched around 1.75 billion. Figure 1.3(b) shows mobile phone penetration by geographic regions.

Some interesting data is shown in Table 1.1.

The mobile phone business is a rapidly growing industry, providing mobile devices, content, and services. As no firm can make everything required for mobile phone devices and network, firms with different resources, capabilities and competences cooperate and form a network to provide products and service to consumers. This is commonly known as a mobile ecosystem, which consists of variety of firms like, network operators (like, Vodafone, Verizon, AT&T), mobile device manufacturers (like Apple, Samsung, Nokia and HP), network infrastructure providers (like Ericsson and Nokia-Siemens), silicon vendors (like Qualcomm, Intel and ST-Ericsson), platform providers (like, Qualcomm and Intel), content providers, system integrators, software providers, application developers, and, of course, consumers. Apart from these players, the growing demand for mobile phone production in recent decades has given rise to so-called original design manufacturers (ODMs) – for example, a company that designs and manufactures a product which is specified and eventually branded by another firm for sale – and original equipment manufacturers (OEMs) – for example, a company that manufactures products or components which are purchased by another company and retailed under that purchasing company's brand name.

Prior to 2010, Nokia was the market leader for mobile device manufacturing and sales. In Q1 2012, based on data from Strategy Analytics, Samsung surpassed Nokia, selling 93.5 million units. In Q3 2014, the top 10 manufacturers were Samsung (20.6%), Nokia (9.5%), Apple Inc. (8.4%), LG (4.2%), Huawei (3.6%), TCL Communication (3.5), Xiaomi (3.5%), Lenovo (3.3%), ZTE (3.0%) and Micromax (2.2%). The top five worldwide mobile phone vendors are shown in Table 1.2.

Table 1.2 Top five worldwide total mobile phone vendors, 2013

Rank	Manufacturer	Source: Gartner (%)	Source: IDC (%)
1	Samsung	24.6	24.5
2	Nokia (now Microsoft)	13.9	13.8
3	Apple Inc.	8.3	8.4
4	LG	3.8	3.8
5	ZTE	3.3	–
6	Huawei	–	3.0
	Others	34.0	46.4

1.4 Past, Present, and Future of Mobile Communication Devices

In the past, the use of a mobile phone was mainly for voice communication, but today there are thousands of applications that a mobile phone offers, including text messaging (SMS), a multimedia messaging service (MMS), Internet access, Web browsing, sending and receiving e-mails, listening to music, reading books, video chat, video recording, location service, time watching, alarm, calendar, calculator. Apart from these, nowadays mobile phones are also used in the field of telemedicine, healthcare, and wearables. In future it has huge potential to be used for watching TV, controlling and tracking remove devices, home automation, object recognition, e-commerce, and so forth.

Further Reading

Arrepim, http://stats.areppim.com/stats/stats_mobile.htm (accessed April 26, 2016).
Das, Sajal Kumar. (2000) *Microwave Signals and Systems Engineering*, Khanna Publishers.
Das, Sajal Kumar. (2010) *Mobile Handset Design*, John Wiley & Sons, Ltd.
Haykin, S. (2005) *Communication Systems*, John Wiley & Sons, Inc.
Proakis, J. G. and Salehi, M. (2005) *Fundamentals of Communication Systems*, Pearson Prentice Hall.
Tse, D. and Viswanath, P. (2005) *Fundamentals of Wireless Communication*, Cambridge University Press.

2

Cellular Systems Modems

2.1 Introduction to Modems

A modem is an electronic device that helps to modulate and demodulate the information at the transmitter and receiver block respectively in order to transmit the information signal reliably through the propagating medium. The word "modem" came from the term "modulator-demodulator." The modulator unit takes a baseband (low-rate / frequency) signal as input and converts it into a high-rate / frequency-modulated signal as output. If the baseband information signal is analog, then analog modulations, like AM, FM, and PM are used, otherwise if the baseband signal is digital then digital modulations like ASK, FSK, and PSK are used in the modulator to produce a low-frequency analog signal, which is later converted to a high-frequency RF signal before transmission through the medium. Initially a modem was also known as "data phone," as it enabled a computer terminal (host) to send and receive information over telephone lines (PSTN) by converting the digital data of a computer terminal into an analog signal used on telephone phone lines and then converting it back to its original form once it was received at the other end. These modems are commonly known as "dialup modems." Wireless modems work in the same way as a dialup analog modem, except they convert digital data into radio signals for transmission through air. The cellular systems modem is also wireless modem used in cellular networks and reside inside a cellular mobile terminal, as shown in Figure 1.2. Today, this modem unit can be integrated inside a mobile phone or it is used in a dongle data card and connected to the host PC device via USB or other interfaces as shown in Figure 2.1. The evolution of the modem over a cellular wireless network has occurred at a much more rapid pace,

Mobile Terminal Receiver Design: LTE and LTE-Advanced, First Edition. Sajal Kumar Das.

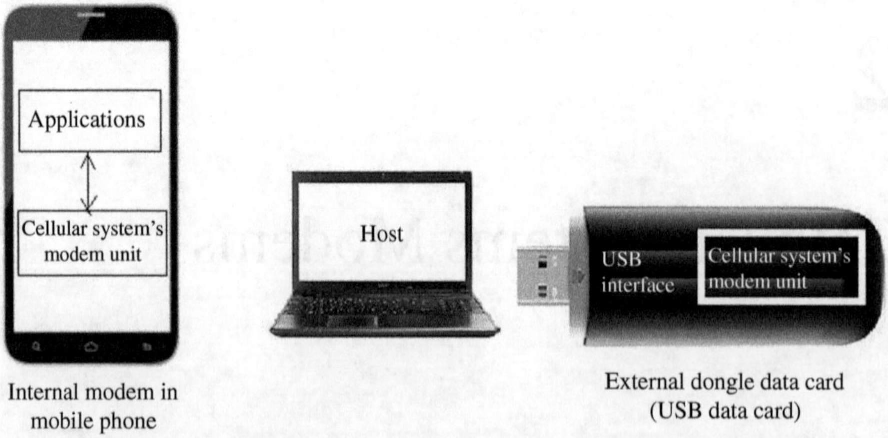

Figure 2.1 Cellular systems modem inside a data card and mobile phone

resulted in the use of these modem devices in a variety of devices (including IoT devices) and achieving data rates of more than 300 mbps. This is expected to increase as the technology evolves.

In this chapter, we will discuss more about how the cellular systems including mobile phone modem system has been evolved over several generations.

2.2 Telecommunication Networks

In recent decades telecommunication has revolutionized the way people communicate. Modern telecommunications networks are result of a long evolution process, which began at the end of the nineteenth century. Electrical telecommunication started in 1838, when Samuel Morse invented his system of dots and dashes for letters of the alphabet, which allowed complex messages to be sent and received. But the history of modern electronic communications began when Alexander Graham Bell invented the telephone in 1876, where speech was converted into an electrical signal, which was transmitted over copper wires and reconstructed at a distant receiver. Thereafter, the nineteenth and twentieth centuries witnessed phenomenal growth in telecommunication networking, mainly through numerous innovations and developments. These unprecedented developments and the synergy of electronics with telecommunications and computing offered a wide range of services and complex applications to corporate and individual users.

In the earliest days there was no concept of a network but only point-to-point links among users. The number of links required in a fully connected system became very large: $n(n-1)/2$ with n entities. To overcome this problem, a switching system or exchange was introduced and users were connected to this. Today, a network is defined as a collection of terminal nodes, links, and intermediate nodes. The nodes are some type of network device and may either be data communication equipment (DCE), such as a modem, hub, bridge, or switch, or data terminal equipment (DTE) such as a digital telephone handset, a host computer, a router, workstation, or server. The links are the means through which the nodes communicate with each other, like copper cables, optical fiber, or radio waves.

Generally, the three main mechanisms through which the communication takes place are (i) transmission, (ii), switching, and (iii) signaling.

- **Transmission** is the process of transporting information between two end terminals in the network. Generally, transmission systems use four basic media for information transfer: copper cables, optical fiber cables, radio waves (air), and free-space optics.
- **Switching** is required to establish the appropriate signal flow path between two communicating terminals. The nodes use circuit switching, message switching, or packet switching to pass the signal through the correct links and nodes to reach the correct destination terminal. In circuit switching the network reserves a dedicated channel (fixed bandwidth) for the entire communication duration as if nodes were physically connected, keeping the bit delay constant. In message switching the message is sent to the nearest directly connected switching node, which then checks for errors, selects the best available route and forwards the message to the next intermediate node. Each node stores the full message, checks for errors and forwards it, so this method is also known as the "store-and-forward" method. Packet switching also uses the store-and-forward mechanism but here the message is broken into small series of packets and then routed between nodes over data links shared with other traffic. Two major packet switching modes are connectionless and connection-oriented packet switching. In connectionless switching each packet has complete addressing or routing information and is routed individually, which sometimes results in out-of-order delivery. In the case of connection-oriented packet switching, a connection is defined and preallocated in the connection setup phase, before any packet is transferred.
- **Signaling** is the mechanism that allows network entities to establish, maintain, and terminate communication sessions in a network.

A logical model that describes how networks are structured or configured and describes how network nodes are interconnected is known as network topology. Various network topologies used today. These are shown in Figure 2.2.

Today, there are several basic types of telecommunications networks in use like, public switched telecommunications networks (PSTNs), cellular networks, computer networks, the Internet network, and the global Telex network. PSTN provides a traditional plain old telephone service (POTS), which relies on circuit switching to connect one phone to another via complex interconnection through a variety of heterogeneous switching systems. A cellular network is a wireless network deployed over cellular structure as explained in detail in section 2.3. A computer network is data network that allows computers to exchange data mainly in the form of packets. It can range from a local area network (LAN) to a wide area network (WAN), based on the size. As there was a need to interconnect these

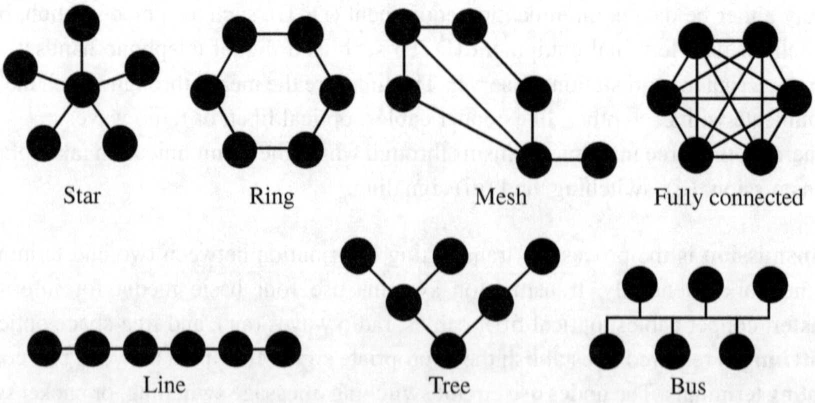

Figure 2.2 Network topology

networks, an internetwork was developed. The Internet network is a global system of interconnected computer networks using a standard Internet protocol suite (TCP/IP).

All kinds of networks are organized in a layering hierarchy, which divides the architectural design into a number of smaller parts, each of which performs a particular subtask and interacts with the other parts of the architecture in a well defined way. However, the different networks do not implement this architecture model in exactly the same way. Of these architectural models, the most widely used layering model is the Open System Interconnection (OSI) model developed by the ISO (International Standard Organization) in 1977. It is an abstract description for layered communications and computer network protocol design. Here, all communication functions are represented in seven layers, where a layer is a collection of conceptually similar functions providing services to the layer above it and receiving service from the layer below it. The functionalities of the seven layers are shown in Figure 2.3.

A set of network layers is also commonly referred to as a protocol stack. The interface between an upper layer and a lower layer is known as service access point (SAP). A protocol data unit (PDU) represents a unit of data specified in the protocol of a given layer, which consists of protocol control information and user data. A PDU is information delivered as a unit among peer entities of networks. A service data unit (SDU) is a unit of data that has been passed down from an OSI layer to a lower layer. The lower layer, n-1, adds headers or trailer, or both, to the SDU, transforming it into a PDU of layer n-1. So, PDU = SDU + optional header or trailer.

Another widely used interoperable network protocol architecture is TCP/IP, which was developed in 1978 by DARPA and driven by Bob Kahn and Vint Cerf. As TCP/IP was designed before the ISO model proposal it has four layers instead of seven but differences between these two models are minor. Figure 2.4 shows the TCP/IP protocol architecture.

The physical and the datalink layers of OSI stack are mapped to a single network interface layer in the Internet (TCP/IP) model. This layer handles the way in which data

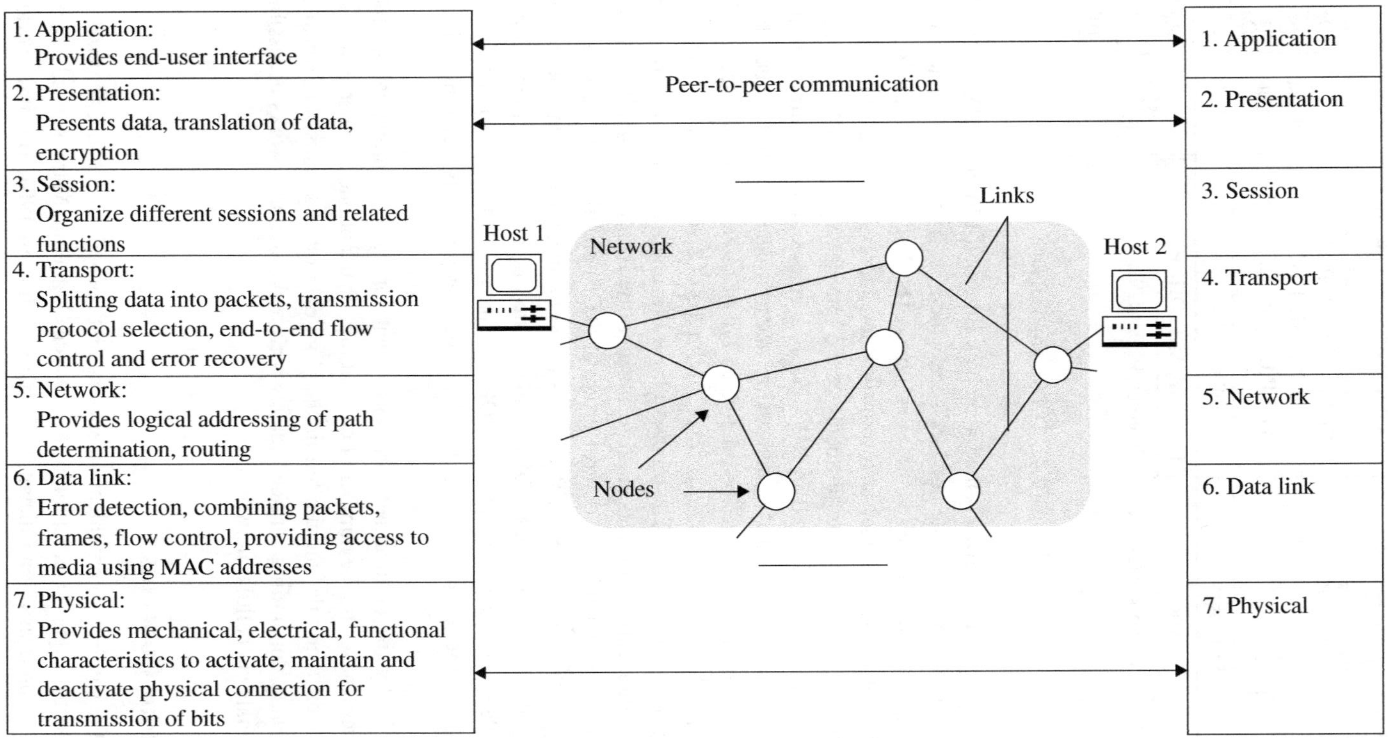

Figure 2.3 OSI seven-layer architecture

OSI model	Internet model (TCP/IP)	Example protocols / services		
Application	Application	HTTP, HTTPS, POP, SSH, SSH, DNS, SSL, FTP, SMTP, IMAP, Telnet, NNTP	Host layers	Data
Presentation				Data
Session				Data
Transport	Transport	TCP, UDP		Segments
Network	Network	IP, ICMP, ARP, DHCP	Media layers	Packets
Datalink	Network interface	Ethernet, PPP, ADSL		Frames
Physical				Bits

Figure 2.4 TCP/IP protocol layer

will be sent over physical network media such as Ethernet, PPP and ADSL. TCP/IP was designed to be independent of the network access method, frame format, and medium, so it can be used to connect differing network types. The Internet layer, analogous to the Network layer of the OSI model is responsible for addressing, packaging, and routing packets on the network. The core protocols of the Internet layer are IP, ARP, ICMP, and IGMP. The IP protocol as defined in RFC 791 is a connectionless, unreliable datagram protocol, primarily responsible for addressing and routing packets between hosts. So, sometimes an IP packet might be lost, delivered out of sequence, duplicated, or delayed, and the IP layer does not attempt to recover. That type of error correction is the responsibility of a higher layer protocol. The transport layer is primarily responsible for session and datagram communication services used to manage the data exchange. This layer's two main protocols are transmission control protocol (TCP) and the user datagram protocol (UDP). As defined in RFC 793, TCP provides a one-to-one, connection-oriented, reliable communications service, whereas UDP provides a one-to-one or one-to-many, connectionless, unreliable communications service. UDP is defined in RFC 768. An application layer provides access to the services of other layers and defines protocols that applications use in order to exchange data.

Though initially the cellular network was meant for voice communication, the rapid growth of Internet use and the number of cellular mobile telephones created a need to bring Internet services to cellular mobile terminals. High data-rate transmission over a cellular network is a very demanding service today, which makes data networks accessible from mobile terminals via cellular networks.

2.3 Cellular Concepts

The door to the wireless communication era was first opened when Clark Maxwell derived a theory of electromagnetic radiation in 1857, which Guglielmo Marconi used as a basis for radio transmission over a long distance via wireless link in 1901. But in a world where users are separated by very long distances, covering such a large geographical area using a

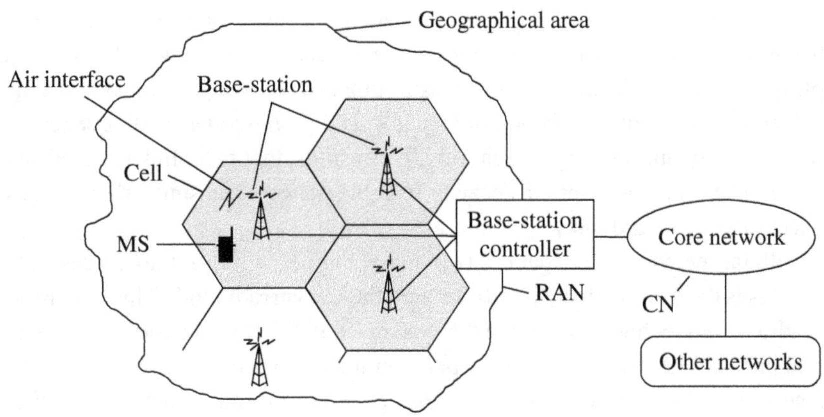

Figure 2.5 Overview of a cellular network

single transmitter transmitting with a huge amount of power was not a real solution. The limitations of such a solution are its waste of transmission power, its poor use of frequency resources, and, above all, the fact that it only covers a particular zone, which means that user mobility is restricted. The ideal solution for this problem was first proposed at AT&T Bell Labs in 1947, which introducing the concept of cell. In 1971, AT&T submitted a proposal to the Federal Communications Commission (FCC) for a cellular mobile concept, where a region is geographically divided into several cells and each cell includes a fixed location transceiver known as base station. This base station wirelessly communicates with the mobile receivers inside that cell area, just like a star-type interconnection topology, and it is also connected to the other base stations and networks via a backbone, which provides global connectivity. Now, the user can roam around among different cells without losing connectivity by means of a handover. When a user moves from one cell to another then a handover from one cell to the other occurs. This provides tremendous mobility for the users. So, a cellular network is a radio network made up of a number of radio cells, each served by one base station. Just as millions of body cells cover our whole body, so a wider geographical area is covered by many such smaller radio cells.

As shown in Figure 2.5, in a cell there will be several user devices, known as user equipment (UE) or mobile station (MSs), and one central base station. The base station and UE communicate via air interface. As air is a public channel, so the air medium is multiplexed among various users (or channels or systems) using different media access technologies. Mobile cellular systems use various techniques like, frequency division multiple access (FDMA), time division multiple access (TDMA), code division multiple access (CDMA), and space division multiple access (SDMA), to allow multiple users to access the same air medium. In fact, many systems employ several such techniques simultaneously. Several radio channels are needed for communication between the network and UEs to carry the user-specific data and control information, and those radio channels are created by using these multiple access techniques. For bidirectional communication, users want to send data as well as receive data, and if this is done simultaneously, then we call it a full duplex. A half duplex is where users either transmit or receive at one time. The technique for multiplexing the available channels for transmitting and receiving is called duplexing, and this is done by time (time division duplexing – TDD) or frequency (frequency division duplexing – FDD) multiplexing. Whenever UE transmits, that radio link is called an uplink (or reverse link), and whenever UE receives (e.g. the network transmits) that link is known as a downlink (or forward link).

The cellular network has mainly two main entities – (i) Radio Access Network (RAN) – this is the front-end and interfaces with the UE via radio link. This mainly depends on the radio access technology used in the system – and (ii) core network (CN) – this is the back-end part and generally does not depend on the radio access technology used. In the whole network, different network entities are connected through different, well defined interfaces, which will be discussed later.

2.4 Evolution of Mobile Cellular Networks

Since the introduction of first-generation cellular mobile networks in the 1970s, cellular networks have undergone tremendous changes. Cellular technology has evolved from being just a voice service and now provides a wide and rich collection of data and multimedia services. Worldwide deployment of cellular networks and the unprecedented growth of the mobile market have enabled global, cost-effective connectivity solutions, which can support a variety of complex applications including many current and emerging healthcare applications. Due to the ever increasing demand for higher data rates, for support for more complex applications, and for a seamless handover between the various networks, the mobile system has evolved over several generations from first generation to fourth generation, and, as a result of these advances in technology, new wireless standards have been developed. The evolution of different cellular systems and standards over several wireless generations is depicted in Figure 2.6.

In this chapter the legacy modems (1G, 2G, and 3G) are briefly discussed and then, in the subsequent chapters, the next-generation modems (LTE, LTE-A, 4G) are discussed in detail.

2.5 First-Generation (1G) Cellular Systems

The first-generation cellular network and mobile phone systems were developed on analog technology. These were characterized by analog modulation schemes (like AM, FM, PM), with FDMA as an air-medium multiple access technique, and were designed primarily for delivering voice services. The first generation cellular system architecture is shown in Figure 2.7. The first automatic analog cellular systems developed by Nippon Telephone and Telegraph (NTT) was deployed in Tokyo in 1979, later spreading to the whole of Japan, and to the Nordic countries in 1981. Next, the Advanced Mobile Phone System (AMPS)

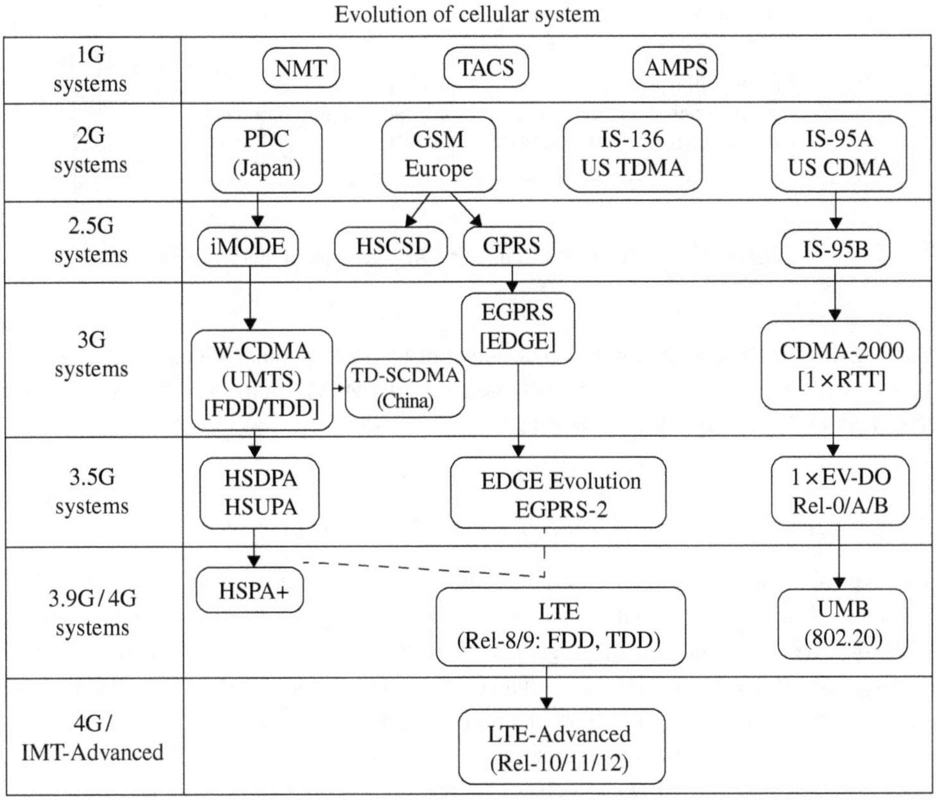

Figure 2.6 Evolution of wireless systems

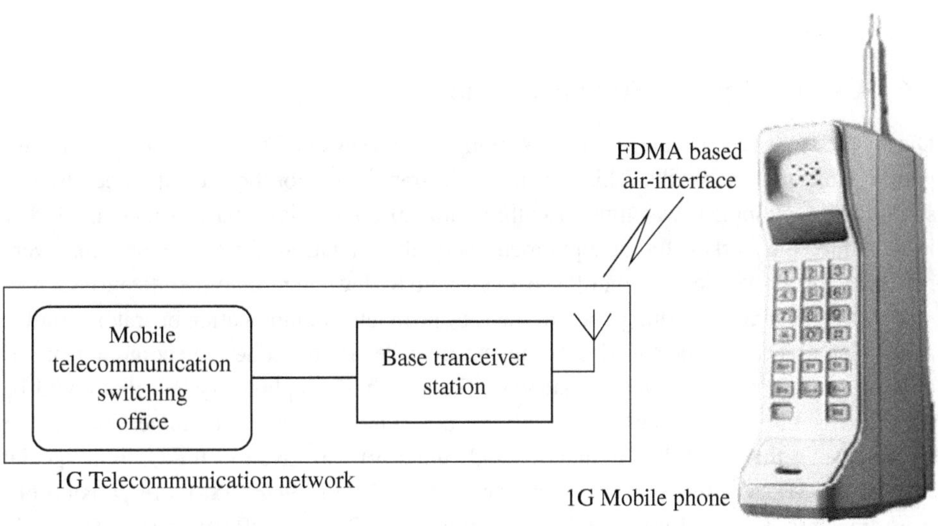

Figure 2.7 First-generation cellular system architecture

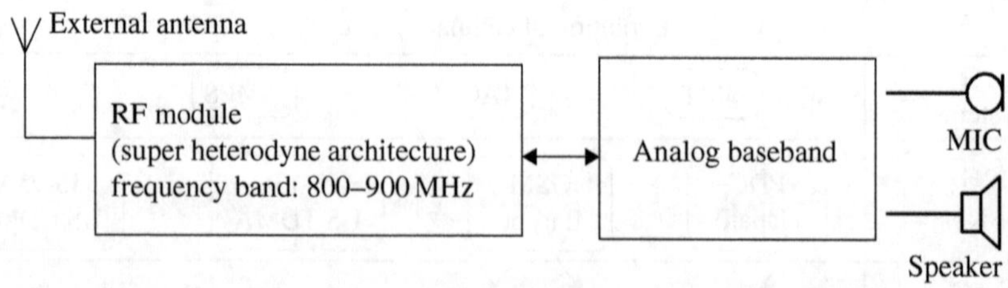

Figure 2.8 System architecture of a first-generation mobile phone

was launched in 1982 in North America. Some of the most popular standards deployed as 1G systems were the Advanced Mobile Phone System (AMPS), Total Access Communication Systems (TACS) and the Nordic Mobile Telephone (NMT).

2.5.1 *First-Generation Mobile Phone Modem Anatomy*

The typical architecture of a first-generation mobile phone is shown in Figure 2.8. It provided analog voice communication using frequency modulation. AMPS used the 800–900 MHz frequency band. Originally 40 MHz of spectrum was separated into two bands of 20 MHz with 30 kHz radio channel bandwidth between mobile station and base stations, and FDMA was used as channel multiplexing technique. The RF receiver was mainly based on super heterodyne architecture. Mobile power level was adjustable. The cellular structure used macro cells of radius around 35 km with frequency reuse and the handoff (handover) concept. Supported features were the ability to dial numbers, talk, and listen, with a talk time of only 35 min.

2.6 Cellular System Standardization

During the 1970s, each country was developing its own system. These systems were incompatible with other networks. This was not a desirable situation because the operation of such mobile equipment was limited within national boundaries, and this incompatibility issue limited the markets for the equipment. Soon the limitations for market potential were realized. This drove the creation of a special group to develop mobile specifications.

In 1982, the main governing body of the European telecommunication operators, known as the Conférence Européenne des Administrations des Postes et des Télécommunications (CEPT), was formed to develop a standard for a mobile telephone system that could be used across Europe. The task of specifying a common mobile communication system for Europe in the 900 MHz frequency band (initially) was given to the Groupe Spécial Mobile (GSM), which was a working group of CEPT. In 1989, GSM's responsibilities were transferred to the European Telecommunication Standards Institute (ETSI), and in 1990 phase I of the GSM standard's specifications were published.

Later, as the cellular market started growing, many organizations such as players in the telecommunications business, network operators, equipment manufacturers, service users, academic experts and approval authorities were interested in the development of new advanced standards to improve capacity, quality, supported features, compatibility issues, and to provide wide-area or even international services. It would be difficult for a single company to develop end-to-end full-system components. It is easier to develop some system entities or components of the end-to-end system. But, in order to interwork among these components developed by various companies together, system interoperability should be guaranteed. So, to form a complete ecosystem for mobile system development, companies felt the need for standardization. Then, due to this growing interest of developing common standard, the 3rd Generation Partnership Project (3GPP) initiative eventually arose. Its original scope was to produce globally applicable technical specifications and technical reports for a 3rd Generation Mobile System based on evolved GSM core networks and radio access technologies with frequency division duplex (FDD) and time division duplex (TDD) modes. It was a global cooperation between six organizational partners – ARIB, CCSA, ETSI, ATIS, TTA and TTC, which was established in December 1998. Rel'99 was the last release specified by ETSI SMG (special mobile group) in summer 2000. After that it was moved to 3GPP. Now 3GPP is actively engaged in developing next-generation mobile standards.

The 3GPP specification work is done in four technical specification groups (TSGs), as shown in Figure 2.9:

- The GSM/EDGE Radio Access Network (GERAN), which consists of three working groups: WG1, WG2, WG3.
- The Radio Access Network (RAN), which specifies the UTRAN and the E-UTRAN and is composed of five working groups: WG1, WG2, WG3, WG4, WG5.
- Service and system aspects (SA), which specifies the service requirements and the overall architecture of the 3GPP system.
- Core network and terminals (CT), which specifies the core network and terminal parts of 3GPP.

The evolution of GSM, WCDMA and LTE systems over different 3GPP releases are captured in the Table 2.1.

2.7 Second-Generation (2G) Cellular Systems

Equipment incompatibility, low traffic-handling capacity, unreliable handover, poor voice quality and poor security issues of first-generation systems created a demand for movement towards second-generation systems. As the number of subscribers grew and demand increased, there was also a need for increased network capacity and wider coverage. So, in early 1990s, a second-generation cellular network was introduced, which uses digital systems and digital modulations to improve channel multiplexing and voice quality.

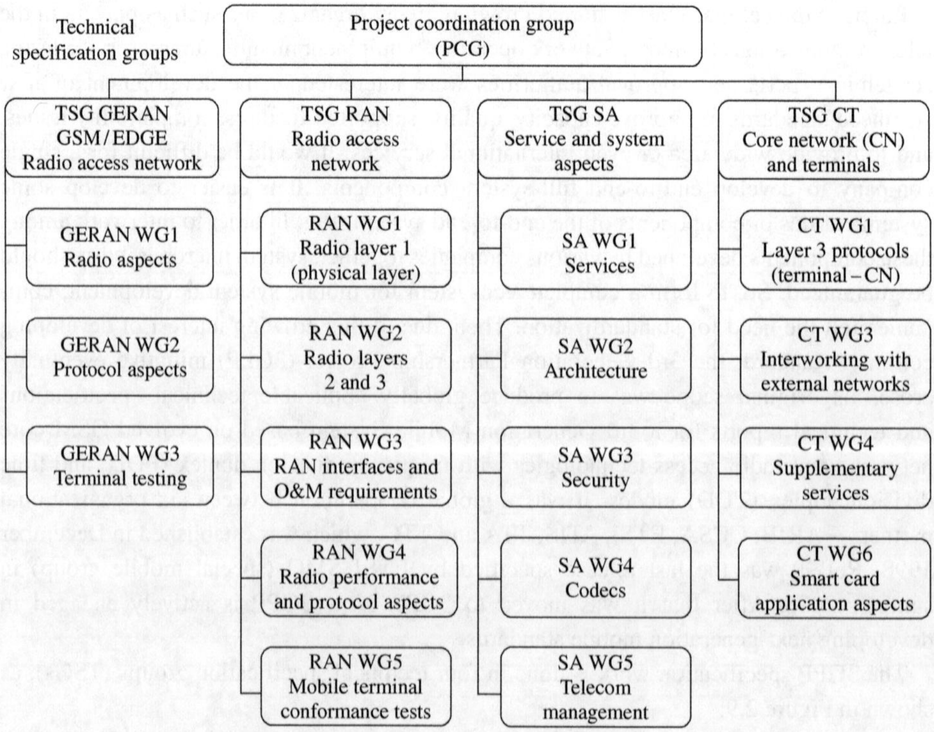

Figure 2.9 3GPP technical specification groups (TSGs)

The two most popular 2G systems are GSM and CDMA one. The CDMA one (also known as IS-95) system is based on Code Division Multiple Access (CDMA) technique. In the next section GSM system is discussed briefly.

2.7.1 GSM System

As discussed in section 2.6, CEPT was formed in 1982 with the task of specifying a common mobile communication system for Europe in the 900 MHz frequency band (initially). In 1989 this responsibility was transferred to the European Telecommunication Standards Institute (ETSI). In 1990 the first phase of GSM (Global System for Mobile Communications) standards specifications was published by ETSI.

2.7.1.1 Overview of GSM System Architecture

As shown in Figure 2.10, the GSM network is composed of several functional entities, whose functions and interfaces are defined in the GSM specification. The interfaces are standardized in order to allow multivendor interoperability, which gives network

Table 2.1 Feature evolutions of GSM, WCDMA, and LTE systems

	Rel'96 (and before)	Rel'97 (1997)	Rel'98	Rel'99 (March 2000)	Rel'4 (March 2001)	Rel'5 (June 2002)	Rel'6 (March 2004)	Rel'7 (2007)	Rel'8 (Dec 2008)	Rel'9	Rel'10	Rel'11	Rel'12	Rel'13
GERAN	GSM HSCSD	GPRS		EDGE			SAIC (DARP-I)	EGPRS-2A, 2B, RTTI, DLDC, MSRD (DARP-II)		VAMOS	TIGHTER		MCDL (downlink multi-carrier)	
UTRAN				WCDMA Data Rate: 2 mbps (indoor), 384 kbps (outdoor) Latency: 150 ms Modulation: QPSK (DL), HPSK (UL)		HSDPA Data rate: 14.4 mbps (DL) Modulations: 16 QAM	HSUPA Data rate: 5.6 mbps (UL	HSPA+ Data rate: 42 mbps (DL) , 11 mbps (UL) Modulations: 64 QAM (DL), 16 QAM (UL) MIMO: 2 × 2	DC-HSDPA Data rate: 42 mbps (10 MHz no MIMO)	DB-HSDPA Data rate: 84 mbps (10 MHz 2 × 2 MIMO)	4 carriers Data rate: 168 mbps (20 MHz 2 × 2 MIMO)	8 carriers Data rate: 336 mbps (40 MHz 2 × 2 MIMO or 20 MHz 4 × 4 MIMO)		
E-UTRAN									LTE Scalable BW: 1.4, 3, 5, 1,15, 20 MHz Modulation: QPSK, 16-QAM, 64-QAM Data rate: UE Cat-5: 300/75 mbps MIMO: 2 × 2 MIMO (UL), 4 × 4 MIMO (DL) Latency: ~10 ms	LTE Femto Cell (HeNB), MBSFN	LTE-Advanced BW: Up to 100 MHz (contiguous or noncontiguous carrier aggregations) Modulations: 64 QAM Data rate: 3000/1500 mbps MIMO: 4 × 4 MIMO (UL), 8 × 8 MIMO (DL)			

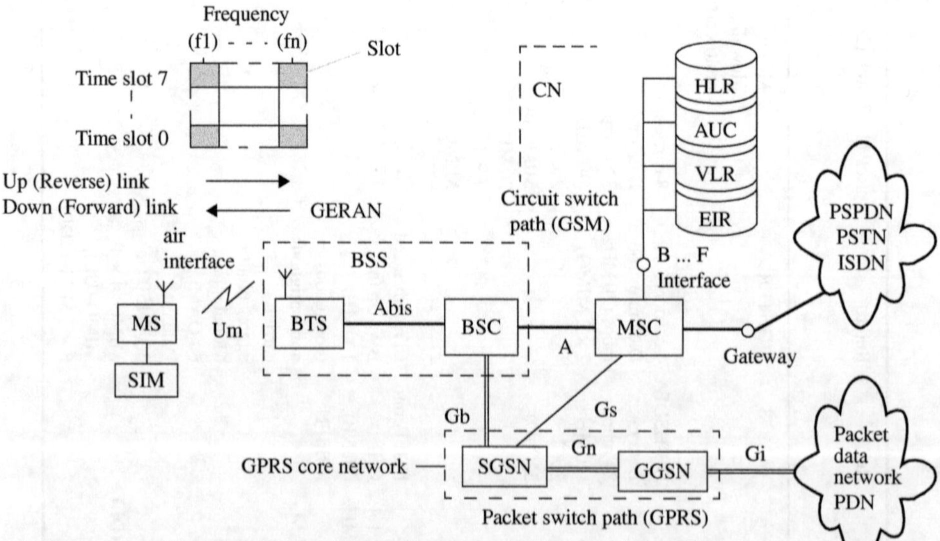

Figure 2.10 GSM/GPRS network architecture

operators the possibility to buy different network elements from different vendors. A network run by one operator in one country is known as a public land mobile network (PLMN), like Vodaphone, AT&T. Different cellular system providers deploy their own GSM networks after buying the frequency licenses from authorities/government. The mobile station (MS), used by the subscribers to access the network, consists of two functional entities: subscriber identity module (SIM) and mobile equipment (ME). A base station (BTS) performs all the transmission and reception functions with MS via air interface. Several BTSes are connected to a base station controller (BSC), which manages the radio channels allocations, handover decisions, power control, and so forth. The GSM radio network part is known as GERAN. Several BSCs are connected to a master switching center (MSC), which is connected to the other MSCs or GMSC or PLMNs or PSTNs or networks. The home location register (HLR) is a database that contains all administrative information for each registered subscriber in that network, including international mobile subscriber identity (IMSI), subscribed services information, service restrictions, and so forth. The visitor location register (VLR) is a database that contains temporary information about a subscriber currently located in a given MSC area but whose HLR is elsewhere. The information includes MSRN, TMSI, MS ISDN, IMSI, location area in which MS has been registered, and supplementary services data. The equipment identity register (EIR) contains a list of valid IMEI numbers to prevent illegal use of equipment. The authentication center (AuC) authenticates users that attempt to connect the network and stores the ciphering keys.

2.7.1.2 Air Interface

As shown in Figure 2.11, the GSM system uses time and frequency division multiple access (TDMA and FDMA) techniques to multiplex the air medium among the users – for example, mobiles. For uplink and downlink separation it uses the FDD technique. The downlink and uplink band is divided into several frequency channels, each having a bandwidth of 200 kHz and each frequency channel is divided into eight time slots, with each slot having a time duration of 577 μs. GSM uses GMSK modulation, where each symbol contains one bit and in one slot duration 156.25 bits are placed for a GSM normal burst transmission, which leads to duration of a bit = 577/156.25 = 3.69 μs.

GSM defines two sets of logical channels – traffic channels and signaling channels (see Figure 2.12). The traffic channels include traffic channel full rate (TCH/FR), half rate (TCH/HR), enhanced full rate TCH/EFR. Signaling channels are divided into (i) broadcast channels – frequency correction (FCCH), synchronization (SCH), broadcast (BCCH); (ii) common control channels – paging (PCH), access grant (GCH), random access (RACH), cell broadcast (CBCH); (iii) dedicated control type – stand-alone dedicated (SDCCH), slow associated control (SACCH), fast associated control (FACCH) channels.

These logical channels are mapped to physical channels, where a physical channel is defined by a bandwidth frequency of 200 kHz and a time-slot duration of 577 μs. The physical layer receives data from a speech encoder or a higher layer (data comes from a protocol stack) at a rate of 20 ms (which is the basic transmit time interval – TTI for GSM) and then it encodes, interleaves, ciphers, and forms the bursts by adding training sequence

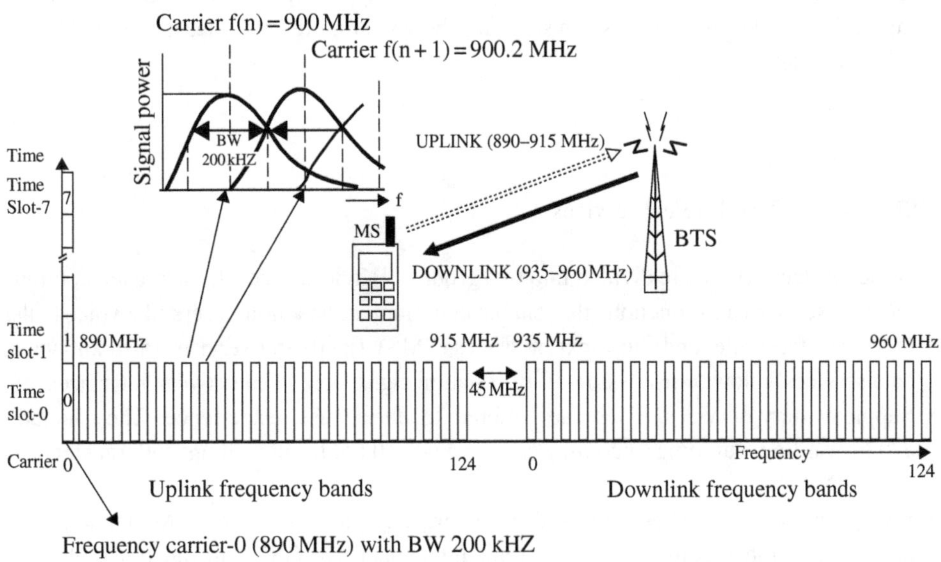

Figure 2.11 GSM uplink and downlink frequency bands

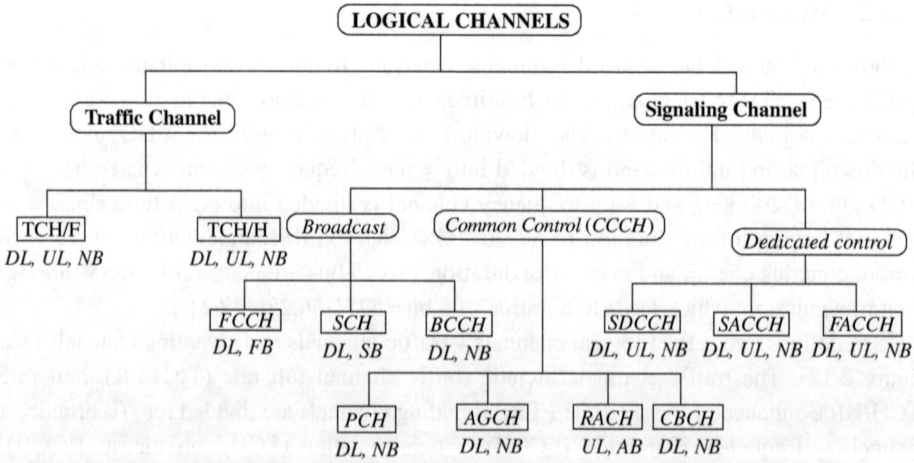

UL – Uplink, DL – Downlink, TCH/F – Traffic Channel (Full Rate), TCH/H – Traffic Channel (Half rate), FCCH – Frequency Correction Channel, SCH – Synchronization Channel, AGCH – Access Grant Channel, CBCH – Cell Broadcast Channel, RACH – Random Access Channel, BCCH – Broadcast Control Channel, PCH – Paging Channel, SDCCH – Standalone Dedicated Control Channel, SACCH – Slow Associated Control Channel, FACCH – Fast Associated Control Channel, NB – Normal Burst, SB – Synchronization Burst, AB – Access Burst, FB – Frequency Correction Burst.

Figure 2.12 GSM logical channels structure

bits, tail bits, and guard bits. Next, it places the burst data in a defined time slot according to the GSM frame structure, which is digital modulated (GMSK), RF upconverted, amplified, and then transmitted. The reverse operations happen in receiver side as shown in Figure 2.13. GSM burst transmission steps are shown in Figure 2.14. Please refer to [1] and [2] for more details.

2.7.1.3 Services

GSM system offers different services.

- Basic services. These are divided into two groups. (i) Teleservices: these are telecommunication services and functions that enable communication among users like voice calls, videotext, facsimile, and short text messages (SMS). (ii) Bearer services (also known as data services) allow transmission of information signals between network interfaces.
- Supplementary services. These are offered to enrich user experience. They include call forwarding, outgoing/incoming call barring, call hold, call waiting, call transfer, and caller ID.
- Emergency services. These are used for emergency purposes. The GSM standard also provides separate facilities for transmitting digital data. The GSM full-rate traffic channel (TCH/FR) allows user data transmission at the primary user data rates of 9.6, 4.8 and ≤2.4 kbps. GSM provides two basic data services, transparent and nontransparent,

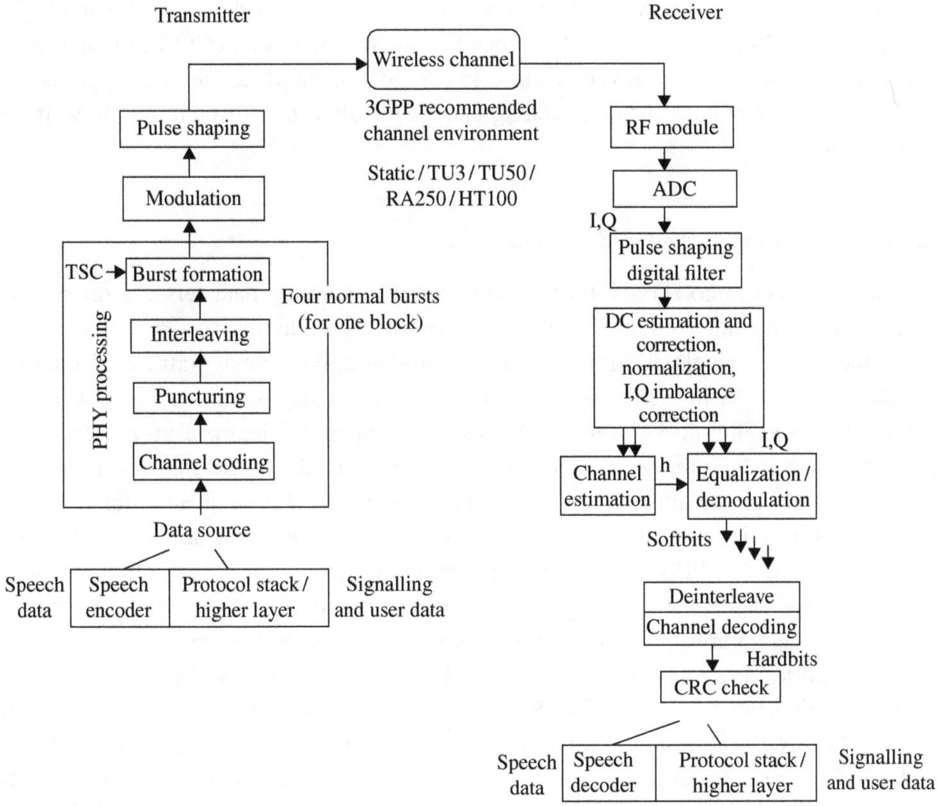

Figure 2.13 Different processes involved in GSM signal transmission and reception

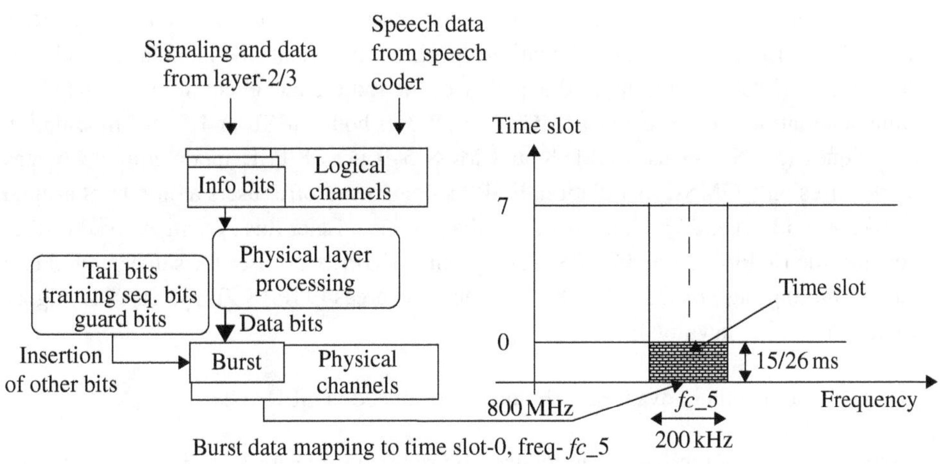

Figure 2.14 GSM burst formation steps

and defines two circuit-switched data protocols, circuit switched data (CSD) and high-speed CSD (HSCSD). CSD was developed for data transmission in GSM system using a single radio time slot to deliver 9.6 kbps. In HSCSD, higher data rates are supported by means of more efficient channel coding and/or the allocation of multiple (up to four) time slots.

2.7.1.4 Evolution of the GSM System

- General packet radio service (GPRS) systems. The wireless data services offered by GSM are based on the circuit switched radio transmission. In this case, a traffic channel is allocated for a single user for the entire call duration. With bursty traffic (like Internet traffic) this results in highly inefficient resource (frequency and time-slot) utilization. So, GPRS was introduced in the GSM system for more efficient packet-switched data transmission, which results in much better utilization of the traffic channels because a channel will only be allocated whenever it is needed and will be released immediately after the transmission of the packets. Using this principle, multiple users can share one physical channel. GPRS improves the utilization of the radio resources, offers volume-based billing, higher data transfer rates, shorter access times, QoS-based service, point-to-point in addition to point-to-multipoint services, and simplifies the access to packet data networks. To support this, GSM network architecture was modified by introducing two new elements the serving GPRS support node (SGSN) and gateway GPRS support node (GGSN), as shown in Figure 2.10. A packet control unit (PCU) is also added into the BSC to control packet channels and separate data flows for circuit and packet-switch services. GPRS employs variable-rate coding schemes (CS) with GMSK modulation and multislot operation but the peak date rate for GPRS is limited to about 115 kbps, which is not sufficient for supporting popular Internet applications.
- E-GPRS systems. Due to higher data rate demand, GPRS system evolved towards EDGE (enhanced data rates for GSM evolution), which is known as E-GPRS (enhanced GPRS). Like GPRS, EDGE uses a rate adaptation algorithm that adapts the modulation and coding scheme (MCS) according to the quality of the radio channel conditions. It supports nine modulation coding schemes (MCS 1–9). It uses both GMSK and 8-PSK modulation techniques (MCS 1–4 uses GMSK and MCS 5–9 uses 8-PSK modulation), whereas GPRS uses only GMSK modulation. EGPRS system can offer users a bit rate of around 250 kbps, with an end-end latency of less than 300 ms. Later this system evolved further towards the EGPRS-2A and EGPRS-2B systems. EGPRS-2A uses the same symbol rate (and sampling rate) as GSM 270.833 ksymb/s, whereas EGPRS-2B systems use higher symbol rate (325 ksymb/s).

There are some other advanced techniques introduced in GERAN:

- Reduced latency – the latency is reduced by reducing transmission time interval from 20 ms (Basic TTI) to 10 ms (RTTI).

- Dual antennas terminals – where two antennas are used in the MS to achieve space diversity or MS receive diversity (MSRD). Similarly, use of multiple antennas (MIMO) and multiple carriers (DLMC) is also proposed for increasing the data rate and GERAN performance. The evolution of GERAN is shown in Table 2.2.

2.8 GSM Mobile Phone Modem Anatomy

As discussed earlier, the modem unit is mainly responsible for user information (e.g. voice/data) and control information transmission and reception of over the channel (air channel). It consists of Radio Frequency (RF) front-end, ADC/DAC, modulation/demodulation, digital baseband processing, and protocol stack unit. The applications sends and receive data via the modem.

2.8.1 Receiver Unit

The internal block diagram of a GSM mobile phone is shown in Figure 2.15.

2.8.1.1 RF Front-End Receiver Unit

The RF receiver module is responsible for signal reception from the air and downconverting it into a baseband signal as discussed in Chapter 6 in more detail.

- Input signal reception. The transmitted electromagnetic (e.m.) signal impinges on the metallic antenna of the mobile receiver and tries to penetrate through it. It is known that the e.m. wave consists of electric field and a magnetic field, which are perpendicular to each other, and also perpendicular to the direction of propagation. From Maxwell's third equation, it can be derived that when the e.m. wave tries to penetrate through the metal (electric conductor), the magnetic field (H) will generate surface current (I) and die down after penetrating the thickness of skin depth of the metal. This current needs to be amplified and sampled. Similarly, the electric field will generate the voltage. As shown in Figure 2.15, the duplexer unit (or tx-rx switch in case of GSM-only phones) separates the transmitter and receiver paths and helps to use the same (single) antenna for transmission as well as reception purposes.
- Band pass filtering. The received signal is then band pass filtered to extract only the desired frequency band from the received input signal.
- Amplification and downconversion. Then the band pass input signal is amplified using a RF low-noise amplifier (RF LNA) circuit, which has very low noise figure (NF). Next, the downconversion is accomplished by employing a mixing process, which produces two converted mixed signal components. Generally, for GSM systems, the receiver RF downconversion architecture is based on simple homodyne receiver architecture. The local oscillator generates a high-frequency signal locally, tuned to the desired

Table 2.2 3GPP GERAN system evolution over different important releases

Rel'96	Rel'97	Rel'99	Rel'6	Rel'7	Rel'9	Rel'10	Rel'12
GSM and HSCSD	GPRS	EDGE (EGPRS)	Single antenna interference cancellation (SAIC - DARP-I)	EGPRS-2A/2B, RTTI, DLDC, MSRD (DARP-II)	VAMOS	TIGHTER	MCDL (downlink multicarrier)
Modulation: GMSK	GMSK	GMSK and 8-PSK		GMSK, 8-PSK, QPSK, 16-QAM, 32-QAM	A-QPSK for VAMOS		
Data rate: GSM: 22.8 kbps CSD: 9.5 kbps HSCSD: 57.6 kbps	CS4: 21.4 kbps/slot For 8 slots theoretical max: 171.2 kbps	MCS-9: 59.2 kbps/slot theoretical max: 473.6 kbps		EGPRS-2A max 98.4 kbps/slot EGPRS-2B: max 118.4 kbps/slot			>2 mbps

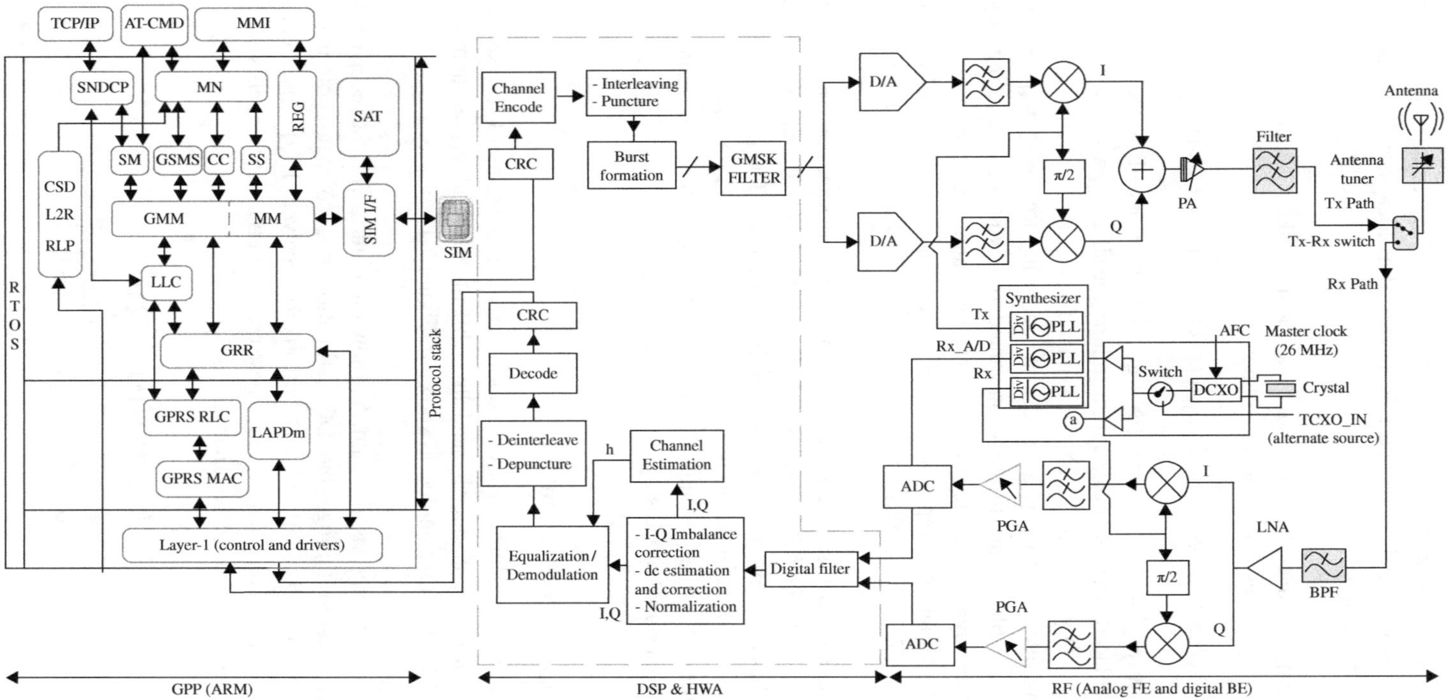

Figure 2.15 Internal block diagram of a GSM/GPRS phone

reception frequency, and passes that to the mixer. In the case of a direct conversion RF receiver, the local oscillator is tuned to the desired receive frequency channel. The mixer circuit produces sum and difference frequencies after local oscillator mixing with the received incoming signal.

If the incoming signal is: $S_{in} = S_s.\cos\omega_s t$ and local oscillator signal is: $S_{LO} = S_L.\cos\omega_L t$, then, after mixing, the resultant signals will be:

$$S_{in}.S_{LO} = S_s.\cos\omega_s t * S_L.\cos\omega_L t = S_s.S_L \left\{\cos\left(\omega_s + \omega_L\right).t/2 + \cos\left(\omega_s - \omega_L\right).t/2\right\} \quad (2.1)$$

where the frequency sum component $(\omega_s + \omega_L)$ will be stopped by the analog low-pass filter, which is placed after the mixer unit. Similarly the sin component will be mixed in the quadrature path (Q path). This analog low-pass filter is also called a channel-select or antialias (A-A) filter, as it helps to reduce the sampling frequency requirements by the ADC circuit as it blocks or stops the unwanted (blocking or interfering) signals. So, the bandwidth and dynamic range requirement reduces and hence the cost of ADC can also be reduced. The variation of input signal strength can cause clipping in ADC, so to avoid that a programmable gain amplifier (PGA) is generally included before the ADC unit.

- Analog to digital conversion (ADC) unit. Next, the mixed and low pass filtered analog signal is sampled by the ADC circuit and generates the [I, Q] digital samples. Each [I, Q] sample pair has a specific value represented by 8 to 16 bits. Then, it is passed for digital baseband processing. Generally, in the baseband, first the [I, Q] samples are digitally filtered (using RRC filter or pulse-shaping filter) for proper signal shaping before the baseband digital signal processing.

2.8.1.2 Baseband Receiver Unit

The digital baseband module deals with digital signal processing of the baseband signal and protocols. In the baseband physical layer bit detection algorithm demodulates and decode this received [I, Q] samples, as shown in Figure 2.13.

- DC estimation and compensation unit. As shown in Figure 2.15, the received [I, Q] samples are first passed through the DC estimation and compensation unit. The DC compensation unit will calculate the DC offsets for received [I, Q] signals in the I and Q paths separately and then it will subtract these estimated DC components separately from each [I, Q] sample in the received burst:

$$\begin{aligned} I_{comp}(n) &= I[k] - \frac{1}{N}\sum_{k=1}^{N} I[k] \\ Q_{comp}(n) &= Q[k] - \frac{1}{N}\sum_{k=1}^{N} Q[k] \end{aligned} \quad (2.2)$$

where, k represent the samples in the [I, Q] buffer and there are N number of [I, Q] sample pairs in a burst.

- Normalization. Then the [I, Q] samples are normalized to unity scale.
- Channel estimation unit. Next, the [I, Q] signal passed to the channel estimation unit. It estimates the channel impulse response (h). There are several methods used for channel estimation. Generally, a set of pilot bits or training-sequence (TSC) bits are inserted in the transmitted signal, which is preknown to the receiver. The channel estimator uses those information bits to estimate the noise and interference signal in the received signal. The channel estimator estimates the channel impulse response based on different channel conditions and passes that to the channel equalizer block.
- Channel equalizer. The equalizer takes the channel impulse response from the channel estimator and the received normalized input [I, Q] samples, then equalizes the received [I, Q] samples using different channel equalization algorithms. Then it demodulates and generates the soft bits for that received burst. The soft bits are scaled using different algorithms.
- Deinterleaver. The scaled soft bits are deinterleaved (by reversing the way that they were interleaved in the transmitter side). Then they are depunctured, if the transmitted bits were punctured (for some of the logical channels puncturing is not used). During the demodulation process, positive soft values (>0) are represented as bit "0," negative soft values are represented as bit "1" and 0 (0×0) soft value is used as neutral. So, most commonly, a soft bit value of 0 is inserted in the punctured bit positions, which has no bias towards the logical 1 or logical 0 bit decision during the hard decoding (final decoding) process.
- Channel decoding. Then the soft bits are passed to the channel decoder unit for making hard decision. Generally, a Viterbi decoder or turbo decoder is used for channel decoding (if convolution coding is used in the transmitter side). The channel decoder helps to correct some bits, if they were received erroneously due to channel or RF impairments, so it is useful for error-correction purposes.
- CRC checking. Once the hard bits are generated the bits are passed to the CRC checking unit to detect errors. So, CRC checking is help for error detection purpose, which means it indicates whether the received block is received correctly or not.
- The protocol stack (PS) unit. The decoded bits from physical layer (after the CRC check) for a data block/frame are sent to the higher layer (protocol stack/application). The signaling data (or control data) is passed to the protocol stack unit's control plane modules and the user data related information is passed to the user plane and to the applications.
- Application unit. There are various applications runs on the mobile phone, such as voice/speech codecs and video codecs, and the received data is passed to the appropriate applications for playing. The PS data could be passed to TCP/IP (based on data service), or AT-CMD (if modem is interfaced for sending commands), or applications like speech decoders, or sent to the MMI for man-machine interface. The complete flow of user data and speech signals through a GSM mobile-phone modem is shown in Figure 2.16.

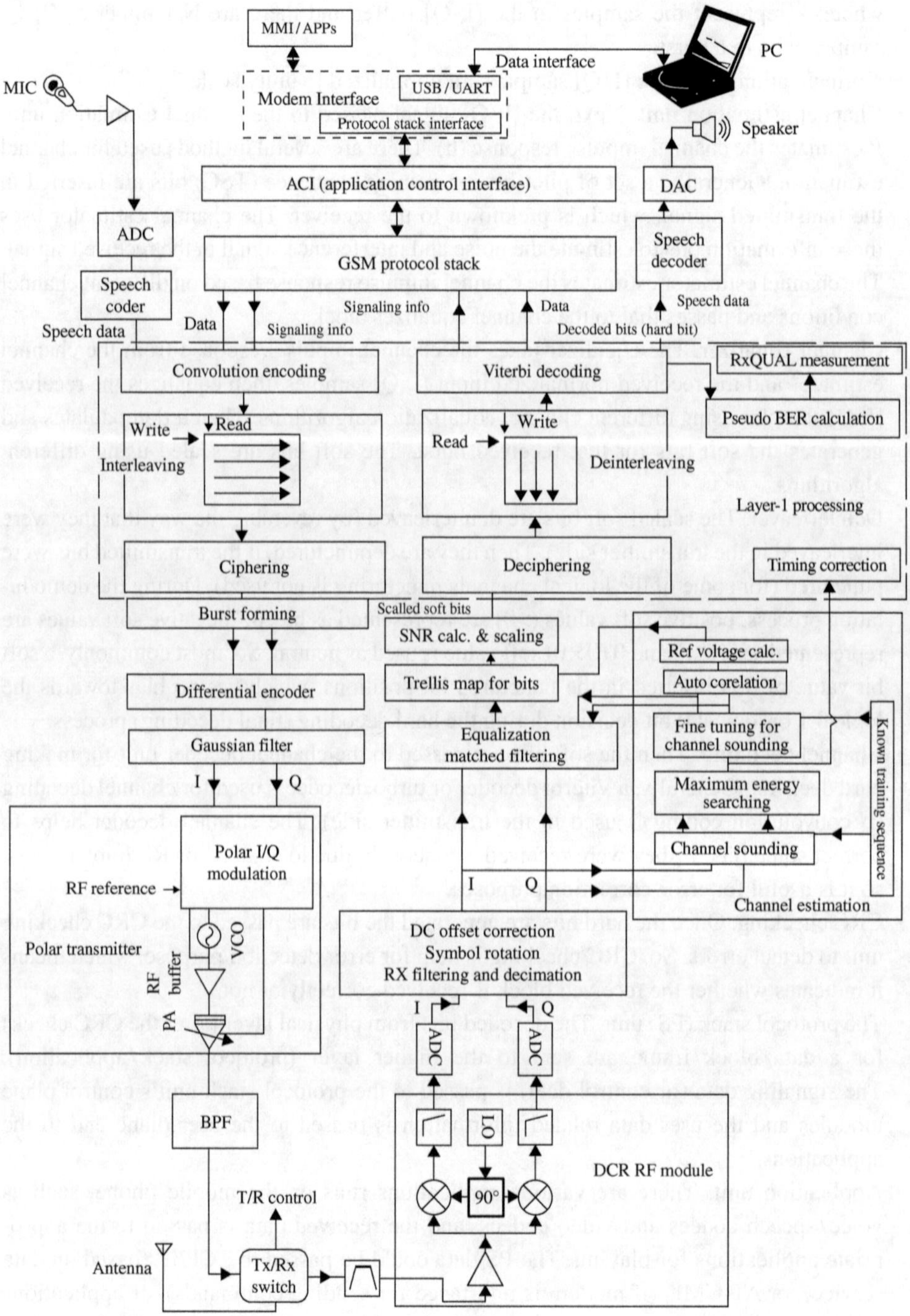

Figure 2.16 Functional block diagram of a GSM mobile phone (modem part)

2.8.2 Transmitter Unit

2.8.2.1 RF Transmitter Unit

The GSM MS RF transmitter unit is shown in Figures 2.15 and 2.16. GMSK modulation is used in GSM, which has constant amplitude and a polar transmitter is mainly used for the RF transmission. This is discussed in detail in Chapter 6.

2.8.2.2 Baseband Transmitter Unit

The physical layer receives the signaling data block (from protocol layers) or user data (frame from speech coder) at every basic TTI interval (20 ms) and processes that block of data and generates four normal bursts as discussed earlier and passes to the RF transmitter for transmission using the allocated frequency channel (ARFCN) and time slot.

2.9 Channel Estimation and Equalization in GSM Mobile Terminals

The wireless channel environment is very complex for several reasons, such as multipath propagation loss, fading, multiuser interference, cochannel interference, adjacent channel interference, and noise signals. The mobile wireless channel becomes more complex due to the user mobility. The Doppler effect becomes prominent, when user becomes mobile. In these scenarios, the air channel plays a pivotal role, as its characteristics mainly influence the signals that propagate through it. Apart from these issues, signal fading makes the mobile wireless channel extremely unpredictable and it varies time to time. On top of fading, noise, interference and attenuation factors, the quality of a wireless link between the transmitter and receiver is highly dependent on the mobile environment, radio propagation parameters, and the air channel's characteristics. Also, intersymbol interference (ISI) plays a significant role, especially, if the symbol duration (T) is shorter than channel delay time. Generally, the multiuser, mobile wireless environments are broadly classified into two categories:

- Sensitivity limited scenarios. In these scenarios, the received signal power at the receiver circuit is very low – for example, the received signal could be very feeble and the signal mainly influenced by the AWGN noise in the receiver circuit and fading characteristics of the propagation channel. Depending on the delay characteristics of the propagation channel, sensitivity-limited scenarios can be classified into two subcategories:
  - Nondelayed channels. In this case, channel delay is less than one symbol period.
  - Delayed channels. In this case, channel delay is more than a symbol period.
- Interference-limited scenarios. In a high-interference scenario, the carrier-to-interference (C/I) ratio reduces. This scenario is different from a sensitivity scenario, as described above. Here, although the desired signal power level might be high or the received signal level (RSSI) might be high, the received input signal mixes with the interference signal (I),

so correct demodulation and decoding of data becomes much difficult. In this case, depending on the nature of the interference, the interference-limited scenarios can be broadly divided into two categories:

- Cochannel interference (CCI). If the interferer signal's frequency is same as the desired carrier's frequency, then the interferer is called as cochannel interferer. Generally this happens because the same frequency channel is reused in a distant cell. This type of interference is called color noise.
- Adjacent channel interference (ACI). If the interferer signal's frequency is next to the desired channel's frequency, then the signals from the nearby frequency channel (adjacent channel) leak into the desired channel. This type of interference is called "adjacent channel interference."

As discussed above, there could be several types of channel conditions based on the type of noise, fading, and interference characteristics. That is why only one type of channel estimation and equalization technique will not be appropriate for all these different types of channel propagation and interference scenarios. Some of these techniques provide better performance gain in cochannel interference conditions but degrade receiver performance in sensitivity-limited conditions. Similarly some techniques enhance the BER performance in some specific channel conditions but degrade the performance in some other channel propagation conditions. Running wrong or inappropriate algorithms in inappropriate channel conditions causes unnecessary processing power wastage and that leads to battery power wastage, without any performance gain. So, to get the best out of everything there is a need for environment or channel condition detectors. These will detect the channel conditions and then enable the appropriate algorithms/solutions to obtain the best performance in that scenario.

2.9.1 Channel Condition Detection Techniques

In GERAN (GSM, GPRS and EDGE systems) mobile phone receivers, the single-antenna interference cancellation (SAIC) algorithm is most commonly used in an interference-limited scenario – when the input signal is dominated by the interferer signal or, the carrier-to-interference ratio (C/I) is lower. The SAIC algorithm uses a whitening process, which improves the CCI and ACI performance but it might reduce the sensitivity performance under low signal conditions by degrading the SNR. So, generally, in sensitivity scenarios such as low signal conditions, another type of channel equalizer is used for processing. That means, when there is a low interference signal present in the received signal – a high carrier-to-interference ratio (CIR) – whitening is bypassed to avoid reducing the SNR (in a sensitivity scenario). So, to take advantage of both, there is a requirement to detect the interference or sensitivity scenario dynamically and select or adjust the equalizer type accordingly. So, in interference-limited channel scenario receiver an algorithm will enable the SAIC equalizer, whereas in a sensitivity-limited scenario the receiver will enable the default equalizer (nonwhitening).

On the other hand, if the detected channel type is adjacent channel interference (ACI), then the input digital [I-Q] samples will be passed through a narrower low-pass filter (with bandwidth <200 kHz for the GSM system) to eliminate the out-of-band signals before the signal is passed for channel estimation and equalization, which will help to enhance the adjacent channel rejection performance. So, first the equalizer switching algorithm has to detect the ACI scenario dynamically and if an ACI condition is detected then only enable the narrow band filter for attenuating the ACI signal.

As discussed earlier, where there is a delayed sensitivity channel environment, generally the channel taps span over more than three taps. In such cases the maximum likelihood sequence estimation (MLSE) equalizer with more channel taps is a better choice to handle multipath delays. So, the channel length or channel type (delayed or nondelayed) has to be detected quickly by the channel condition detection algorithm and then, based on that detected channel type, the equalizer switching will select the channel tap lengths (three or seven taps) and MLSE taps accordingly. First, the channel-type detection switch has to dynamically detect the channel type and after that it has to adjust or select the channel tap length and enable MLSE for a greater number of taps, instead of keeping more taps in the feedback path in the decision feedback sequence estimation (DFSE) type of equalizer.

As discussed above, the algorithm should be doing two tasks: (i) channel environment detection (sensitivity, CCI, ACI … detection) and (ii) adaptive filtering and equalizer processing.

2.9.1.1 Dynamic Channel Environment Detection Method

Depending on the received signal strength (RSSI), the presence of interference signals (CCI, ACI) and channel delay characteristics (three taps, seven taps), the wireless channel can be broadly classified into four categories: cochannel (CCI), adjacent channel (ACI), sensitivity delayed, and sensitivity nondelayed. The channel environment detection module has two parts: (i) channel type detection: this submodule will detect whether the channel is sensitivity limited (AWGN) or ACI or CCI dominated, and (ii) delayed or nondelayed channel detection: if the detected channel type is sensitivity limited, then this submodule helps to detect the channel length to indicate whether it is a delayed (> three taps) or nondelayed channel.

2.9.1.2 Detection of Channel Type

A simple method can be used to dynamically detect the channel condition. The received complex-valued, baseband, symbol-spaced signal can be modeled as:

$$r(n) = \sum_{k=0}^{N} \mathrm{h}(\mathrm{k})\mathrm{s}(\mathrm{n}-\mathrm{k}) + \mathrm{I}(\mathrm{n}) \tag{2.3}$$

where, s(n) is the transmitted symbol, h(k) is the channel response, h(k)s are the coefficients of the baseband channel, and I(n) is undesired signal. I(n) includes white noise (AWGN) as well as colored (correlated) noise sequences, including CCI, ACI, and multipath components.

Based on the received sequence and known transmitted bit sequence, which is actually the training sequence code (TSC) in GSM, the channel impulse response h(k) is computed. The autocorrelation of the signal is defined as:

$$\rho_{vv}(k) = E\{I(n)I^*(n-k)\}.$$

Let us consider that the undesired/unwanted signal I(n) is white, which means the autocorrelation of I(n), $\rho_{vv}(k) = \delta(k)$, the ML estimate (which is the optimal estimate) of h(k)s is the least-squares estimate (LSE). But, when the noise I(n) is not white (i.e. $\rho_{vv}(k) \neq \delta(k)$) the least-squares estimate is not the maximum likelihood (ML) estimate of h(k).

In any typical cellular mobile receiver system, the undesired signal I(n) can be modeled as the sum of three signals (CCI, ACI and AWGN), as below, which is passed through the received filter:

$$I(t) = \left[I_{CCI}(t) + I_{ACI}(t) + I_{WN}(t)\right] * p(t) \tag{2.4}$$

and

$$I(n) = I(n \times T_{symbol})$$

Where, p(t) is the analog received filter, $I_{ACI}(t)$ is the analog adjacent channel interferer (ACI) before the received filter, $I_{CCI}(t)$ is the analog cochannel interferer (CCI) signal before the received filter and $I_{WN}(t)$ is the additive Gaussian thermal noise (AWGN) before the received filter.

From the above composite signal, I(n) is obtained by sampling I(t) at every T_{symbol} seconds. $I_{CCI}(t)$ or $I_{ACI}(t)$ can be colored so I(n) might become colored. On top of that, if p(t) is not a Nyquist filter then I(n) might become colored. Generally, I(n) can be colored and the color of the disturbance might change from one received burst to another. In case of colored noise, if the ML estimate is used then it will not be appropriate as the ML estimate of the channel coefficients is not the least-squares estimate in case of colored disturbances. In this work, it is assumed that the autocorrelation of the disturbance belongs to a finite set of candidate autocorrelations. It is also assumed that this autocorrelations set, and the whitening filter corresponding to each of these autocorrelations, is known a priori. Now, let us enumerate these candidate autocorrelations by $\{\rho_{vv}^i(k)\}_{i=1}^N$, and represent the corresponding whitening filters as $\{h^i(k)\}_{i=1}^N$. Then we find out the channel estimation for each autocorrelation and can be represented as $h^i(k)$ which minimizes the maximum-likelihood criteria considering that this autocorrelation is the right one. Then we need to select the channel estimate and autocorrelation estimate among these N pairs of channel estimates and autocorrelation estimates, which minimizes the ML criteria in the above equation. In the receiver, it is assumed that two-pass channel estimation is used and, in the first pass channel estimation, a fixed whitening filter is selected to cater for ACI, CCI, or AWGN scenarios, depending on the power of the residual noise remaining once the training sequence (e.g. pilot bits) signal part is filtered with each of these three types of filters. According to above discussion, three precalculated whitening filter taps in three noise models could be:

- white noise (AWGN);
- cochannel interference (CCI);
- adjacent channel interference (ACI).

So, during the processing the following steps are performed in sequence to detect which noise model is most suitable (or most appropriate) for the received input signal's processing:

1. The DC offset is estimated and then compensated. Next, the I,Q samples are normalized and then channel estimation is performed on the received DC compensated normalized [I-Q] samples. At the first stage, the simple channel estimation is performed using only three channel taps.
2. Next, the reference synchronization (pilot or training) sequence (s′) and channel estimate (h) are convolved to get x^, estimated synchronization (training) sequence. The noise samples (ns) are computed by subtracting x^ from the received synchronization samples (r) as given in the below equation.

$$\begin{aligned} x\wedge &= s' * h \\ ns &= r - x\wedge; \end{aligned} \tag{2.5}$$

3. The receiver should have three predesigned filters, made for ACI, CCI, and AWGN. Upon filtering using the precalculated whitening filter, the receiver computes the power residues of noise samples – which will provide three power residues. The minimum out of these three values will indicate whether it is a sensitivity scenario, a CCI scenario, or an ACI scenario. Sensitivity is indicated as 0, CCI as 1 and ACI as 2.
4. The output from the above detection will be an index value, which indicates whether the current burst experiences the channel as white noise (=0), CCI (=1), or ACI (=2) type disturbances. Depending on the detected index value, the present channel type will be set to sensitivity, or CCI, or ACI.
5. The wireless channel conditions or channel environment may change dynamically from one burst to another burst, so some sort of averaging might be required. The current detected channel type requires to be stored for detecting the long-term averaging of channel type value consideration. This is done by introducing a forgetting factor variable and the average channel type is selected using the forgetting factor. Then that average channel type will indicate the channel environment at any given point of time.

2.9.1.3 Detection of Channel Length

In the previous section we have discussed the mechanism for channel-type detection. After the channel-type detection, the receiver algorithm requires to detect the channel length (L), which indicates whether it is a time-delayed or nondelayed channel. For channel length estimation the following steps will be executed:

1. To find the scaled squared error, perform a "four-tap" channel estimation where the scaled error is the product of the estimated error and the modified Akaike information criterion (AIC) factor:

   Scaled_sqerror1 = 4-tap channel estimation error* Akaike information criterion (AIC) factor

2. To find the scaled squared error, perform a "seven-tap" channel estimation; where the scaled error is computed by multiplying the error by the modified Akaike information criterion (AIC) factor.

   Scaled_sqerror2 = 7-tap channel estimation error* Akaike Information Criterion (AIC) factor

3. Next, the minimum errors from both the above computations are compared and it is decided whether the channel length will be four tap or seven tap, based on the one which is has the minimum squared error.

   Channel_length = four tap if Scaled_sqerror1 = min (Scaled_sqerror1, Scaled_sqerror2)

   Channel_length = seven tap if Scaled_sqerror2 = min (Scaled_sqerror1, Scaled_sqerror2)

4. A variable is used to indicate the channel length. The channel length is set to "1" or "0" based on whether the current detected channel length is seven or not. From this, an average channel length for channel length seven is derived using an exponential averaging.
5. Then the final channel length is decided by comparing the averaged variable with a threshold.

Once the channel condition is detected then a switching technique can be employed to select the appropriate equalizer according to the channel type detected.

2.9.2 *Protocol Stack of GSM Mobile*

Different protocol layers inside the MS, BTS, BSC, and MSC of a GSM system are shown in Figure 2.17. MS, being the end entity, has all the protocol layers (of ISO), whereas on the network side, the protocol layers are spread among different entities like BTS, BSC, and MSC. During the operation, the protocol layers in the mobile station (MS) (as shown in Figure 2.15) interact with their counterparts, which are spread across those network entities.

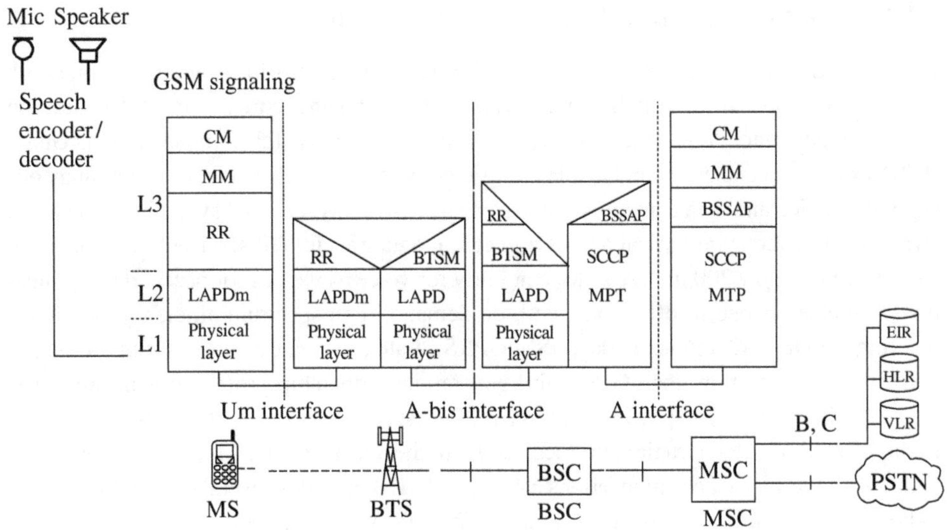

Figure 2.17 GSM protocol layers

At every transit time interval (TTI) (GSM basic TTI=20 ms), the speech encoder/decoder application inside the GSM mobile handset directly provides the user's speech data (traffic data) to the physical layer for processing. Based on the interface, the GSM signaling protocol is assembled into three general layers:

- Physical layer (L1). This uses the channel structures over the air interface based on TDMA and FDMA multiplexing. It is responsible for channel encoding/decoding, interleaving, ciphering, burst forming, and so forth.
- Data link layer (L2). This is a modified version of the link access protocol for the D channel (LAP-D), used in ISDN, and it is called the link access protocol on the Dm channel (LAP-Dm – m stands for "modified"). Its functions are the organization of L3 information into frames, peer-to-peer transmission of signaling data as defined in frame formats, and the establishment of data links on signaling channels.
- Layer 3. This is divided into three sublayers: radio resource management (RR), mobility management (MM), and connection management (CM). Mobility management is responsible for location management and security. It updates the location information of the mobile station as the mobile station changes its location, performs the authentication procedure, assigns the TMSI (temporary mobile system identification) to the mobile station, and controls the attach and detach function.

The connection management (CM) function includes the setup and releases of the circuit switched connections in mobile originating and terminating calls and assistance to the SMS connection.

RR manages the quality of the radio link, assigns the radio channels, performs frequency hopping, performs the handover procedure and power control of mobile station.

2.10 Third-Generation (3G) Cellular Systems

Due to several limitations of second-generation (2G) systems – for example low network capacity, low data-rate support, higher latency, and weaker data security – the work to develop third-generation mobile systems was started by the International Telecommunications Union (ITU) using frequencies around 2 GHz with emphasis on a higher data rate, simultaneous support of voice and data capability, improved speech quality, channel switching and packet switching transfer, symmetrical and asymmetrical data transfer (IP services), low roundtrip packet delay (below 200 ms), seamless mobility for voice as well as for packet data applications, variable bit rate to offer bandwidth on demand, high spectrum efficiency, and interworking with the existing networks (GSM/GPRS). Later, to create a single forum for mobile systems standardization, the 3GPP (third-generation partnership project) organization was formed, which develops specifications for a 3G system based on the Universal Terrestrial Radio Access (UTRA) radio interface, and on the enhanced GSM core network. The Universal Mobile Telecommunications System (UMTS) provides several different terrestrial radio accesses, the most popular of which is WCDMA. In parallel, there was another 3G system development (3GPP2) going on based on IS-95 systems, which focused on the development of cdma2000 and the multicarrier mode of cdma2000.

2.10.1 Overview of UMTS System Architecture

Operators spent plenty of money to deploy the GSM network. For this reason they wanted the GSM network to coexist with the new 3G network, so that they could also do business using the already deployed legacy GSM network. To support this, the UMTS network architecture is built on top of the existing GSM network. The UMTS network architecture is shown in Figure 2.18, which can be broadly divided into three parts:

- User equipment (UE). This interfaces with the user. It has two parts – mobile equipment (ME), which is the single or multimode terminal used for radio communication, and the UMTS subscriber identity module (USIM), a smart card that contains the subscriber identity, subscribed services, and authentication and encryption keys. The UE interfaces with the network via a W-CDMA air interface, which is known as Uu.
- The UMTS Terrestrial Radio Access Network (UTRAN) handles all radio-related functionality and this is specific to UMTS system. It has two main entities. The first is node B, which is equivalent to BTS in GSM/GPRS, and performs the air-interface processing (channel coding, rate adaptation, spreading, synchronization, power control, etc.). It consists of transmitter-receiver and antenna subsystems. The other, the radio network controller (RNC), is equivalent to GSM BSC and is responsible for radio resource management and control of the node Bs, handoff decisions, congestion control, power control, encryption, admission control, protocol conversion, and so forth.
- The core network (CN) is responsible for transport functions such as switching and routing calls and data, tracking users, and so forth. The CN entities are more or less

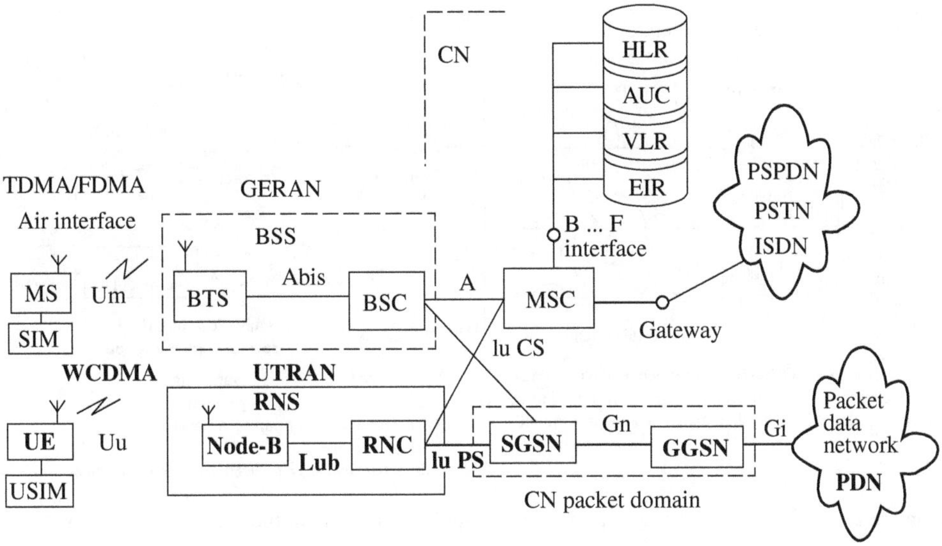

Figure 2.18 UMTS network architecture

remains similar to legacy GSM system. Interface to CN (Iu) are separated into two logical domains: circuit switched domain (IuCS), which handles the circuit switched service, including signaling and the packet switched domain (IuPS), which handles all packet data services.

2.10.2 UMTS Air Interface

The UMTS uses wideband code division multiple access (W-CDMA) radio access technology to offer greater spectral efficiency and capacity to mobile network operators. In CDMA, each data symbol is mapped to several chips, where a chip is a bit in a code word or sequence, which is used to modulate the information signal. In the time domain, each chip bit has a smaller duration than a data symbol – chip rate is more than data rate (one data duration has several chips). As shown in Figure 2.19(a), in the time domain every chip has a smaller duration (~0.26 μs) than data so, in the frequency domain (converted using Fourier transform), each chip signal will take more bandwidth than a data signal. So, when each data symbol is mapped to several chips, then it occupies more bandwidth, which is why this technique is known as spread-spectrum transmission – see Figure 2.19(a). The ratio between the chip rate and the data rate is called the spreading factor (SF), which is also represented as processing gain – the ratio of transmission bandwidth and the original data bandwidth. To support a higher data rate and increased capacity, the chip rate has to be increased, and that requires more bandwidth. For this reason, in the UMTS system, wide-band-CDMA (5 MHz BW) is used.

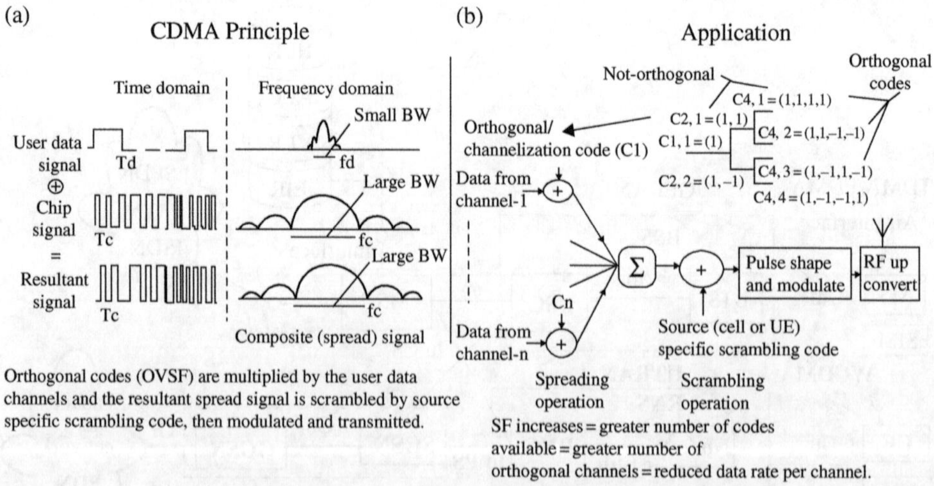

Figure 2.19 (a) CDMA working principle and application for multiple access scheme. (b) OVSF code tree

Generally, three types of codes are used in UMTS: orthogonal codes, scrambling codes and synchronization codes. Orthogonal codes have good orthogonality (e.g. ideally they do not interfere with each other), so these are used for creating several simultaneous channels from the same transmitting source. This code is also known as channelization or OVSF codes. The user data is XORed with orthogonal code sequences or data is converted to NRZ (nonreturn to zero, 0 to +1 and 1 to −1) signal and then multiplied with the OVSF code signal. This operation is known as spreading. From one source (UE or NB), there is a need to transmit/receive many simultaneous channels (traffic, signaling, etc.). Each source has a set of OVSF codes and these are used for separating different channels' data from that source. As OVSF code displays orthogonality, so when different channels' data are multiplied with different OVSF codes and summed together, as shown in Figure 2.19(b), they will not mix up in the transmitter – and they can be easily separated again at the receiver by multiplying the respective OVSF codes for each channels. After the spreading operation, all these spread signals from the transmitting source are added to form a composite signal. On the receiver side, each channel's data can be recovered by multiplying the composite spread signal with the individual orthogonal code used for spreading that channel on the transmitter side. But, these orthogonal codes have very poor correlation, so they exhibit bad orthogonality when these codes are not time aligned, which happens in a multipath channel environment due to the overlapping of code symbols as a result of propagation channel delay spread. So to overcome that issue, a scrambling operation is performed on top of the spreading operation by multiplying the spread signal with PN sequence code (scrambling code), which helps to preserve the orthogonality among the spread signals. Each source (either node B or UE) has a unique PN sequence code or scrambling code, which is used to scramble the composite spread signal from that particular source by multiplying the composite spread data (as explained earlier) with that source's specific scrambling code.

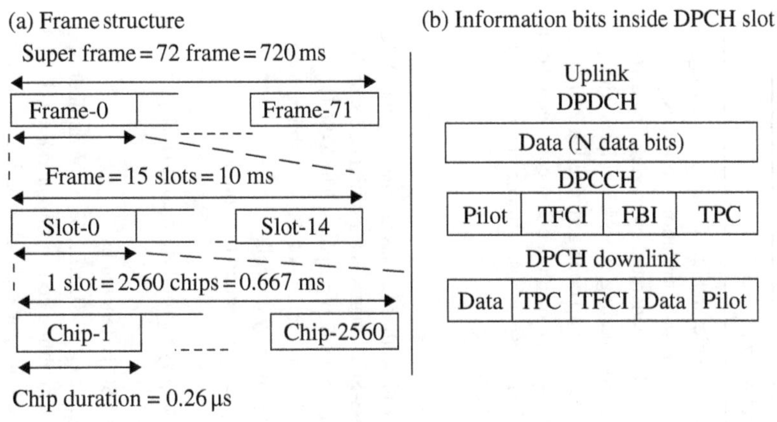

Figure 2.20 (a) UMTS (FDD) radio frame structure. (b) Information bits inside slots of DPCH

A synchronization code is used by the synchronization channels (PSCH and SSCH) for initial synchronization purposes. See the details of three types of codes in table 2.3(a). The UTRA system encompasses two modes: frequency division duplex (FDD) and time division duplex (TDD). In the TDD mode, the same carrier is used for both the uplink and the downlink direction. Each time slot in a TDD frame can be allocated between the uplink and downlink directions. In the FDD mode the uplink and downlink are separated by two different frequency bands and, in each band, there are several carriers, each of bandwidth 5 MHz. As some channels are transmitted continuously and some are not, so, for activation/deactivation of different channels at different times, each frequency carrier is divided into 10 ms radio frames, and each radio frame is dived into 15 time slots, each of 667 μs duration (Figure 2.20(a)). The UTRAN uses a fixed chip rate of 3.84 Mcps – in each slot duration there are always 2560 chips present but the number of information (user data, control) bits will vary based on the SF used. The spreading factor ranges from 256 to 4 in the uplink and from 512 to 4 in the downlink. Again, based on the physical channel type (Table 2.3(b)), in each slot there is a different quantity of user data, and control bits (pilot, transport format indicator (TFCI), feedback indicator (FBI), transmit power control (TPC)) are present – see Figure 2.20(b).

2.10.2.1 Separation of Cells (NBs), UEs, Channels

In the UMTS network, cells (and sectors) are separated by different cell-specific downlink primary scrambling codes. UEs are separated by uplink scrambling codes, which are dynamically assigned to it. Different physical channels from the same source (UE or node B) are separated by different channelization (OVSF) codes. Uplink and downlink paths (channels) are separated by different frequency bands in the FDD mode and by different time slots in the TDD mode. The spreading factor (SF) is derived from OVSF code tree, as shown in Figure 2.19(b), for channel separation. At any given point there could be several channels transmitted from the same source, so several OVSF codes would be used. In the tree, at any

Table 2.3(a) Different codes used in WCDMA system (3GPP TS 25.213)

Code	Channelization code	Scrambling code	Synchronization code
Property	Has good orthogonal properties, but does not have good correlation properties. These are Walsh codes. This is derived from the OVSF code and denoted as $C_{ch,SF,k}$, where *ch* is the channel, *SF* is the spreading factor of the channel and *k* is the code number in the tree	These are PN sequence codes and basically Gold codes (derived from two M sequences). They have very good correlation properties.	These are hierarchical Golay codes. They have good autocorrelation properties. The PSC is generated by modulating a 16-chip code running at 3.84 Mchips/s with another 16-chip code generated at 240 kchips/s. The result is a 256-chip sequence at 3.84 Mchips/s whose autocorrelation function can be found rapidly.
Usage	Downlink: separation of different DL channels to all the connected UEs from the same NB Uplink: separation of DPDCH and DPCCH from the same UE	Downlink: separation of different sectors/cells Uplink: separation of different UEs in a cell/sector	Downlink: these codes are used by the P-SCH and S-SCH channels for initial cell search and synchronization purposes. The S-SCH enables the identity of the scrambling code group used by the cell. They are not used in the uplink.
Length	Uplink: 4–256 Downlink: 4–512	Uplink: can have a long or short code. Long code: the scrambling code period is truncated to 10 ms and repeated every 10 ms. That means it has a period of 38400 chips and repeats every 10 ms. RAKE receiver are used for this. The short code has a period of 256 chips and repeats 150 times in 10 ms. Multiuser detection (MUD) receiver is used for this. Downlink: only long codes, 10 ms 38400 chips	256 chips Primary synchronization code (PSC) denoted acp The SSC is denoted by $ac_{si,k}$, where I = 0, 1,…, 63 is the number of the scrambling code group, and k = 0, 1,…, 14 is the slot number. Each SSC is chosen from a set of 16 different codes with a length of 256.

Number of codes	Number of ch. codes under one scrambling code = spreading factor	Uplink: 2^{24-1} Downlink: a total of (2^{18-1}) number of scrambling codes (numbered from 0 to 262142) can be generated and only 8192 codes are allocated. The 8192 downlink scrambling codes are divided into 512 sets, each set consists of one primary scrambling code and 15 secondary scrambling codes. Total = 512 * (15 + 1) = 8192. Then 512 primary scrambling codes are further divided into 64 scrambling code groups (512 = 64 * 8), each group consisting of 8 primary scrambling codes. The 64 groups have a one-to-one mapping to the sequence of secondary synchronization codes (in SSCH). Each cell is allocated only one primary scrambling code to uniquely identify it.	Every cell across the system (regardless of network operator) transmits the same primary synchronization code in the P-SCH channel. There are 64 sets from which the secondary synchronization channels will be selected
Code family	OVSF	Long codes: 10 ms Gold codes, Short codes: extended S (2) code family. The long scrambling codes are 38 400 chip segments of Gold codes and last one frame of 10 ms. The codes are formed by a bitwise addition of two m-sequences. Short scrambling codes are 256 chips in length.	The primary synchronization code is constructed from a generalized hierarchical Golay sequence. The secondary synchronization codewords require the Hadamard sequence.
Effect on BW	Increases bandwidth	Does not increase transmission bandwidth	Increases bandwidth but predefined

Table 2.3(b) Different channels in the UMTS (FDD) system (Rel-99) and their mapping

Logical channels	Transport channels	Physical channels
Broadcast control channel (BCCH)	Broadcast channel (BCH)	Primary common control physical channel (PCCPCH) (DL ch)
Dedicated control channel (DCCH)	Forward access channel (FACH)	Secondary common control physical channel (SCCPCH) (DL ch)
Paging control channel (PCCH)	Paging Channel (PCH)	
Common traffic channel (CTCH)	Random access channel (RACH)	Physical random access channel (RACH) (UL ch)
Dedicated traffic channel (DTCH)	Common packet channel (CPCH)	Physical CPCH (PCPCH) (UL ch)
Common control channel (CCCH)	Downlink shared channel (DSCH)	Physical DSCH (PDSCH) (DL ch)
	Dedicated channel (DCH)	Dedicated physical data channel (DPDCH) (DL, UL) Dedicated physical control channel (DPCCH) (DL, UL) (in DL DPDCH and DPCCH are time multiplexed in DPCH) Primary synchronization channel (PSCH) (DL) Secondary synchronization channel (SSCH) (DL) Common pilot channel (CPICH) (DL) Acquisition indication channel (AICH) (DL) Paging indication channel (PICH) (DL) CPCH status indication channel (CSICH) (DL)

particular stage, all the derived codes are orthogonal, but in any branch, the parent code and the derived child codes are not orthogonal. So, the OVSF codes will be used or reserved accordingly as they will be used for different broadcast, control, and user-specific channels.

2.10.2.2 UMTS Channel Structure

According to the information carried by the channel, the channels are categorized into three different levels: (i) logical-level channels; (ii) transport-level channels; and (iii) physical-level channels. These are mentioned in Table 2.3(b). More details can be found in [1] and [3].

2.10.3 Physical Channel Transmission

2.10.3.1 Downlink Channel Transmission (From Node B)

The transmission of different downlink channels from a node B is shown in Table 2.3(b). PSCH and SSCH are multiplied PSC and SSC codes (synchronization codes) respectively, and summed then transmitted only for the first 256 chip duration (256 * 0.26 μs = 66.7 μs) at the beginning of every slot (every slot is 2560 chips e.g. 667 μs). These channels are not scrambled as all UEs have to detect the PSCH and SSCH without any prior knowledge of system parameters. The scrambling code employed in UTRA FDD is a 38 400 chip segment of a 2^18 − 1 length Gold code. So, a total of 2^18 − 1 scrambling codes, numbered 0 to 262 142, can be generated; however, not all are used to keep the cell search procedure in UE simple. The allocated downlink scramble codes are divided into 512 sets and each set consists of a primary scrambling code and 15 secondary scrambling codes. So, 512 * 16 (16 = 1 primary + 15 secondary) = 8192 scrambling codes are used. The 512 primary codes can indicate 512 cells uniquely; these 512 are further divided into 64 scrambling code groups to make cell search operation faster. So, each group will contain eight primary scrambling codes (512 = 64 * 8). The 64 code groups have a one-to-one mapping to the sequence of secondary synchronization codes repetition over a radio frame. Each cell is allocated only one primary scrambling code. PCCPCH, PCPICH is always transmitted scrambled using primary scrambling codes.

The common pilot channel (CPICH) is always transmitted with the spreading code $C_{ch,256,0}$ and the channel data is all logical 1. This is scrambled with the cell-specific primary scramble code. BCH transport channel data (containing system parameters) is mapped to PCCPCH and continuously transmitted (constant rate 30 kbps, fixed SF = 256, TTI is fixed to a 20 ms, no TFCI bit) by spreading with fixed OVSF code $C_{ch,256,1}$. The spread PCCPCH channel's data is scrambled using cell specific scrambling code and transmitted only in the remaining (2560 − 256 chips) duration in a slot (see Figure 2.21, first 256 chip duration is

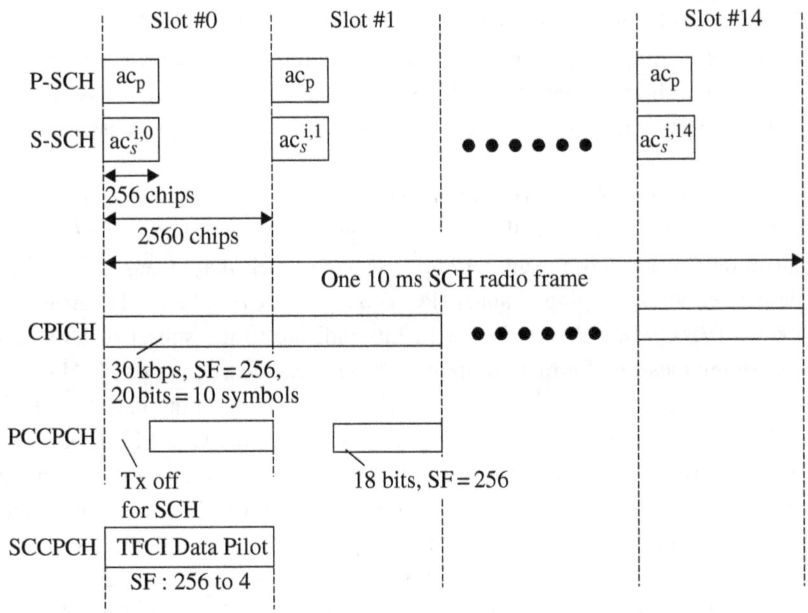

Figure 2.21 Different downlink (DL) physical channels

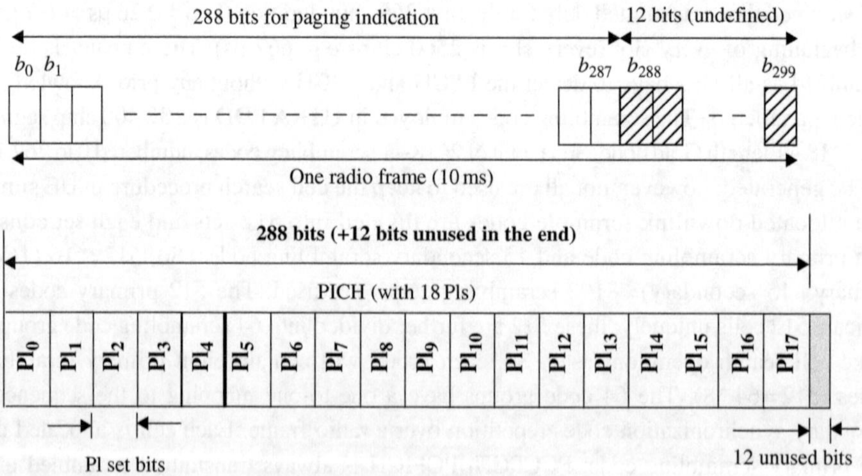

Figure 2.22 Structure of PICH

used for SCH transmission). The SCCPCH carries FACH and PCH data and these are of two types – one includes TFCI and another one does not. This channel's data rate varies from 30 kbps to 1920 kbps as SF can vary from 256 to 4. This channel is only transmitted when required. A downlink dedicated physical channel (downlink DPCH) contains dedicated data generated at Layer 2 and above – the dedicated transport channel (DCH) – and transmitted in time multiplex with control information generated at Layer 1 (known pilot bits, TPC commands, and an optional TFCI). There could be several such channels and each one is spread using a separate OVSF code and scrambled as shown in Figure 2.19(b). Physical downlink shared channel (PDSCH) is a shared channel used to carry DSCH transport channel and supports variable data rates. It is shared among various users using different channelization code.

The UE moves to an URA_PCH state to save power (see Figure 2.29) but remains accessible to the network by periodically checking paging indicator channel (PICH). PICH is also transmitted for a specific paging group (a group of UEs that means group of USIMs) and is associated with a paging channel (PCH) on S-CCPCH. The PICH carries the page indicator bits (PIs), where N number of PI bits indicate to the subset of UEs about the incoming paging message for that group of UEs. The frame structure of PICH is shown in Figure 2.22. Each group of UE is associated with a particular paging indicator (PI). PICH uses SF 256, so, there are 20 * 15 = 300 bits per 10 ms frame. One PICH radio frame of length 10 ms consists of 300 bits (b0, b1, …, b299) and, out of these, 288 bits (b0, b1, …, b287) are used to carry paging indicators (PIs) for different UEs in a paging group. The remaining 12 bits (b288, b289, …, b299) are unused. Then out of 288 bits, N bits together (N = 18, 36, 72, 144) are used to indicate PI for a particular UE. If there are m UE groups and N bits per PI, for which the PICH channel is transmitted, then N * m = 288. If received PI indicates that there is a paging message for that group of UEs, then all UEs in that group need to read the SCCPCH (PCH is mapped to SCCPCH), which appears after three slots after that PICH frame. But, if these bits are set to 0, then no paging is indicated. If a paging message is there, then the UE sends the channel request to the network using the RACH uplink channel. Acquisition Indication channel (AICH) is used to respond RACH and carries acquisition indicator (AI). AI value +1 indicates positive acknowledgement (ack), −1 negative acknowledgement (nack) and 0 Null (transmit with more power until it gets ack or nack). Its frame structure is based on 2 radio frames similar to RACH. So, 15 access indication slots occupy 20 ms.

The transmission mechanism for different downlink channels from Node B is described in Figure 2.23.

2.10.3.2 Uplink Channel Transmission (from UE)

There are two types of uplink dedicated physical channels: (i) the uplink dedicated physical data channel (uplink DPDCH) and (ii) the uplink dedicated physical control channel (uplink DPCCH). These are transmitted in parallel with the I/Q code multiplexed within each radio frame. The uplink DPDCH is used to carry the DCH transport channel and the DPCCH is used to carry control information generated at the physical layer, which includes pilot bits to support channel estimation, transmit power control (TPC) commands, feedback information (FBI), and an optional transport-format combination indicator (TFCI). At any time there may be zero, one, or several (maximum six) uplink DPDCHs and only one uplink DPCCH on each radio link. The uplink transmission scheme is shown in Figure 2.24(a).

The physical random access channel (PRACH) is an uplink channel used by the UE for connection request purposes. The random-access transmission is based on a slotted ALOHA approach with fast acquisition indication. The UE can start the random-access transmission at the beginning of a number of well defined time intervals, denoted as access slots. There are 15 access slots per two frames (20 ms). The random-access transmission consists of one or several preambles with a length of 4096 chips formed by 256 repetitions of a signature with a length of 16 chips (Walsh code), followed by a message with a length of 10 ms or 20 ms. A maximum of 16 signatures are available. The UE decodes the BCH (SIB) of the target cell to find out the cell-specific spreading codes available for preamble and message parts, the signatures and access slots available in the cell, the spreading factor allowed for the message part and the PCCPCH transmit power level. Then, to access the network, the mobile randomly selects the signature and access slots to be used for the RACH burst. The mobile estimates the downlink path loss and calculates the required uplink transmit power to be used for the random access burst. Then a 1 ms preamble is sent with the selected signature and waits for the response in the AICH channel from the network. The terminal decodes the AICH to see whether the base station has detected the preamble. If no AICH is detected, the terminal increases the preamble transmission power by a step given by the network (in a system information message) as a multiple of 1 dB and transmits in the next available access slot. If the AICH is received with the signature S of the PRACH, then the message part is sent. For message part transmission, each slot consists of two parts: a data part to which the

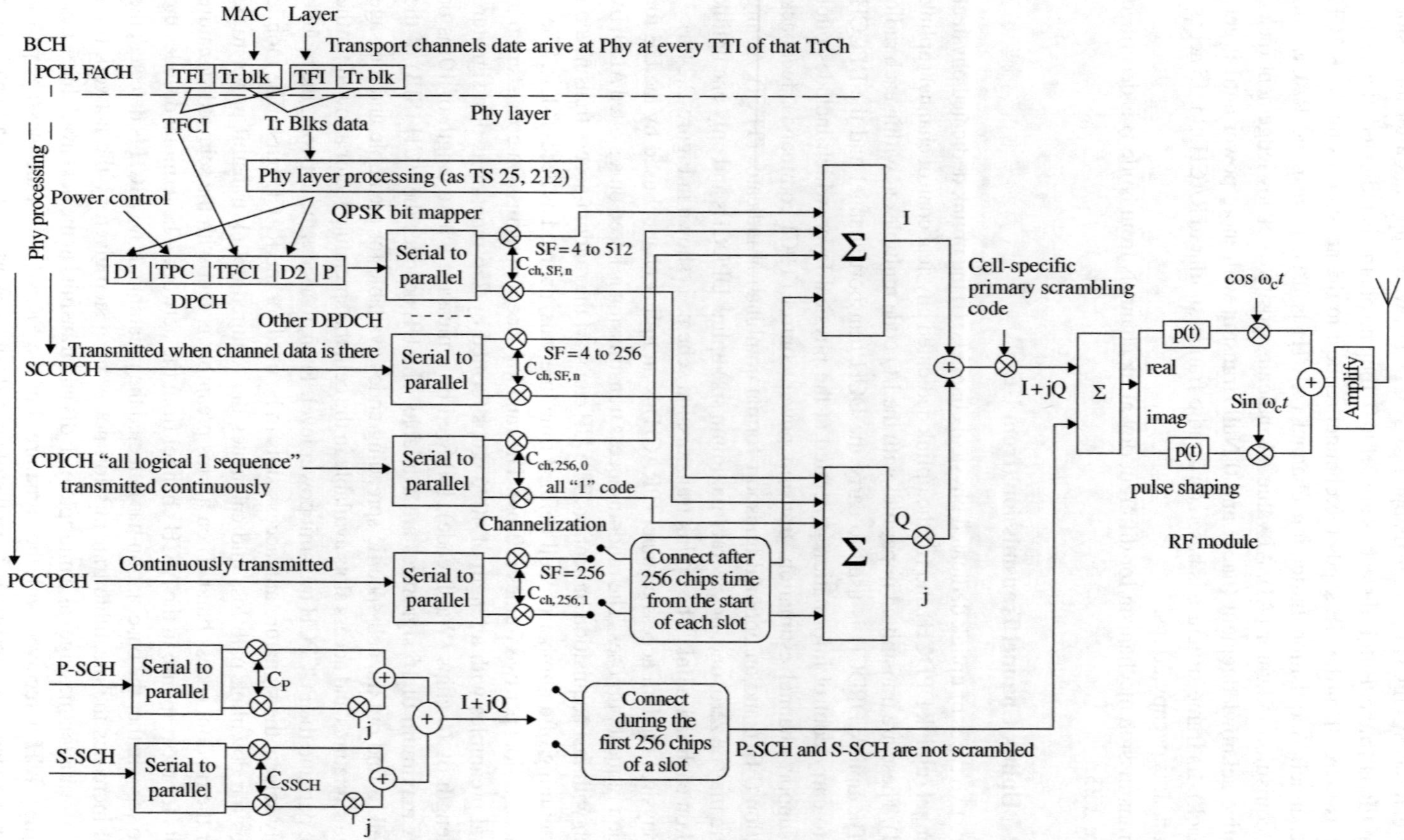

Figure 2.23 Downlink physical channels transmission blocks (when only primary scrambling is used on the network side)

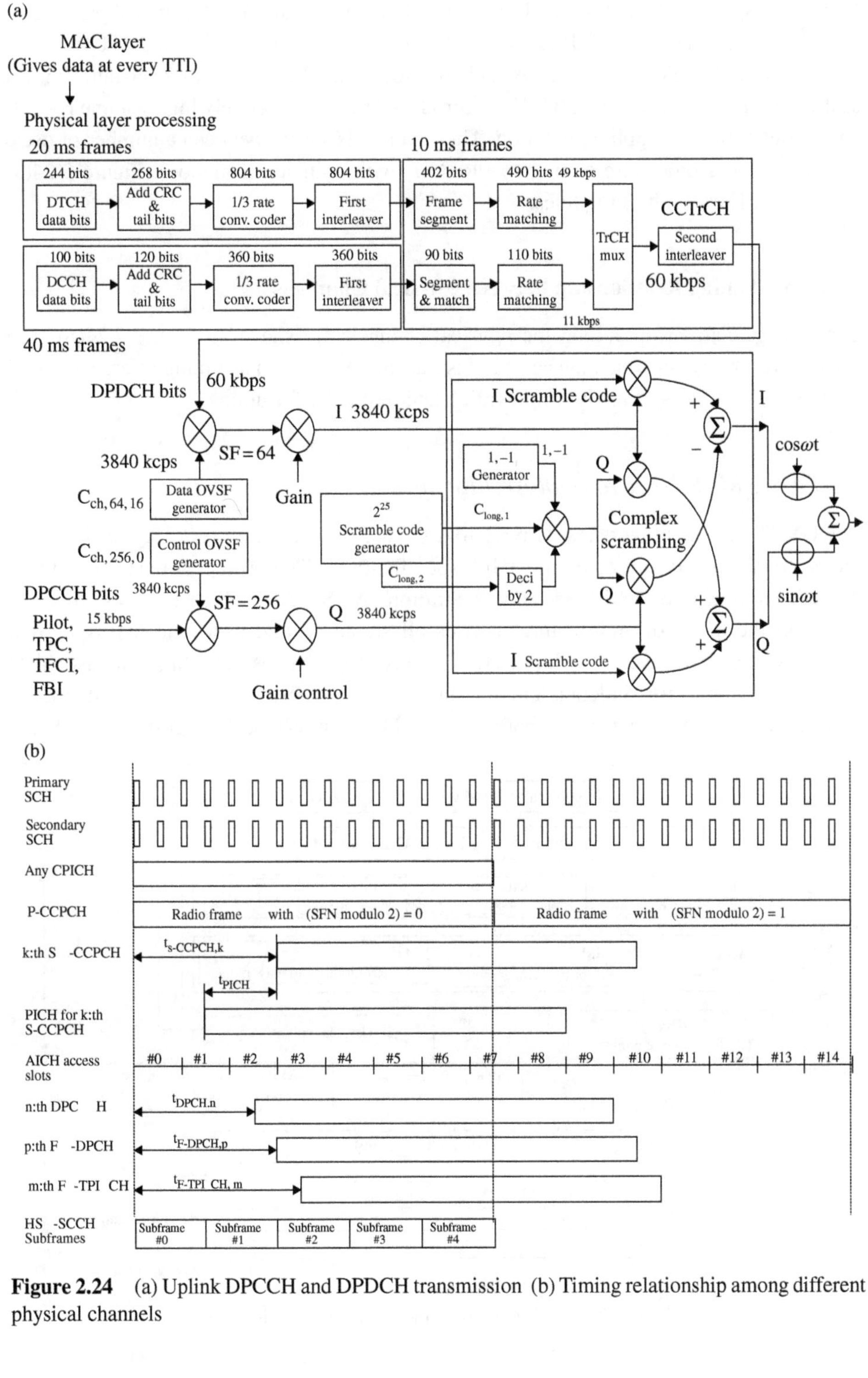

Figure 2.24 (a) Uplink DPCCH and DPDCH transmission (b) Timing relationship among different physical channels

RACH transport channel is mapped (transmitted in the I channel) and a control part (transmitted in the Q channel), which carries layer 1 control information (pilot and TFCI).

Apart from the RACH, like the FACH in the downlink, there is one more channel known as the common packet channel (CPCH), introduced to carry relatively high volume packet-based user data in the uplink direction. This channel is shared between a number of users and uses access procedure similar to PRACH (with collision detection). Figure 2.24(a) shows the UE transmitter modules.

2.10.3.3 Timing Relationship between Physical Channels

The timing relationship among the physical channels are shown in Figure 2.24(b). The P-CCPCH (BCH), on which the cell SFN is transmitted, is used as timing reference for all the physical channels, directly for downlink and indirectly for uplink.

2.10.4 UMTS UE Protocol Architecture

The UMTS protocol layers inside WCDMA UE are shown in Figure 2.25. The design of the protocol stack is guided by the 3GPP specifications. The protocol are separated into Access Stratum (AS) and Non-Access Stratum (NAS). AS (which consists of lower layers in the protocol architecture) carries all signaling and user data messages that relates to the radio access technology, whereas NAS carries signaling and user data messages which are independent of underlying access mechanism that means independent of radio access technology (RAT), air-interface like GSM, UMTS are

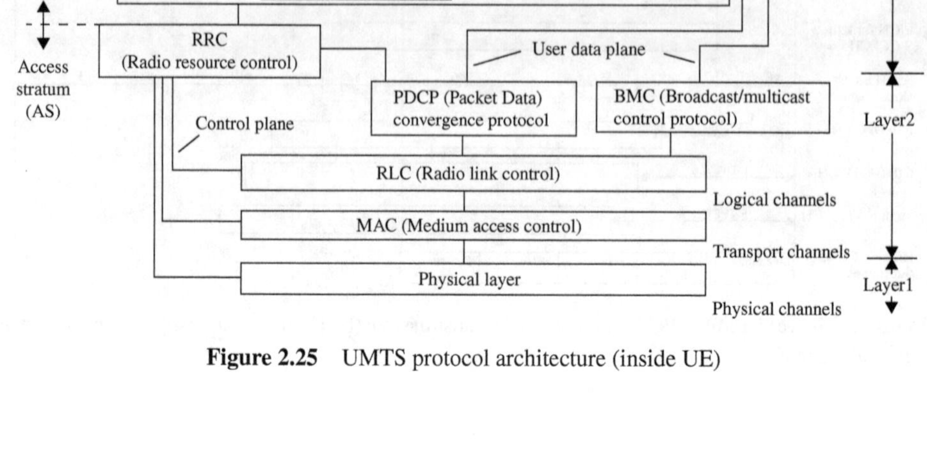

Figure 2.25 UMTS protocol architecture (inside UE)

used. Different protocol layers are briefly described below and more can be found in the 3GPP technical specifications as mentioned below.

- Physical layer (L1). The main functions of the physical layer are : (i) FEC encoding/ decoding and error control of transport channels; (ii) physical layer measurements and indications to higher layers, for example about received signal quality, channel decoding quality; (iii) macrodiversity distribution/combining and soft hand-over execution; (iv) multiplexing of transport channels and demultiplexing of coded composite transport channels; (v) rate matching; (vi) mapping of coded composite transport channels on physical channels; (vii) Modulation and spreading/demodulation and despreading of physical channels; (viii) frequency and time synchronization; (ix) closed-loop power control (x) Power weighting and combining of physical channels (xi) RF processing. (Refer to 3GPP TS 25.211, 25.212, 25.213, 25.214.)
- MAC Sublayer (L2). The main functions of MAC sublayer (part of L2) are: (i) mapping between logical channels and transport channels; (ii) selection of the appropriate transport formats for each transport channel depending on instantaneous source rate; (iii) priority handling between data flows of a UE; (iv) multiplexing/demultiplexing of higher layer PDUs into/from transport blocks delivered to/from the physical layer on common transport channels; (v) multiplexing/demultiplexing of the higher layer PDUs into/from transport block sets delivered to/from the physical layer on dedicated transport channels; (vi) traffic volume monitoring; (vii) maintenance of a MAC signaling connection between peer MAC entities; (viii) dynamic transport channel type switching; (ix) ciphering (in transparent RLC mode). (Refer to 3GPP TS 25.321.)
- RLC sublayer (L2). The RLC sublayer performs several functions to deliver the layer 2 services. (i) Connection control – this function performs the establishment, release, and maintenance of a RLC connection (ii) Segmentation and reassembly. This function performs segmentation/reassembly of variable-length higher layer PDUs into/from smaller RLC payload units (PUs). One RLC PDU carries one PU, except in the case where header compression is applied, where there are several RLC PUs. The size of the smallest retransmission unit is determined by the smallest possible bit rate. The RLC PDU size is adjustable to the actual set of transport formats. (iii) Header compression. This feature compresses several payload units into one RLC PDU and is referred to as RLC header compression. RLC header compression should be applied for an acknowledged data transfer service. Its applicability is negotiable between UTRAN and UE. (iv) Concatenation. If the content of an RLC SDU does not fill an integer number of RLC PUs, the first segment of the next RLC SDU is put into the RLC PU in concatenation with the last segment of the previous RLC SDU. (v) Padding. When concatenation is not applicable and the remaining data to be transmitted does not fill an entire RLC PDU of a given size, the remainder of the data field is filled with padding bits. (vi) Transfer of user data. This function is used for the conveyance of data between users of RLC services. The RLC supports acknowledged, unacknowledged, and transparent data transfer. QoS setting controls the transfer of user data. (vii) Error correction. This function provide

error correction by retransmission, for example Selective Repeat, Go Back N, or a Stop-and-Wait ARQ, in acknowledged data transfer mode. (viii) In-sequence delivery of higher layer PDUs. This function preserves the order of higher layer PDUs that were submitted for transfer by RLC using the acknowledged data transfer service. If this function is not used, out-of-sequence delivery is provided. (ix) Duplicate detection. This function detects received RLC PDUs that are duplicated and ensures the resultant higher layer PDU is delivered only once to the higher layer. (x) Flow control. This function allows an RLC receiver to control the rate at which the peer RLC transmitting entity sends information. (xi) Sequence numbering check (unacknowledged data transfer mode). This function guarantees the integrity of reassembled PDUs and provides a mechanism for the detection of corrupted RLC SDUs through checking the sequence number in RLC PDUs when they are reassembled into a RLC SDU. A corrupted RLC SDU is discarded. (xii) Protocol error detection and recovery. This function detects, and attempts to recover from, errors in the operation of the RLC protocol. (xiii) Ciphering. (Refer to 3G TS 25.322.)

- Packet data convergence protocol (PDCP) (L2). This uses the services provided by the RLC sublayer. This sublayer is responsible for header compression and the decompression of IP data streams, transfer of user data, maintenance of PDCP sequence numbers, and so forth. (Refer to TS 25.323.)
- Broadcast/multicast control (BMC) (L2). This protocol adapts broadcast and multicast services on the radio interface. It is responsible for scheduling of BMC messages, transmission of BMC messages to UE, and delivery of cell broadcast messages to the upper layer. (Refer to TS 25.324.)
- RRC sublayer (L3). The radio resource control (RRC) layer handles the control plane signaling of Layer 3 between the UEs and the UTRAN. The RRC perform several functions. (i) Reception of broadcast information provided by the nonaccess stratum (core network). (ii) Reception of broadcast information related to the access stratum. (iii) Establishment, maintenance, and release of an RRC connection between the UE and UTRAN. (iv) Establishment, reconfiguration, and release of radio-access bearers. (v) Assignment, reconfiguration and release of radio resources for the RRC connection. (vi) RRC connection mobility functions. The RRC layer performs evaluations, makes decisions and executes actions related to RRC connection mobility during an established RRC connection, such as handover, cell reselection, and cell/paging area update procedures. These functions shall be based on measurements from the lower layers. (vii) Paging/notification. The RRC layer shall handle broadcast paging information from the UTRAN addressed to the UE. The RRC layer shall also handle paging during an established RRC connection. (viii) Routing of higher layer PDUs (ix) Control of requested QoS (x) UE measurement reporting and control of the reporting (xi) Outer loop power control. The RRC layer shall control setting of the target of the closed loop power control. (xii) Control of ciphering. The RRC layer shall provide procedures for setting of ciphering (on/off) between the UE and UTRAN. (xiii) Initial cell selection and reselection in idle mode. The RRC shall select the most suitable cell based on

idle mode measurements and cell selection criteria. (xiv) Congestion control. The RRC manages the internal data buffer during information transfer. (Refer to 3GPP TS 25.331.)

- Call control (CC). This is one of the protocols in the communication management (CM) sublayer. Every UE supports the call-control protocol. If a UE does not support any bearer capability at all, then it responds to a SETUP message with a RELEASE COMPLETE message. In the call control protocol, it is possible to define more than one CC entity. Each CC entity is independent from the others and communicates with the correspondent peer entity using its own MM connection. Different CC entities use different transaction identifiers. The elementary procedures can be grouped into the following classes: (i) call establishment procedures. (ii) Call clearing procedures. (iii) Call information-phase procedures. (iv) Miscellaneous procedures. The terms "mobile originating" or "mobile originated" (MO) are used to describe a call initiated by the UE. The terms "mobile terminating" or "mobile terminated" (MT) are used to describe a call initiated by the UTRAN.

  Call establishment procedures. Establishment of a call is initiated by request from a higher layer in either the UE or UTRAN. It shall consist of: (i) the establishment of a CC connection between the UE and UTRAN. (ii) The activation of the codec or interworking function. The UE shall support the following types of call establishment – (i) Mobile originating call establishment. There are two kinds of a mobile originating call, a basic call and an emergency call. The request to establish a MM connection shall contain a parameter to specify whether the call is a basic or an emergency call. (ii) Mobile terminating call establishment. It is possible to terminate a call from a UE, provided that a MM connection is already established by the UTRAN.
- Session management (SM). The session management (SM) provides management services to the GPRS point-to-point data services at the UE radio interface. The SM supports PDP context handling of the UE. The SM procedures for identified access are performed only if a GMM context has been established between UE and UTRAN. For anonymous access, the SM procedures are performed without a GMM context being established. The SM procedures are: (i) PDP context activation – this procedure is used to establish a PDP context between UE and UTRAN for specific QoS on a specific NSAPI. The PDP context is initiated by the UE or upon request, by the network. (ii) PDP context modification. This procedure is used to change the QoS negotiated during the PDP context activation procedure or a previously performed PDP context-modification procedure. The network initiates the procedure at any time when a PDP context is active. (iii) PDP context deactivation. This procedure is used to deactivate any existing PDP context between the UE and the network. The context deactivation is initiated by the UE or the network. (iv) Anonymous PDP context activation. This procedure is used to establish a PDP context anonymously between the UE and network for a specific QoS on a specific NSAPI. The procedure is initiated by UE only. (v) Anonymous PDP context deactivation. This procedure is used to deactivate any anonymous PDP context that exists between the UE and the network. The context deactivation is initiated by the UE or network.

Mobility Management (MM)

The main function of the mobility management sublayer shall be to support the mobility of UEs, for example by informing the UTRAN of its present location and providing user identity confidentiality. Other functions of the MM sublayer is to provide connection management services to the different entities of the higher connection management (CM) sublayer. There are two sets of procedures defined for the MM: (i) MM procedures for non-GPRS services, performed by the MM entity of the MM sublayer, and (ii) GMM procedures for GPRS services, performed by the GMM entity and GMM-AA entity of the MM sublayer.

Depending on how they are initiated, there are three types of MM procedures. (i) Common procedures. It is always possible to initiate a MM common procedure. Procedures that belong to this category are: TMSI reallocation, authentication, identification, IMSI detach, MM information procedure. (ii) Specific procedures. It is possible to initiate a MM specific procedure only if no other MM specific procedure is running, or no MM connection exists. The procedures belonging to this type are: normal location updating, periodic updating, the IMSI attach procedure. (iii) MM connection management procedures. These procedures are used to establish, maintain, and release a MM connection between the UE and the UTRAN, over which an entity of the higher CM layer can exchange information with its peer. It is possible to perform a MM connection establishment only if no MM specific procedure is running. It is possible for multiple MM connections to be active at the same time.

- GMM common procedures. There are four types of GMM common procedure: P-TMSI reallocation, GPRS authentication and ciphering, GPRS identification, and GPRS information.
- GMM specific procedures. Two types of GMM specific procedures shall be supported in the UE in the GMM context. One shall be initiated by UE and other shall be initiated by UTRAN.

Universal Subscriber Identity Module (USIM) Interface

The USIM interface provides the transmission protocol for retrieving information elements that are stored in the USIM for 3GPP network operations. The transmission protocol is in accordance with ISO/IEC 7816-3 standards. The USIM interface retrieves the following USIM related information upon request from the UE: (i) administration Information. Mode of operation of USIM – for example, normal, type approval; (ii) USIM service table. Optional services provided by the USIM; (iii) IMUI; (iv) language indication; (v) location information; (vi) cipher key, Kc, and cipher key sequence number; (vii) access control class(es); (viii) forbidden PLMN; (ix) phase identification; (x) ciphering key for GPRS; (xi) GPRS location information; (xii) cell broadcast-related information; (xiii) emergency call codes; (xiv) capability and related parameters; (xv) HPLMN search period; (xvi) BCCH information, list of carrier frequencies to be used for cell selection; (xvii) phone numbers – abbreviated dialing numbers and fixed dialing numbers.

In addition the USIM interface, via directions from the UE, provides the functions to manage and provide storage for the following information: PIN, PIN enabled/disabled indicator, PIN error counter, unblocked PIN, unblocked PIN error counter, data integrity keys, subscriber authentication keys.

Man Machine Interface (MMI)
The MMI interfaces with the user and provides user procedures for call control, physical input and output, such as indications and displayed information. The MMI is positioned above the protocol stack and has interfaces with keypad, display, and USIM. For all the features mentioned, the MMI uses the services of the protocol stack, keypad drivers, and LCD drivers. The following features are supported by the MMI: (i) called number display; (ii) indication of call progress signals; (iii) country/PLMN indication; (iv) country/PLMN selection; (v) basic key pad entry – physical means of entry of 0–9, +, * and #; (vi) service indicator; (vii) call control. SEND and END function keys for call initiation and termination respectively; (viii) call acceptance – the call is accepted when the user presses the SEND function key; (ix) off-hook call initiation; (x) call termination.

2.10.5 UMTS Addressing Mechanism

The addressing mechanism used in UMTS is similar to GSM. Some new addresses are introduced apart from IMSI, TMSI, and packet-TMSI:

- s-RNTI: this radio network temporary identifier (RNTI) is assigned by serving RNC. It uniquely identifies UE within the SRNS. It is 20 bit.
- u-RNTI: this is assigned by the SRNC and uniquely identifies UE within the UTRAN. It is 32-bit UTRAN identity.
- c-RNTI: this is allocated by the controlling RNC when UE accesses a new cell. It is valid only in the cell to which it is allocated.
- UE ID dedicated channels: when UE is in dedicated mode it is addressed explicitly by frequency, channelization code, scrambling code, and so on. So, UE is addressed purely through physical layer but not through u-RNTI or c-RNTI.

2.10.5.1 URA, LA, RA, CGI

In UMTS the highest level in hierarchy is the PLMN. It is defined as a telecom network that provides mobile cellular services. The location area is defined as an area in which a UE may move freely without updating its current location at VLR. If the UE moves out of that area then it sends the location update message. Routing area (RA) is used in the PS domain; it is defined as an area in which UE might move freely without updating its current location at the SGSN.

The UTRAN registration area (URA) is defined as an area covered by several cells. A routing area generally, contains one or more URAs. A URA contains one or more cells.

A URA is used to track the location of a UE within UTRAN. A URA is uniquely identified by URA identity.

Each cell has an identity. To identify cells uniquely across PLMNs, an identity called "cell global identity" (CGI) is defined.

2.10.6 *Radio Links, Radio Bearers, and Signal Radio Bearers*

Radio links (RL) represents the physical links between the UE and some access point in UTRAN. Each radio link is defined by its frequency, channelization, and scrambling code.

The radio bearer (RB) is a layer 2 connection between UE and RNC. It is used for both control signaling and user data. RBs used for signaling is called SRBs.

To provide user plane data transfer we have a higher layer entity, known as the radio access bearer (RAB). The RAB comprises an RB and a connection from the SRNC to the SGSN, which is known as an Iu bearer.

2.11 UMTS UE System Operations

In the WCDMA system, once the UE is powered ON, the UE first does the booting and system initialization. Next it starts up the modem operation.

2.11.1 *Carrier RSSI Scan*

If the UE supports WCDMA, then the protocol layer (higher layer) will indicate to the RF block the supported frequency bands that need to be scanned. The RF block will be tuned to different RF frequencies, one after another, and the carrier RSSI will be measured for different frequencies and carriers will be ranked based on the signal strength. Next, the RF will be tuned to the carrier, which has the highest signal strength (provided that carrier is allowed to camp according to the USIM data), and the cell search operation is performed.

2.11.2 *Cell Search*

The process of searching for the best suitable cell and achieving synchronization with that is known as the cell search process. The goal of the cell search procedure in the W-CDMA subsystem is to find new W-CDMA cell candidates. If the W-CDMA is configured as the active RAT then, upon power ON, the UE tries to find a suitable cell to camp on (for initial cell selection to enter in RRC idle mode), or find another better cell to camp on (during cell reselection in RRC connected mode to find cells for reselection and handover candidates).

The cell search functionality provides the capability to search for new cells in many cases like: initial cell search, PLMN cell search, background cell search, cell search during interfrequency measurements, cell search during intrafrequency measurements in idle, cell search during interfrequency measurements in idle, cell search during CPC measurements, cell search during passive measurements, cell search at blind activation, and so forth.

The cell-search algorithm is divided into three stages: (i) slot boundary detection; (ii) code group identification; (iii) cell / sector primary scrambling code identification.

Figure 2.21 shows the slot-and-frame structure of the synchronization channels used in cell search. The primary-synchronization channel (P-SCH), secondary-synchronization channel (S-SCH) and the common pilot channel (CPICH) are used for the cell search procedure. Each slot contains 2560 number of chip sequences. The P-SCH and S-SCH are transmitted simultaneously and transmitted all the time only for the 256 chips duration at the beginning of each slots. That means it occupies only 10% of each slot at the beginning. One frame is 15 slots.

Slot Synchronization (CS1)

The same P-SCH sequence is used by all the node Bs and same sequence is transmitted in every slot. So, the P-SCH sequence is identical in all slots and in all WCDMA cells. As the same sequence is used by all the transmitting stations, only one matched filter is sufficient to detect the slot boundary value. So, at the first stage (CS1), the circuit only detects the slot boundary. Generally, correlation with the locally stored P-SCH sequence, of 256 chips length, will give the start position of slot.

Conventional detection of the slot boundary entails:

1. Correlating the received data over 256 chips with the PSC.
2. Then performing this correlation over N_t slots, which is set to 15 slots (=1 frame).
3. Then accumulating (integrating and sum over the symbol period) all the N_t correlation values.
4. Finally, selecting the hypothesis that corresponds to the maximum correlation value.

The simplest method of detecting P-SCH is to use matched filter. There are some computationally efficient methods like, Hierarchical matched filter, efficient Golay correlator etc. are used those exploits the special characteristics of the P-SCH.

Frame Synchronization and Code Group Identification (CS2)

The S-SCH sequences vary slot by slot, based on 16 varieties of SSC sequences. There are 16 SSC sequences available for 15 slot positions in a frame. So, out of 16 S-SCH sequences, 15 sequences are selected for each different code groups and these 15-sequence numbers are arranged according to the code group number (#0 to #64). Then these 15 SSC sequences are placed in 15 slots of a frame for transmission. As shown in Table 2.4, to create 64 different scrambling code groups (in each code group there are eight cells), the S-SCH sequence numbers (0 to 16) are arranged in different ways in 15 available slots over a frame. For example, for Group 0, the S-SCH sequences transmitted over slot#0 to slot#14 are: 1,1,2,8,9,10,15,8,10,16,2,7,15,7,16. Knowing these sequences over a frame, UE can easily detect the group number.

Table 2.4 Scrambling code groups (refer to 3GPP TS 25.213)

Scrambling Code Group	Slot number														
	#0	#1	#2	#3	#4	#5	#6	#7	#8	#9	#10	#11	#12	#13	#14
Group 0	1	1	2	8	9	10	15	8	10	16	2	7	15	7	16
Group 1	1	1	5	16	7	3	14	16	3	10	5	12	14	12	10
_	_	_	_	_	_	_	_	_	_	_	_	_	_	_	_
Group 62	9	11	12	15	12	9	13	13	11	14	10	16	15	14	16
Group 63	9	12	10	15	13	14	9	14	15	11	11	13	12	16	10

The UE uses the SCH's secondary synchronization code (SSC) to achieve frame synchronization and identify the code group of the cell found in stage 1. This is done by correlating the received signal with all possible SSC sequences (out of 16 possible sequences each of length 256 chips) and identifying the maximum correlation value. Since the cyclic shifts of the sequences are unique, the code group, as well as the frame synchronization, is determined.

Scrambling Code Identification (CS3)

In each code group there are eight cells (primary scramble code). The UE needs to identify the cell (each cell has unique primary scrambling code). The UE knows the code group; now it needs to find out the cell that means the primary scrambling code. The scrambling code is identified by correlating the symbols in the received CPICH channel (CPICH data is preknown and all logical "1"s) with all possible eight scrambling codes in that identified scrambling code group.

Now, as cell the primary scrambling code is known, UE is ready to descramble and despread the primary CCPCH (the spreading code of PCCPCH is fixed and preknown to the system), which is mapped from BCH transport channel and contains the system and cell specific broadcast information (SIB).

2.11.3 System Information Reception

The UE should read the system information (SI) transmitted over the BCH (normally through PCCPCH or SCCPCH in case of DRAC) after the first cell search or periodically afterwards. The UE needs to locate and read the SI prior to starting any radio connection to UTRAN. For BCH, 20 ms fixed TTI is used (BCH transport block is 246 bits fixed). The system information block (SIB) forms system information message (RRC PDU) that makes BCH transport block and it is divided into two frames (two SFNs) as shown in Figure 2.26(a). The cell SFN counts the radio frames from 0 to 4096 (SFN spans over total 12 bits) and is used for scheduling SIBs. System information is organized with a treelike hierarchy and the master information block (MIB) contains the scheduling information of SIBs directly or other scheduling blocks (which again contains scheduling information for SIBs). The

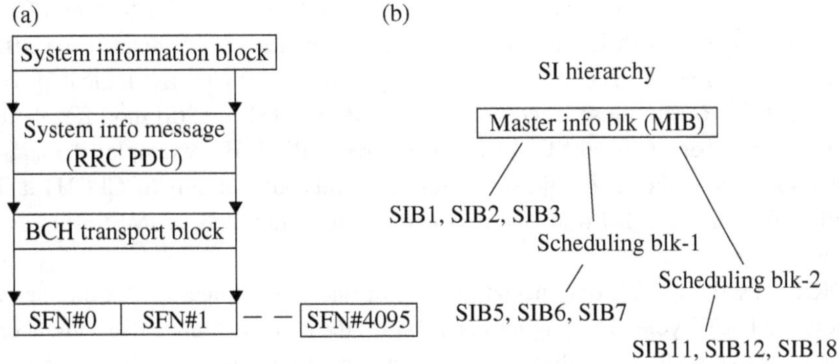

Figure 2.26 Information structure of SIBs – (a) SIB transmitted in two frames. (b) SI hierarchy

MIB is repeated over eight radio frames. The UE first needs to locate the MIB and find out scheduling blocks (see Figure 2.26(b)). Each SIB is scheduled independently to allow different rate for SIBs transmission. The value tag inside a MIB or scheduling blocks indicates whether the corresponding SIB information has changed or not. The UE does not need to read all SIBs constantly so it can employ DRX (discontinuous reception) procedures. Also, a paging type 1 message contains a value tag for MIB that indicates the change about MIB and SIBs information. That helps to prolong UE sleep duration because now UE can read value tag when it wakes up for paging reception and based on that it decides whether to read any further SIB information or not. (Please refer to technical standard 3GPP TS 25.331 for more details.)

2.11.4 Paging Reception and DRX

Most of the wireless mobile networks (including GSM) employ discontinuous reception (DRX) to conserve the battery power of UEs. DRX allows an idle UE (when the UE has nothing to transmit) to power off the radio receiver for a predefined period (called the DRX cycle t_D) instead of continuously listening to the radio channel in the downlink. DRX allows UE to move to sleep mode (where it can shut down many of its functions to save power). The longer the DRX cycle length, the longer the UE is in sleep state. For UMTS, it is defined by DRX cycle-length coefficient (k) as: DRX cycle length $= 2^k$ frames for FDD mode.

The k value might change based on the current UE state. When UE is in DRX (sleep) mode, it has to periodically wake up and read incoming paging message from network (CN or UTRAN). The paging information is transmitted to select UEs in idle, CELL_PCH or URA_PCH state using PCCH channel on an appropriate paging occasion by transmitting a PAGING TYPE 1 message. The CN may request paging or UTRAN may initiate paging for UEs in CELL_PCH or URA_PCH state to trigger a cell update procedure, for UEs in idle mode, CELL_PCH and URA_PCH state to trigger reading of updated system information.

First, UE needs to find out which of the SCCPCHs it has to use for reading PCH. In a cell, a single or several PCHs may be established, which is communicated via SIB 5 of BCH. For each defined PCH there is one uniquely associated PICH, also indicated. The UE selects a SCCPCH from those listed in SIB 5 based on IMSI as IMSI mod K, where K is equal to the number of listed SCCPCHs that carry a PCH. For example, if there are 4 SCCPCHs available, then using IMSI relation, UE finds out that only SCCPCH1 it should use. When there is no IMSI (the USIM is not inserted) then default IMSI=0 is used for emergency calls.

If DRX is used, then UE only needs to monitor one page indicator (PI), on one paging occasion per DRX cycle. The paging occasion defines which frames (SFN number) UE must monitor to read PICH to check whether any incoming message is there or not.

Paging occasion (in SFN)=(IMSI div K) mod (DRX cycle length)+n*DRX cycle length.

Here, n=0,1,2 …4095 and K is the number of SCCPCHs that carry a PCH. So, if the DRX coefficient K is 6, DRX cycle length will be 64 frames and if the paging occasion is 5, 69 … then, at every 64-frame interval, the UE should wake up and check the 5th, 69th … frames for PI. A PI is used as short indicator that is transmitted on PICH to indicate to a UE that there is a paging message on an associated paging channel carried by SCCPCH. As discussed in section 2.10.3, for FDD mode, the number of PIs per frame (Np) can be 18, 36, 72 or 144. Using the equation, PI=DRX index mod Np, the UE finds out which PIs it should monitor. DRX index=IMSI div 8192. For example, PI is computed as 18. So, now, the UE has to check whether in SFN 5's PICH's PI 18 bits are set or not. If these bits are set then the UE should get ready to read the actual paging message (PCH) that will appear on SCCPCH (here SCCPCH1- as discussed earlier) after a defined offset time (which is 3*2560 number of chips). The paging message can include eight paging records and it also contains a BCCH field to indicate the MIB value tag and modification information.

Similarly, UE in connected mode CELL_DCH or CELL_FACH state can also receive the paging message and UTRN initiates the procedure by transmitting a PAGING TYPE 2 message on the DCCH using AM RLC. (More can be found in TS 25.211, 25.304.)

2.11.5 RRC Connection Establishment

At the start, the UE is in idle mode, then, for a RRC connection request, the UE sends RACH for RRC_Connection_Request with the structure as defined in SIB5. The UE uses the connection frame number (CFN) based on the SFN for the common channel with the relations: CFN=SFN mod 256. One element in the RRC_Connection_Request message is the establishment cause, which is used to inform UTRAN about the nature of the RRC connection required. The network sends the RRC connection setup message. The UE has to listen to the SIB5 message to know the structure of common channels. The UE must discover the SCCPCH that carries FACH for connection set up message reading. The initial UE identity is the identity sent by UE during the RRC connection request message, and in

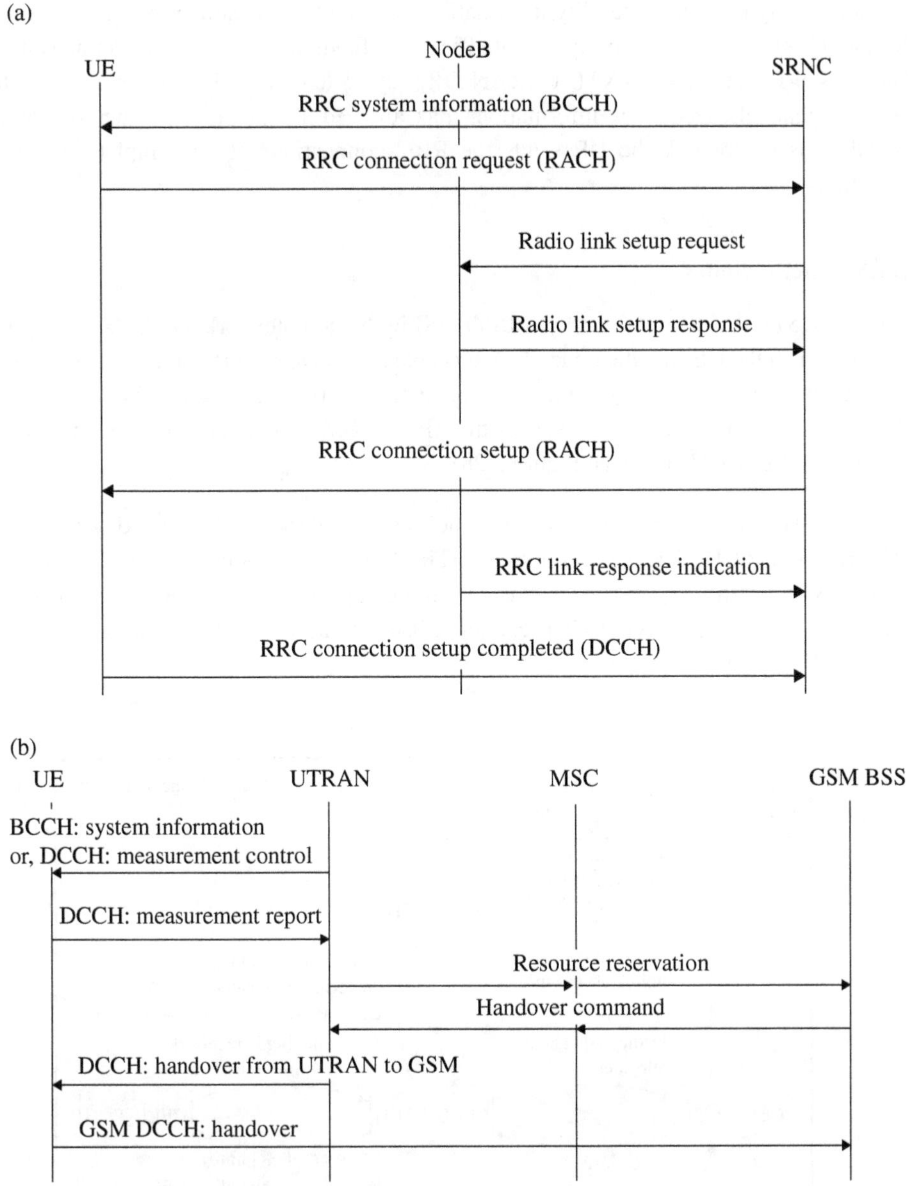

Figure 2.27 (a) Message sequence for RRC connection establishment, (b) Message sequence for inter-RAT handover

the RRC connection setup UTRAN uses the same. This is required for the first exchange of information to allow the network to identify the UE prior to a temporary UTRAN identity allocation. The RRC state indicator field in setup message indicates to which RRC connection state the UE should move (as shown in Figure 2.27). The UE in CELL_FACH state assigned a common channel (SCCPCH), whereas in CELL_DCH state a dedicated physical channel is assigned. The capability information field in set up message indicates whether UE should intimate the capability or not. The other fields in this message are: activation time, new u-RNTI, new c-RNTI, UTRAN DRX cycle length, SRB info set up, UL-DL transport channels, frequency information, max allowed TX power etc. Once the setup procedure is completed, the UE sends the RRC connection set up complete message to UTRAN.

2.11.5.1 RRC States

In idle mode the UE is identified by IMSI/TMSI in the core network, but UTRAN has no information on the UE. In idle mode, the UE is able to receive system and cell broadcasting information. The establishment of a RRC connection is triggered due to a higher layer of UE or paging. In such cases, UE receives the RR connection setup and goes to the CELL_FACH or CELL_DCH state (see Figure 2.28).

- URA_PCH: in this state no dedicated channel is assigned to the UE. In the downlink, the UE receives PICH. There is no uplink. UTRAN is aware about the UE's location at UTRAN registration level. The UE executes the cell update procedure only if the UTARN registration area is changed. DCCH cannot be used in this state; all activities are initiated by PCCH or RACH.

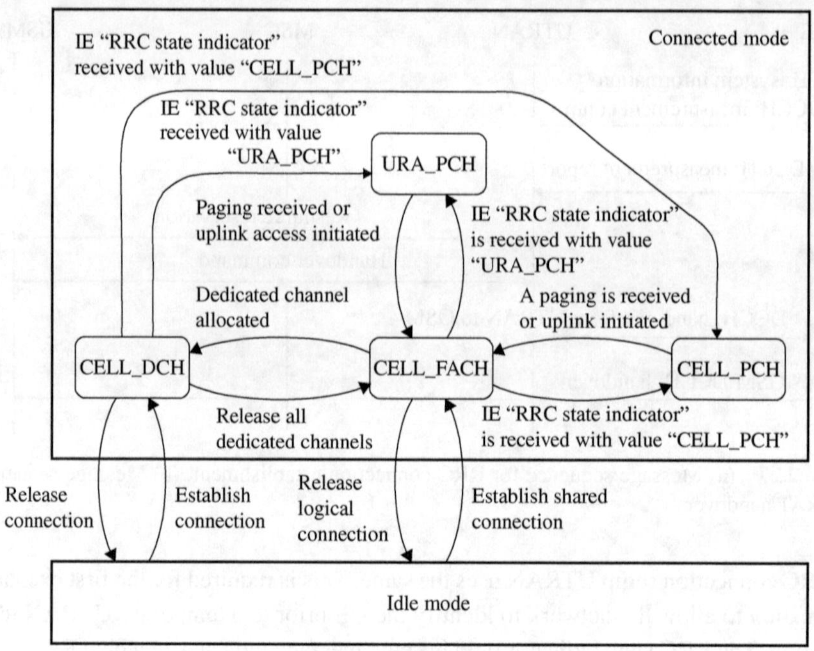

Figure 2.28 RRC connection states

- CELL_PCH has no dedicated channel. In the downlink the UE receives PICH. There is no uplink. UTRAN knows the UE's location up to cell level. The UE can be reached via PCH. The UE listens to BCH. In the event of cell reselection the UE automatically moves to Cell_FACH state.
- CELL_FACH: no DPCH is assigned, the UE receives FACH in downlink, and in UL it can use a common channel (RACH) for data transmission. The UE listens to BCH. Cell reselection is performed. UTRAN knows the UE's location up to cell level.
- CELL_DCH: Entered from idle mode or by establishing a DCH from the Cell_FACH state. DPCH is assigned to the UE and the UE's location up to cell level is known. There is active set updating, and measurement and reporting is on.

2.12 WCDMA UE Transmitter Anatomy

A generic block diagram of a WCDMA UE transmitter is shown in Figure 2.29. The steps followed in the UE transmitter are described below.

- Higher layer data. The MAC layer (L2) generates new transport block every 10 ms (or a multiple of that), fills it with the necessary information, and sends it to the physical layer

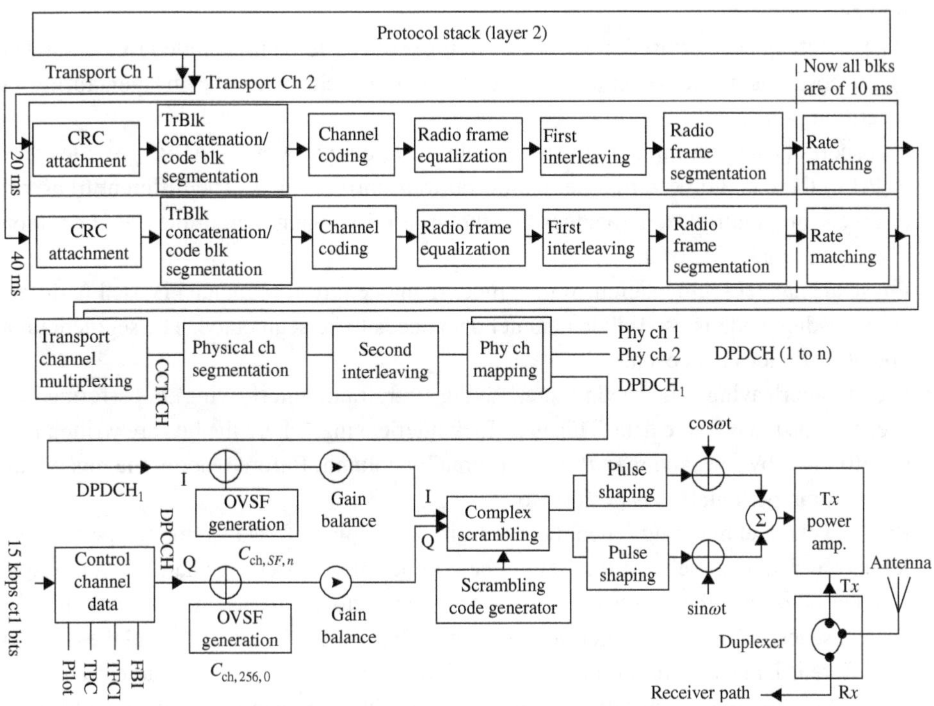

Figure 2.29 WCDMA UE transmission block

(L1). It is possible to send several transport blocks (transport block set) via the same transport channel within one radio frame in parallel.

- CRC attachment. Then CRC bits are inserted for error detection purpose. There are five CRC polynomial lengths in use (0, 8, 12, 16, and 24 bits), and higher layers will indicate which should be used for a given transport channel.
- Transport block concatenation. All transport blocks on a transport channel within a TTI are serially concatenated. If the resulting block size is larger than the maximum size of a code block (depends on the channel coding method to be used for the TrCH), then additional code block segmentation is performed.
- Channel coding. Next, channel coding takes place for helping error correction purpose. Turbo encoding is effective for high-quality data (rate = 1/3, constraint length = 4) and convolution coding is effective for speed and other low-rate data. The scheme used is based on the QoS requirements for the channel.
- Radio frame equalization. Here, data is divided into equal sized blocks when transmitted over more than a single radio frame (10 ms).
- First interleaving. This is used for interframe (among several 10 ms radio frames based on TTI) interleaving. It is used when the delay budget allows for more than 10 ms. The length of interleaver is defined as 20, 40 and 80 ms.
- Radio frame segmentation. If the first interleaving is used, the frame segmentation will distribute data coming from the first interleaving over two, four or eight consecutive frames.
- Rate matching. This is used to match the number of bits to be transmitted to the number of bit positions available on a single frame. This is achieved either by puncturing or repetition of bits.
- Transport channel multiplexing. At any point (TTI boundary when MAC provides data to PHY) there could be zero, one, or several transport channels. The different transport channels are multiplexed together by the transport channel multiplexing operation (TFCI) and make CCTrCH.
- Physical channel segmentation. Where more than one physical channel is used (a different spreading code is used) then channel data needs to be segmented. The segmentation operation divides data evenly.
- Second interleaving. The second interleaving (intraframe interleaving) is performed on every 10 ms radio frame data. This is a block interleaving, where the bits are written into a matrix row by row, and read from it column by column. Before reading the bits out an intercolumn permutation is performed.
- Physical channel mapping. Then the bits from the second interleaver are mapped on different physical channels. At this stage, the number of bits accommodated in a physical channel is exactly the number that the spreading factor of the frame can transmit. As shown, several physical channels are generated (Phy ch-1, Phy ch-2,…) and these could be different $DPDCH_n$. In the uplink, the DPDCH data may vary on a radio frame-by-frame basis. Some channels data originates in the physical layer itself, like the DPCCH, CPICH and SCH. The control data (pilot bits for channel estimation, TPC bits for power

control, TFCI bits and FBI bits) is transmitted through the DPCCH. The DPCCH is transmitted continuously and DPDCH rate and combination information is sent in the TFCI of the DPCCH. After the bit multiplexing, spreading, complex scrambling and data modulation (HPSK) takes place, where the data are mapped to I-phase and Q-phase components. As shown in Figure 2.29, the DPCCH is always mapped to the Q channel and DPDCHs are either mapped to the I or the Q channel, as discussed in section 2.10.3. The symbol of physical channels before spreading is mapped to value +1, 0, −1 (1 mapped to −1, 0 mapped to +1 and 0 value is used to indicate for discontinuous transmission).

- Spreading. The DPCCH is always spread to the chip rate (3.84 Mcps) by the channelization code $C_{ch,256,0}$ (SF = 256), whereas the nth $DPDCH_n$ ($0 <= n <= 6$) is spread to the chip rate by the channelization code $C_{ch,SF,n}$. The spreading operation is explained in Chapters 5 and 13 of reference [1], where each symbol will be multiplied by the chip sequence to spread the data. Also, see Figure 2.24(a) for UE transmission and Figure 2.23 for NB transmission blocks.
- Gain balance. Next, the actual gain value is set for the DPCCH and DPDCH channels.
- Summing. Then the complex valued chip sequence is generated through the summation of complex valued chips. As the values from several spread channels are summed, so each summed value in the resultant sequence might vary from the maximum positive value to the minimum negative value.
- Scrambling. The complex scrambling is performed on the complex data sequence ($\Sigma I + jQ$) with the use of complex scrambling code. The scrambling code used was assigned by the network to the UE. The complex scrambling code is generated by the time-shifting real sequence. The merit of complex scrambling is that it helps to reduce the peak power.
- Pulse shaping. The resulting chip data sequence is restricted to a 5 MHz band by a pulse shaping filter (roll off factor 0.22).
- RF transmission. Then this is converted into an analog signal by multiplying $\cos\omega t$ and $\sin\omega t$ in the I and Q branches. The orthogonally modulated IF signals are converted to RF signals in the 2 GHz band and are subjected to power amplification, and then sent via the antenna.

2.13 WCDMA UE Receiver Anatomy

The typical architecture of a WCDMA UE receiver is shown in Figure 2.30.

2.13.1 Baseband Architecture

The baseband modules of the WCDMA receiver are shown in Figure 2.31.

The mobile phone needs to execute spreading code synchronization, which consists of two processes – acquisition and tracking. Tracking maintains the sync timing within +/– 1 chip of acquisition. The despreader may be a sliding correlator or a matched filter with high-speed synchronization capabilities.

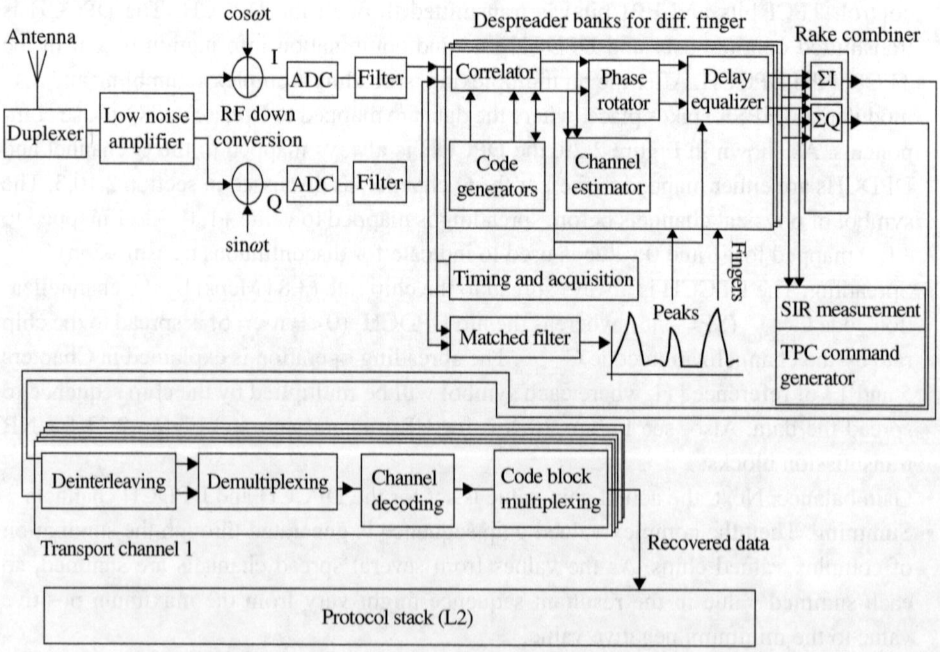

Figure 2.30 WCDMA UE receiver internal architecture

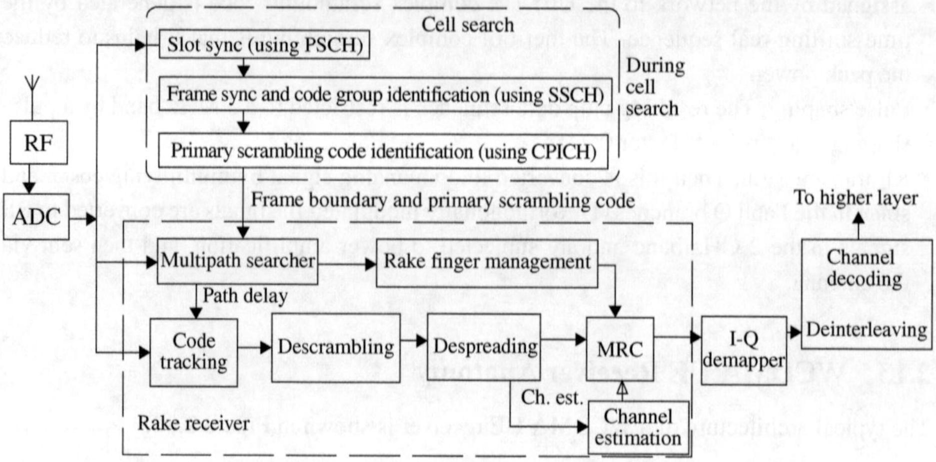

Figure 2.31 WCDMA UE bit detection method

As explained in Chapters 3, 5 and 13 of reference [1], in a typical WCDMA rake receiver the channel estimates are used to combine the various multipath signals. A receiver model is shown in Figure 2.31, which shows how different paths are combined using maximum ratio combining (MRC). Let us consider, a_k is the complex transmitted symbol, p_k is the complex spreading (OVSF) and scrambling code combined, N is the spreading factor, f(t) is the pulse shaping filter and T is the chip duration, then transmitted signal u(t) can be represented as:

$$u(t)=\sum_{k=0}^{k-1} a_k \sum_{n=0}^{N-1} p_k(n) f(t-nT-kNT)$$

If the channel is modeled as a filter with complex taps given by c_j and delays of d_j, where $\{j=0.J\text{-}1\}$ for $J\text{-}1$ different paths and $g(t)$ is the AWGN, then the received signal $y(t)$ at the mobile unit is:

$$y(t)=\sum_{j=0}^{J-1} c_j u(t-d_j)+g(t)$$

At the receiver end the received data is first passed through a matched filter and the received signal $r(t)$ can be represented as:

$$r(t)=\int f^*(\tau-t)y(\tau)d\tau$$

A path searcher estimates the delay for the different multipath signals. Then the received signal is delayed by the amount estimated by the path searcher and multiplied by the scrambling and spreading code as used for transmission. The descrambled and despread data are then summed over one symbol period. Different estimates over different paths of the same symbol are generated using the equation below:

$$x_k(t)=\frac{1}{N}\sum_{m=0}^{N-1} p_k^*(n) r(t+mT-kNT)$$

Then these are combined by the rake receiver with the corresponding channel estimate as:

$$\hat{a}=det\left(\sum_{j=0}^{\hat{J}-1} \hat{c_j}^* x_k(\hat{d}_j)\right)$$

where, C_j is the channel estimates, d_j is the estimated path delays, J is the estimated number of strong paths, *det( )* is a simple decision device, and â is the estimated bit/symbol obtained at the output of the rake receiver.

Generally, as shown in Figures 2.30 and 2.31, the following steps are performed in a WCDMA receiver:

1. **Multipath searcher and finger detector**. This detects the different multipath signals based on the peak and find out the delay for different paths and their relative signal strength. Received signals are multiplied by the scrambling code and delayed versions of the scrambling code. A path searcher determines the delays prior to descrambling. Each delay corresponds to a separate multipath.

2. **Rake finger management**. This tracks and manages the detected fingers.
3. **Descrambling**. This descrambles the data by multiplying the received I,Q vectors with the locally generated version of the respective scrambling code (and delayed version of it, as the delay amount is supplied by the path searcher).
4. **Despreading**. The descrambled data of each path is despread simply by multiplying the descrambled data by the appropriate spreading code.
5. **Channel estimation**. The purpose of the channel estimation is to estimate the channel phase and amplitude for each of the identified paths. Then, once this information is known, it can be used for combining each path of the received signal in maximum ratio combiner (MRC). In the WCDMA system, the channel estimation can be performed using a common pilot channel (CPICH), which is continuously transmitted, or the time-multiplexed pilot bits in a dedicated traffic channel (DPCH has two to eight pilot symbols). But, the advantage of using CPICH for channel estimation is that all the data in the frame can be used in compared to only a few symbols in the DPCCH/DPDCH. Also, since the CPICH is transmitted with higher power than in the traffic channel, so the reception at the mobile is better. One situation in which the time-multiplexed pilot bits become useful is when the mobile is at the cell edge. This is because the dedicated channels are power controlled, whereas the CPICH is not power controlled.
6. **Integration**. The despread data is integrated over one symbol period, giving one complex sample output per QPSK symbol. This process is carried out for all paths to be combined by the rake receiver.
7. **Symbol combining**. The same symbols obtained via different paths are then combined using the corresponding channel information and a combining scheme such as MRC. Generally, there are three basic techniques used for diversity combining: (i) selection diversity – here the best signal is selected from several input diversity signals based on some criteria; (ii) equal-gain diversity combining – the multiple input signals are cophased and them summed with equal gain weight; (iii) maximal ratio combiner – multiple input signals are cophased and then scaled according to their received signal quality and then summed in proportion to their weight. This gives optimum performance, and it is most commonly used in the rake receiver.
8. **I, Q demapper**: The combined output is sent to a simple decision device to decide on the transmitted bits.
9. **Deinterleaving and decoding stage.**

For every slot, the UE receiver:

- estimates the channel from the pilot bits on the DPCH or from the CPICH as explained in step 5. above;
- estimates the SIR from the pilot bits for each slot and accordingly sends the TPC command in the uplink direction to the node B to control its downlink transmission power; Received Signal Code Power (RSCP) = RSSI * (Ec / N0).
- decodes the TPC bit in each slot and adjusts the downlink power of that connection accordingly.

For every 10 ms frame:

- The TFCI information is decoded from the DPCH frame to obtain the bit rate and channel decoding parameters for the DPCH.

For transmission time intervals (TTI, interleaving period) of 10, 20, 40 or 80 ms the DPCH data is decoded. The uplink receiver (node B) typically needs to perform the following tasks when receiving the transmission from UE. The node B receiver starts receiving the frame and dispreading the DPCCH and buffering the DPDCH according to the maximum bit rate, corresponding to the smallest spreading factor.
For every slot:

- the channel is estimated from the pilot bits on the DPCCH;
- the SIR is estimated from the pilot bits for each slot and the TPC command is sent in the downlink direction to the terminal to control its uplink transmission power;
- the TPC bit is decoded in each slot and the downlink power of that connection is adjusted accordingly.

For every second or fourth slot:

- the FBI bits are decoded, if present, over two or four slots and the diversity antenna phases, or phases and amplitudes are adjusted, depending on the transmission diversity mode.

For every 10 ms frame:

- the TFCI information is decoded from the DPCCH frame to obtain the bit rate and channel decoding parameters for the DPDCH.

For transmission time intervals (TTIs, interleaving periods) of 10, 20, 40 or 80 ms:

- the DPDCH data are decoded.

2.14 Evolution of the UMTS System

Due to continuous demand for a higher data rate and lower round trip delay or latency, the UMTS system has evolved continuously. A high-speed downlink packet access (HSDPA) was introduced as part of 3GPP Release 5 to improve downlink data rate and latency. High-speed uplink packet access (HSUPA), also known as enhanced uplink or EUL, was introduced as a part of 3GPP Release 6. HSDPA and HSUPA, are together called "high-speed packet access," (HSPA) (refer to TS 25.848). HSPA is deployed on top of the WCDMA network either on the same carrier or using another carrier. To achieve low delays in the link control, the MAC functionality for the HS-DSCH has been moved to the node-B from

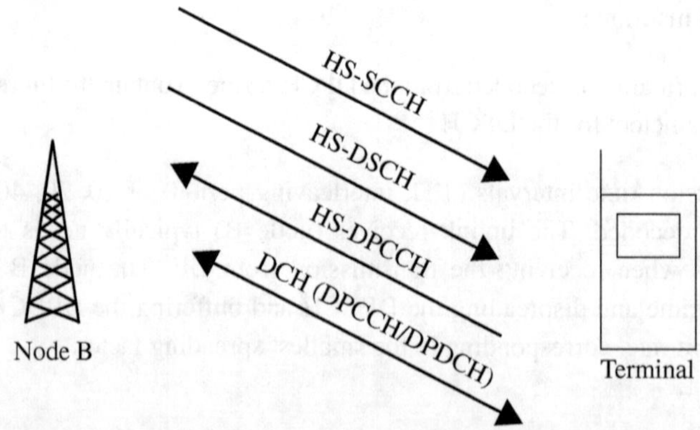

Figure 2.32 Channel requirement for Release 5 HSDPA operation

the RNC. The retransmission is handled in node-B, and there is no separate drift and serving RNC, so no downlink soft handover concept in the HSDPA network architecture.

2.14.1 HSDPA

HSPA and WCDMA can share all the network elements in the core network (CN) and in the radio network (RAN). The channel needed for HSDPA operation is shown in Figure 2.32.

2.14.1.1 Channel Structure

In Release'99 WCDMA specifications, the three different channels that can be used for downlink packet data transmission are: a dedicated channel (DCH), a downlink-shared channel (DSCH), and a forward access channel (FACH). The DSCH is always used together with a DCH. The spreading of DCH in DL does not vary from frame to frame, whereas the DSCH has a dynamically varying SF informed on a 10 ms frame-by-frame basis, with the transport format combination indicator (TFCI) signaling carried on the associated DCH. The DSCH code resources can be shared between several users and the channel may employ either single-code or multicode transmission.

HS-DSCH
In HSDPA, the DSCH is replaced by the high-speed DSCH (HS-DSCH) transport channel for the user data transmission. In release 5, the HSDPA is always operated along with the DCH in parallel, so, the same DCH channel exists. In case of only packet data service, DCH is used for carrying signaling radio bearer, whereas a circuit-switched service runs on the DCH. Release 6 also allows signaling to be carried without the DCH. HS-DSCH applies a channelization code resource shared between users. Dynamic sharing between users is

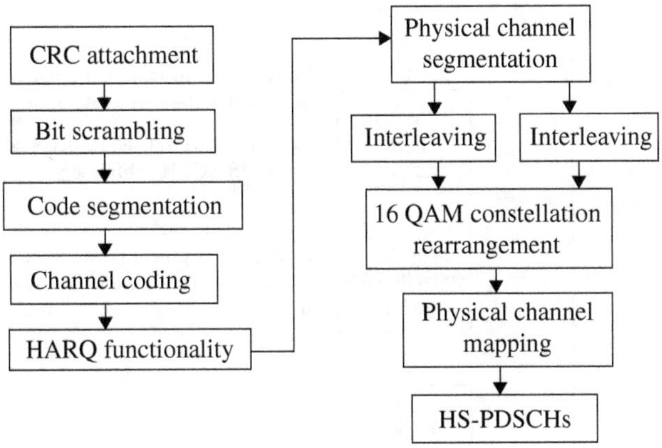

Figure 2.33 HSDSCH processing flow

done in the time domain every 2 ms. This channel supports fast link adaptation, H-ARQ, and scheduling.

It can use a number of multicodes with fixed spreading factor 16. Theoretically, the maximum number of codes available is 16 but the common channels and associated DCHs need some room, so the maximum feasible number is 15 and one code tree branch is reserved for control channels and DCHs. A shorter TTI length of 2 ms is used. User allocation occurs with node B-based scheduling every 2 ms. Data arrives from the MAC every 2 ms for processing (if data needs to be sent), then, as shown in Figure 2.33, first CRC is appended to it. Next, bit scrambling is performed to reduce the possibility of repeating the same symbol, which ensures good signal properties for demodulation. The physical channel segmentation maps the data to the physical channel, to two interleavers. The HS-DSCH transport channel is mapped on the high-speed physical downlink shared channel (HS-PDSCH) in the physical layer. Here, QPSK and 16QAM (in later release 64 QAM) are used for data modulation. The process flow is shown in Figure 2.33. No discontinuous transmission (DTX) is used on the slot level. The HS-PDSCH is either fully transmitted or not transmitted at all during the 2 ms TTI. Adaptive modulation and coding (AMC) is used to select the most suitable modulation technique and coding scheme based on received channel state information from the UE.

HS-SCCH

For the associated signaling needs in the DL, there is the high-speed shared control channel (HS-SCCH). This needs to be available before decoding HS-DSCH, so it has two slots offset from the HS-DSCH. This enables the HS-SCCH to carry time-critical signaling information, which allows the terminal to demodulate the correct codes for HS-DSCH dispreading (as shown in Figure 2.34). There are no pilots or power control bits on the HS-SCCH and, thus, the phase reference is always the same as for the HS-DSCH. The spreading factor of 128 allows 40 bits per slot to be carried. It is divided into

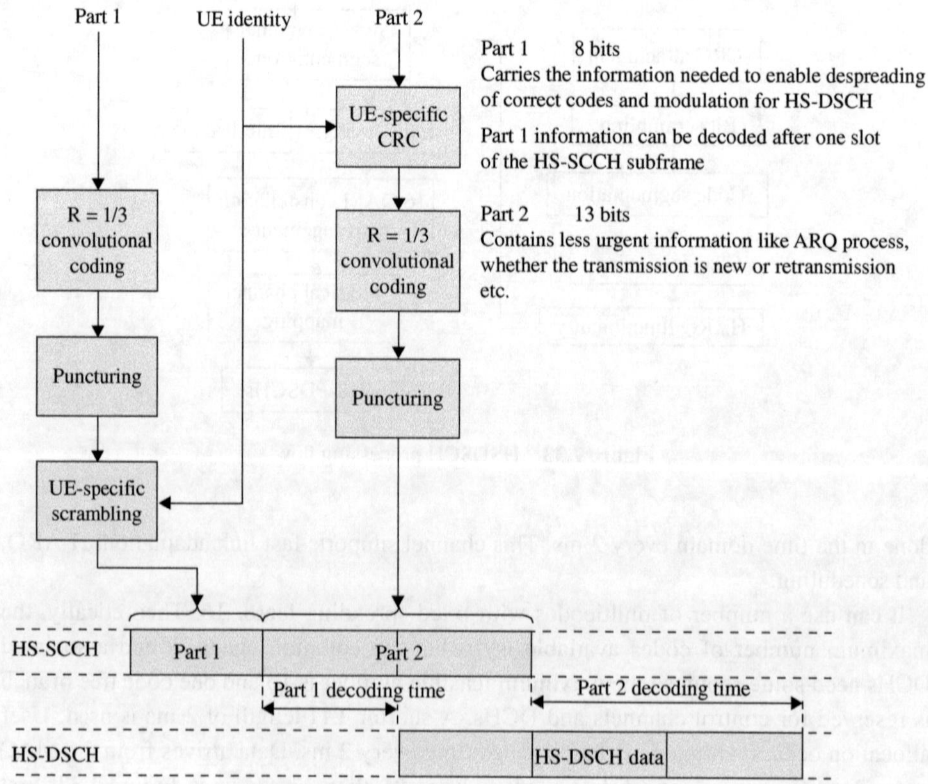

Figure 2.34 HS-SCCH and HS-DSCH slot offset

two parts: the first part carries the information (code, modulation type and so forth) required for HS-DSCH despreading. The second part contains less urgent information, such as the ARQ scheme. The HS-DSCH transport format consists of 7 bits channelization code, 1 bit modulation scheme indication, 6 bits transport block size indication. The HARQ related information consists of 3 bits HARQ process number, 3 bits redundancy version, 1 bit new-data indicator. A UE id (16 bits) identifies the UE for which HS-SCCH information is intended. UE id is not explicitly transmitted rather implicitly included in the CRC calculation. When there is a need to have code multiplexing, then more than one HS-SCCH needs to be included. A single terminal may consider at most four HS-SCCHs. The system itself could configure even more. Here, one-third convolutional coding is used.

HS-DPCCH

For the associated signaling needs in the UL, there is the high-speed dedicated physical control channel (HS-DPCCH). The HS-DPCCH uses the fixed spreading factor of 256 and has a 2 ms/three-slot structure. The first slot is used for HARQ feedback, which informs the base station whether the packet in DL was decoded correctly or not. The coding for the HARQ is simple, with a sequence of "1s" sent for the ACK and "0s" for the NACK. The two remaining slots are used for channel quality information (CQI), which informs the base station scheduler of the data rate the terminal is expected to be able to receive at a given moment. For CQI, (20.5) coding is applied – similar to TFCI coding.

All the Release'99 transport channels are terminated at the RNC. So, the retransmission procedure for the packet data is located in the serving RNC. But, that imposes more delay in the case of repeat transmission schemes. So, in HSDPA, retransmissions can be controlled directly by the node B, leading to faster retransmission and thus shorter delay with packet data operation when retransmissions are needed. To support that new entity MAC-hs is introduced. As scheduling has been moved to the BTS, there is now a change in the overall RRM architecture. The SRNC will still retain control of handovers and is the one that will decide the suitable mapping for quality of service (QoS) parameters. RRC states are the same with HSDPA and HSUPA, as in Release 99.

H-ARQ

Retransmission requests are managed by the Node B instead of the RNC as in Release'99. If the decoding of the initial transmission fails, a retransmission is sent that can be combined with the initial transmission or is self-decodable. The UE does not discard a failed retransmission rather stores it and later combines it with retransmissions to increase the probability of successful decoding. Combining of different transmissions provides improved decoding efficiency while minimizing the need for additional repeat requests over the air interface. The HSDPA supports chase combining (CC) and incremental redundancy (IR).

In CC, if a UE detects that it has received an erroneous packet, it sends an NAQ to node B. Node B transmits the packet again with the same coding scheme. If this is also received in error, this packet is combined with previously received packet in an attempt to recover from errors. Eventually this packet with be received without error or a retransmission limit will be reached and error recovery will be left to higher layers (such as RLC and TCP for TCP-based applications).

IR is similar to CC but retransmitted data is coded with additional redundant information to improve the chances that the packet will be received either without errors or with enough errors removed to allow combining with previous packets to allow error recorrection. In order to better utilize the waiting time between acknowledgements, multiple processes are allowed to run for the same UE using different TTIs.

This mechanism is called N-channel stop-and-wait protocol. If one channel is waiting for an acknowledgement, the remaining N-1 channels continue to transmit. N is up to six for advanced node B implementations.

MAC-hs deals with the functions critical to delay and performance and is located at node B. For non-HSDPA support, node B stations are connected to the RNC, which provides scheduling, coding parameters, and retransmission services to UE devices. To support HSDPA, these parameters are determined based on the instantaneous channel conditions as reported by the UEs. The delays associated with the forwarding of channel data to the RNC for processing coupled with the burden of an RNC having to service multiple node B stations requires that the node B, and not the RNC, perform these services for HSDPA. Support for the HS-DPA feature is optional for mobile terminals.

Adaptive Scheduling

Release 5 moves the scheduling decision to the node B. The base station uses terminal feedback information about channel quality, terminal capabilities, QoS needs, and air-interface resource availability to take the best decision in real time. The proportional fair scheduler is one example of a scheduler that prioritizes applications in the best channel conditions while also prioritizing users that have received lower throughout than other users. Reduction of TTI to 2 ms for HSDPA leads to significantly lower latencies than Release 99.

HSDPA in Release 5 does not support soft handover. The UE continuously monitors all the node Bs in its active set and reports to UTRAN when a change in the best cell occurs. UTRAN then reconfigures the serving HS-DSCH cell using either synchronous or asynchronous reconfigurations. Both internode B and intranode B handovers are supported. There is no separate DRNC here. Data are sent from one serving HS-DSCH cell only. Only serving FS-DSCH cell will send HS-DSCH and HS-SCCH. Only serving HS-DSCH cells needs to decode uplink feedback.

UE Capabilities

There are several different categories of UEs defined for HSDPA that specify the following parameters (3GPP TS 25.211):

- maximum number of HS-DSCH codes that a UE can simultaneously receive;
- minimum inter-TTI time, which is defined as the minimum time between the beginning of two consecutive transmissions to the UE;
- supportf modulations (QPSK, 16QAM);
- maximum number of transport block bits received within an HS-DSCH TTI;
- maximum number of soft-channel bits over all the HARQ processes;

Based on that, the HSDPA UE categories are given in Table 2.5.

2.14.2 HSUPA

As discussed earlier, high-speed uplink packet access (HSUPA), also known as enhanced uplink or EUL, was introduced as a part of 3GPP Release 6 (see 3GPP TR 25.823 for more details). The newly introduced transport channel, E-DCH, supports fast node B-based scheduling, fast physical layer HARQ with incremental redundancy, and, optionally, a shorter 2 ms transmission time interval (TTI). Unlike HSDPA, however, the support of this

Table 2.5 HSDPA UE categories (see 3GPP TS 25.306)

Category	Codes	Inter-TTI	TB size	Modulations	Total number of soft bits	Data rates (mbps)
1	5	3	7300	QPSK/16QAM	19 200	1.2
2	5	3	7300	QPSK/16QAM	28 800	1.2
3	5	2	7300	QPSK/16QAM	28 800	1.8
4	5	2	7300	QPSK/16QAM	38 400	1.8
5	5	1	7300	QPSK/16QAM	57 600	3.6
6	5	1	7300	QPSK/16QAM	67 200	3.6
7	10	1	14600	QPSK/16QAM	115 200	7.2
8	10	1	14600	QPSK/16QAM	134 400	7.2
9	15	1	20432	QPSK/16QAM	172 800	10.2
10	15	1	28776	QPSK/16QAM	172 800	14.4
11	5	2	3650	QPSK	14 400	0.9
12	5	1	3650	QPSK		1.8

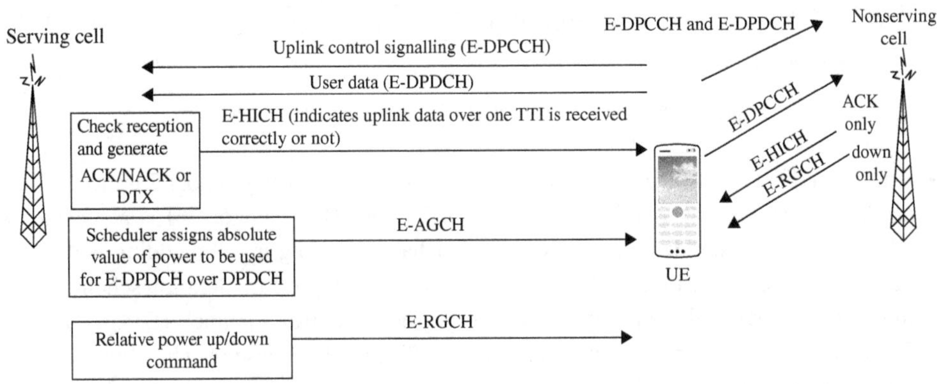

Figure 2.35 Newly introduced HSUPA channels

2 ms TTI in the UE is not mandatory; rather, it depends on the UE capability. It is configured at call setup whether 2 ms TTI or 10 ms TTI is to be used for HSUPA transmission. The advantages of 2 ms TTI is the delay between retransmissions is shorter compared with the 10-ms. But, when UE is closer to the cell edge, then signaling using a 2-ms period will consume lot of transmission power. The uplink scheduler is located in the node B close to the air interface. Task of the uplink scheduler is to control the uplink resources of the UEs in the cell. The scheduling mechanism is based on absolute and relative grants. Node B contains a new medium access control entity called MAC-e, and the RNC contains a new medium access control entity called MAC-es.

The enhanced dedicated channel (E-DCH) is introduced as a new transport channel (see Figure 2.35) for carrying user data on the uplink and which translates into two new uplink physical channels:

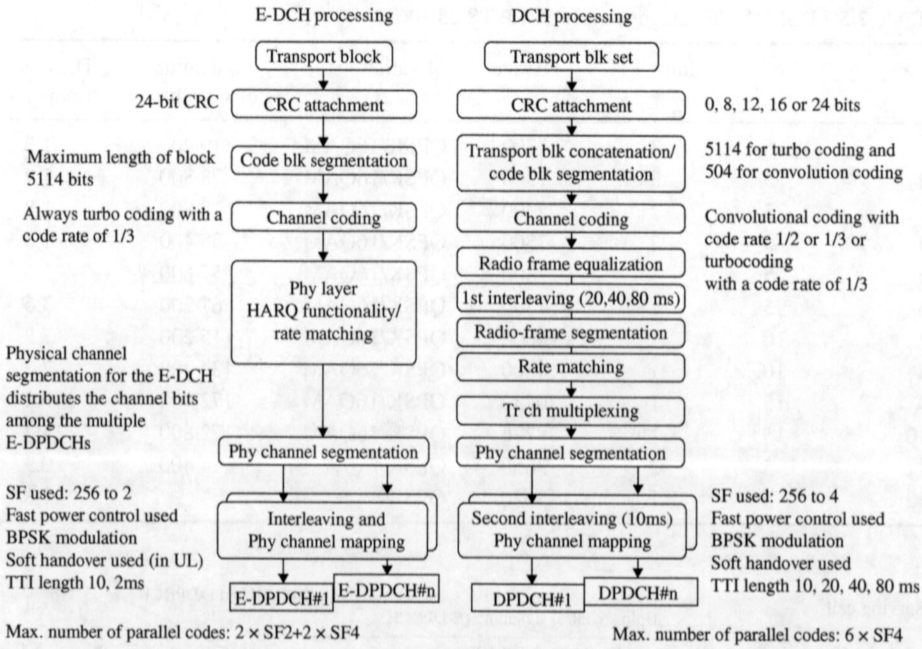

Figure 2.36 E-DCH and DCH physical layer processing steps

- E-DCH dedicated physical data channel (E-DPDCH): spreading factors 2 up to 256 are used. The maximum possible data rate of 5.76 mbps (UE category 6) is achieved by allocating two channelization codes of SF 2. Channel coding is turbo coding with code rate 1/3 and the BPSK modulation scheme is used (no adaptive modulation scheme). It is time aligned with the uplink dedicated physical control channel (DPCCH). The E-DPDCH has a very similar structure to the DPDCH of Release 99.They both support orthogonal variable spreading factors to adjust the number of channel bits to the amount of data actually being transmitted and supports multiple channels transmission in parallel. E-DPDCH supports fast physical layer level HARQ and fast Node B based scheduling which DPDCH does not. The physical layer processing steps involved for E-DPDCH and DPDCH are shown in Figure 2.36.
- E-DCH dedicated physical control channel (E-DPCCH): the E-DPCCH carries control information associated with the E-DPDCH. E-DPCCH exists along with the E-DPDCH. It is transmitted simultaneously using a separate channelization code with spreading factor 256.

This carries all the necessary information needed in order to decode E-DPDCH. It contains – (a) 2-bit RSN (Retransmission Sequence Number) informing the HARQ sequence number of the transport block currently being sent on E-DPDCHs. (Initial TB sent with RSN = 0, first one with RSN = 1, second with RSN = 2, and all subsequent transmissions uses RSN = 3), (b) 7-bit E-TFCI which indicates the transport format of E-DPDCHs from

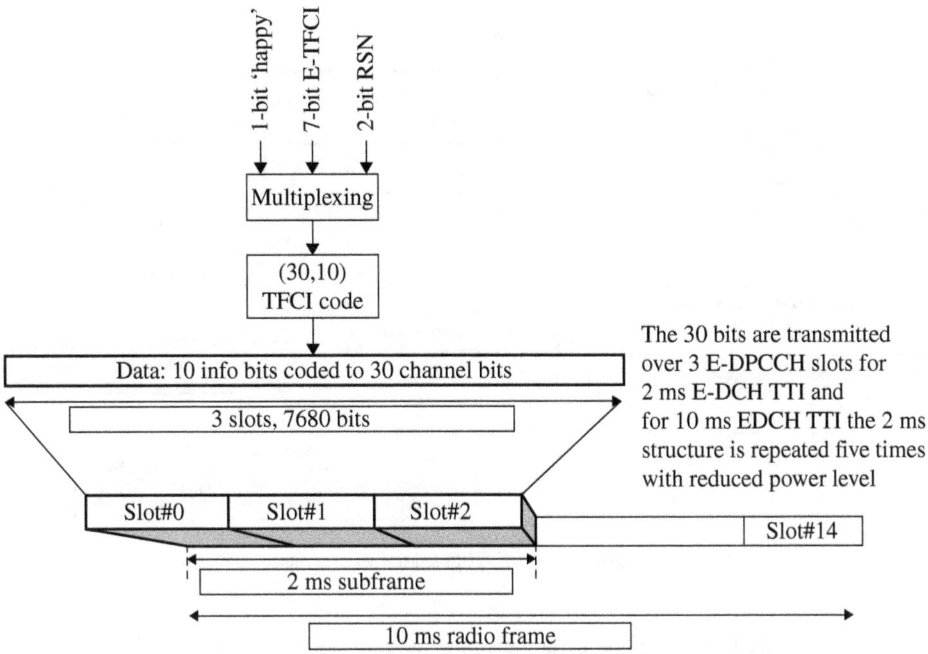

Figure 2.37 E-DPCCH transmission

that receiver can derive how many E-DPDCHs are transmitted in parallel and the spreading factor, (c) 1-bit rate request ('happy bit') indicating whether the UE is happy with the current data rate or the relative power allowed to be used for E-DPDCHs or whether it want to use higher power allocation. Figure 2.37 shows the E-DPDCH channel processing.

In the downlink, three new channels are introduced for control purposes:

- E-AGCH: E-DCH absolute grant channel carrying absolute grants – it is only transmitted from the serving cell. It sets the maximum allowed E-DPDCH power to a specific level indicating the exact power level the E-DPDCH may use in relation to the DPCCH. Thus effectively telling the UE the maximum transmission data rate it may use. It uses a fixed spreading factor of 256 and QPSK modulation. It uses convolutional coding of code rate 1/3. This channel contains- (a) Absolute Grant value, which is 5-bit integer number ranging from 0 to 31 (mapped to specific E-DPDCH/DPCCH power ratio that UE may use), (b) Absolute grant scope, which is 1 bit and can be used to activate/deactivate a particular HARQ process (identified by the E-AGCH timing) or all HARQ processes. It can only be used with a 2-ms E-DCH TTI. In addition to this the E-AGCH uses a primary and a secondary UE-id for identifying the intended receiver and delivering one additional bit of information. A 16-bit CRC is calculated over the 6 information bits and masked with either a primary or a secondary UE-id. Next it is coded and rate-matched to fit in three-slot-long (2-ms) SF 256 channel. If a 10-ms E-DCH TTI is used then the three slots are repeated five times to fill the whole radio frame.

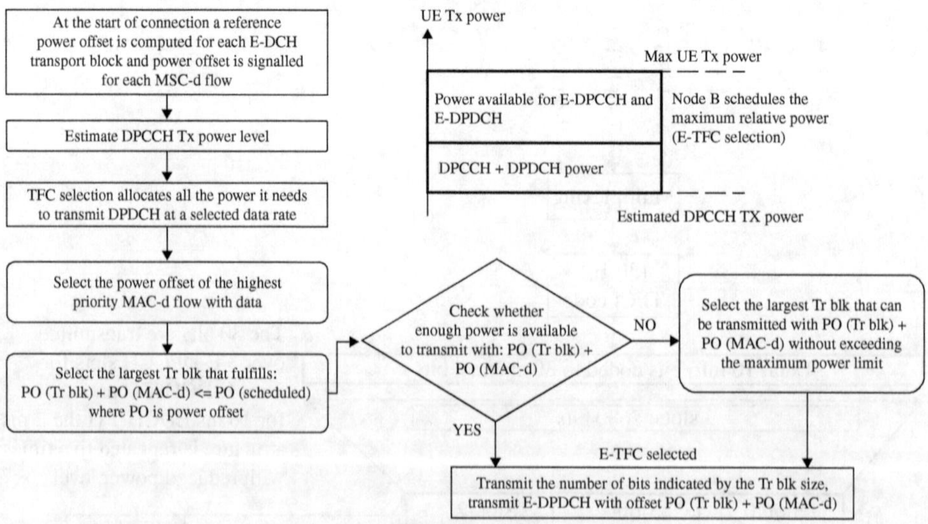

Figure 2.38 E-TFC selection process

- E-RGCH: E-DCH relative grant channel, carrying relative grants to step up/down E-DCH power relative to the current transmission. The cells belonging to the serving E-DCH radio link set of a UE by definition transmit the same E-RGCH content, thus, enable the UE to soft-combine these Channels. The cells not belonging to the E-DCH serving radio link set may only transmit down (DTX) and thus only the serving cell and the other cells belonging to the same E-DCH radio link set can increase the UE's allowed maximum relative transmission power of the data channel.
- E-HICH: E-DCH hybrid ARQ indicator channel, carrying ACK/NACK. If the Node B received the transmitted E-DPDCH correctly- it will respond with a positive acknowledgment (ACK), if received incorrectly it will respond with a negative acknowledgment (NACK). The serving E-DCH radio link (serving E-DCH cell), can transmit both ACKs and NACKs. Node Bs that do not contain the serving E-DCH cell only transmit ACKs and if such a cell does not receive the E-DPDCH TTI correctly, then it does nothing (DTX). The UE will continue retransmitting until at least one cell responds with an ACK. If one of the cell indicates Positive ACK, then UE knows UL data transmission is successful.

E-RGCH and E-HICH are transmitted from radio links that are part of the serving radio link set and from nonserving radio links. Both the E-HICH and the E-RGCH use the same structure, same channelization code and scrambling code. Thus, with length 40 signature sequences, 20 users, each with 1 E-RGCH and 1 E-HICH, can share a single channelization code The power for different users' E-HICH and E-RGCH can be set differently and information is BPSK-modulated with on/off keying.

Node B Scheduler commands to UE for selection of transport format (number of bits to be transmitted in a TTI) for the E-DCH transport channel and UE MAC layer selects the

E-TFC as described in Figure 2.38. The basic principle behind HARQ for HSUPA is the same as that for HSDPA. In uplink also, several UE categories are defined based on-terminal's multi-code capability, support of a 2-ms TTI (10 ms default), minimum total terminal RLC/MAC buffer size (see 3GPP TS 25.306).

In addition to HSDPA, Release 5 introduces the IP multimedia system (IMS) architecture that promises to greatly enhance the end-user experience for integrated multimedia applications and offers wireless operators an efficient means of providing such IP-based multimedia services. Release 5 also introduces the IP UTRAN concept to reduce network costs and to increase network efficiencies. Release 6 features include the following: multimedia broadcast multicast service (MBMS), enhanced dedicated channels (E-DCH), advanced receiver performance specifications (receive diversity at the terminal), IMS enhancements, enhancements to support interworking with WLAN, wideband AMR speech codec, IP flow-based bearer-level charging, push-to-talk over cellular (PoC), and support for emergency services. Release 6 IMS allows provision of a CS domainlike service via the PS domain.

2.14.3 HSPA+

HSPA+ is an evolution of HSPA with the introduction of MIMO, multicarrier, higher order modulation schemes and other advanced features. The evolution of UMTS specification is shown in Table 2.6. To achieve a higher data rate in the downlink, 3GPP Release-8 introduced dual-cell (DC-HSDPA) operation for two adjacent carrier cells (5 + 5 = 10 MHz) operating in the same frequency band using intraband continuous carrier aggregation (CA). From the higher layer perspective, each component carrier appears as a separate cell with its physical cell identifier. A cell is characterized by a combination of scrambling code and carrier frequency. Now, in that sense, two carriers, along with a scrambling code, can form dual cells. Dual-cell (DC) HSDPA is the natural evolution of HSPA by means of carrier aggregation in the downlink. Here, the two cells belong to the same node-B and are on different carriers:

- Anchor carrier. This is also known as a primary carrier (primary serving cell – PSC) and it has all the physical channels (DPCH/F-DPCH, E-HICH, E-AGCH, and E-RGCH).
- Supplementary carrier. This is also, known as secondary carrier (secondary serving cell – SSC).

Using this dual carrier technique, the peak-data rate is doubled from 21 mbps to 42 mbps even without the use of MIMO. Again, Release 9 introduces DC-HSDPA in combination with MIMO on both carriers. This will allow a theoretical speed of up to 84 Mbit/s. Often UMTS licenses are issued in a paired spectrum of either 10 MHz or 15 MHz blocks – for example two or three carriers, for uplink and downlink. So, the DC-HSPDA implementation becomes easy for operators, and network and UE vendors, by using two adjacent carriers. But, in many cases, operators have different frequency bands so they wanted to use carrier aggregation among different frequency bands. So, Release 9 allows the paired cells

Table 2.6 WCDMA HSPA evolution

Rel'99	Rel'5	Rel'6	Rel'7	Rel'8	Rel'9	Rel'10	Rel'11
WCDMA Modulation: QPSK (DL), HPSK (UL)	HSDPA DL: added 16 QAM	HSUPA	HSPA+ 64 QAM (DL) 16 QAM (UL)	DC-HSDPA	DB-HSDPA	4 carriers	8 carriers
Data rate: 2 mbps (indoor) 384 kbps (outdoor)	14.4 mbps (DL)	5.6 mbps (UL)	BW 5 MHz, 2×2 MIMO: 28 mbps (DL) 11 mbps (UL)	Dual carrier: 10 MHz, no MIMO: 42 mbps	BW 10 MHz , 2×2 MIMO: 84 mbps	BW 20 MHz, 2×2 MIMO: 168 mbps	BW 40 MHz, 2×2 MIMO or BW 20 MHz, 4×4 MIMO: 336 mbps

operation on two different frequency bands. This is known as dual-band dual-carrier (DB-DC) HSDPA and this is interband CA. Support for this optional feature is signaled to the network via UE capability signaling.

2.14.4 Receiver Architecture (RAKE and G-RAKE) Evolution for WCDMA

A receiver is considered to be optimum in the sense that it minimizes the error rate of the signal that is detected over the channel. In WCDMA, one method of implementing an optimum receiver is based on correlation and integration. The correlator demodulator was defined as an optimum demodulator for the AWGN channel but for a radio channel that has severe multipath delays the demodulated output includes significant energy in the components that are delayed. So, the consequence is that the receiver is no longer optimum.

For single-path reception, the received signal could be multiplied by a locally generated or stored dispreading waveform and the resultant signal is integrated over the duration of the transmitted symbol or bit period. At the end of the integration period the output is sampled and the integrator is reset. In case of a multipath environment, however, reflected signals over different delay paths carry some energy from the same original transmitted chip sequence, so, using the maximum ratio combining (MRC) technique, multipath input signals could be co-phased (by compensating the delay) and then scaled according to their received signal quality. For that reason several other correlators are added in the receiver structure, which will be used to tap the energy from the reflected paths. As shown in Figure 2.39, such correlators ("subreceivers") or multiplier pairs are called "fingers" and this type of receiver is known as a RAKE receiver, where each correlator is assigned to a different multipath component to track and extract the energy. Please refer to [1] for more details. There are two basic approaches: The first is symbol-level

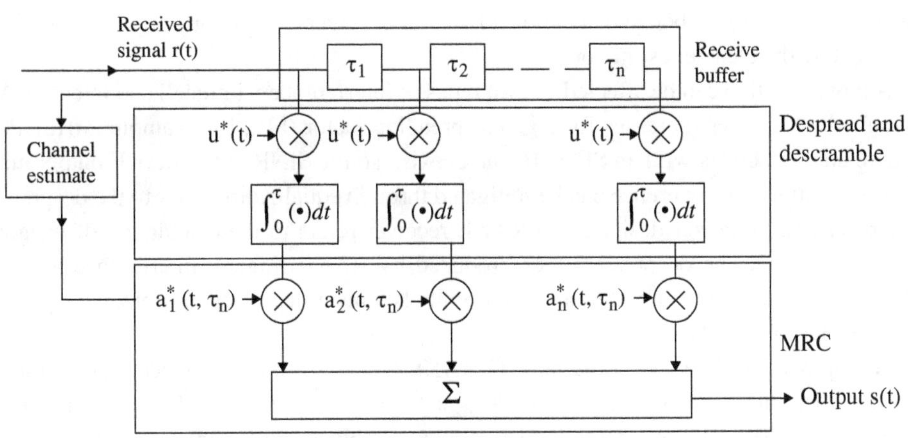

Figure 2.39 RAKE receiver (symbol-level combining)

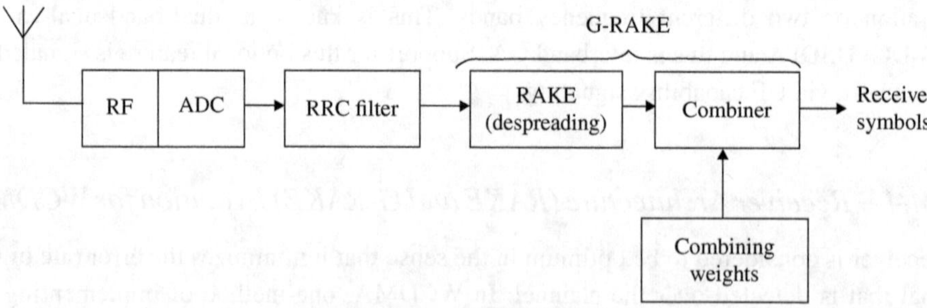

Figure 2.40 G-RAKE receiver

combining, where the received signals are despread and descrambled prior to being combined in MRC. Here, the multiplication of the signals by the channel impulse response is performed at the symbol rate. This provides a faster processing speed. The second is chip-level combining, where received signal scaling and combining occurs at the chip level. Then dispreading, descrambling and integration are performed. Symbol rate combining is more computationally efficient, except where there is a low spreading-factor scenario (SF < number of fingers).

As shown in Figure 2.39, the first multiplier multiplies the received signal by u*(t), which is a complex combination of dispreading and descrambling codes. This is then multiplied by $a_n^*(t, \tau_n)$, which is the complex conjugate of the estimated channel impulse response for that path. By multiplying these two factors, it co-phases the signal and scale according to the amplitude of the received signal. Next, the signals from different fingers are summed to produce a maximum ratio combined signal. The number of fingers to be used in a RAKE receiver depends upon the amount of multipath energy present in the received signal and it is not explicitly stated in the WCDMA system specification. However, the maximum number of cells in the active set is eight, which implicitly indicates that the UE requires eight fingers in the RAKE receiver. A searcher block is responsible for making decisions on the number and location of the rake fingers based on the measurements reported by the channel estimator.

A mobile station can be affected by two types of interference – intracell and intercell. As the number of codes increases for a given spreading factor (SF), for example in HSDPA, where up to 15 codes with an SF of 16 can coexist, so the SINR at the RAKE output may degrade. Intracell interference can be mitigated through equalization prior to the despreading receivers. Equalizer design for a RAKE receiver refers to the calculation of weights used to combine the despread values produced by RAKE fingers. Interference at each RAKE finger consists of interference from symbols transmitted by other users, intersymbol interference and AWGN noise.

To improve interference suppression, G-RAKE (generalized RAKE) receivers are introduced, where the combiner weights are designed to maximize the signal while minimizing interference. GRAKE includes a RAKE plus a combiner (RACOM) (see Figure 2.40). There are two important methods of calculating the channel weights: (i) MMSE and (ii) the

ML method. The ML method uses the noise covariance matrix R_u but the MMSE method uses the data correlation matrix R_y. In nonparametric MMSE-based equalization, the signal can be modeled as: $w = (R_y)^{-1}h$, where R_y is the data correlation matrix, which is computed by finding the correlation of despread data symbols observed across fingers; "h" is the channel impulse response and "w" is the weight gain. In practical implementation the data correlation matrix is filtered across slots to smooth out noise. The MMSE method finds the weight (w) that minimizes the mean-squared difference between the combined weighted despread symbols ($w^H y$) and the transmitted symbols.

ML-based equalization can be modeled as: $w = (R_u)^{-1}h$. In the ML method we can find R_u (i) *parametrically,* where we express R_u as a function of different parameters and we substitute the values of the parameters to find R_u, or (ii) *nonparametrically,* where we estimate R_u from the despread pilot symbols. Pilot symbols are known at the receiver so we can use them to estimate the noise correlation matrix. If R_u is found parametrically, such an equalizer implementation is called a parametric GRAKE (P-GRAKE) and if R_u is found nonparametrically, the equalizer implementation is called a nonparametric G-RAKE (NP-GRAKE).

References

[1] Das, Sajal Kumar. (2010) *Mobile Handset Design*, John Wiley & Sons, Ltd.

[2] Holma, H. and Toskala, A. (eds.) (2006) *HSDPA/HSUPA for UMTS: High Speed Radio Access for Mobile Communications*, John Wiley & Sons, Ltd.

[3] 3GPP TS 25.211 (2010) *Physical channels and mapping of transport channels onto physical channels (FDD)* (Release 8) V8.7.0 (2010–09). 3rd Generation Partnership Project.

Further Reading

3GPP (2016a) *3GPP Specification Series*, http://www.3gpp.org/DynaReport/45-series.htm (accessed April 26, 2016).

3GPP (2016b) *3GPP Specification Series*, http://www.3gpp.org/ftp/Specs/html-info/25-series.htm (accessed April 26, 2016). 3rd Generation Partnership Project, Sophia Antipolis.

3GPP TR 25.823 (2008) *Feasibility study on synchronized E-DCH for UTRA FDD*. 3rd Generation Partnership Project.

3GPP TR 25.825 (2008) *Dual-cell HSDPA operation*. 3rd Generation Partnership Project.

3GPP TR 25.950 (2005) *UTRA high speed downlink packet access*. 3rd Generation Partnership Project.

3GPP TS 25.212 (2009) *Multiplexing and channel coding (FDD)* (Release 8) V8.6.0 (2009-09). 3rd Generation Partnership Project.

3GPP TS 25.213 (2009) *Spreading and modulation (FDD)* (Release 8) V8.5.0 (2009-12). 3rd Generation Partnership Project.

3GPP TS 25.214 (2009) *Physical layer procedures (FDD)* (Release 8). 3rd Generation Partnership Project.

3GPP TS 25.306 (2010) *UE radio access capabilities* (Release 8) V8.10.0 (2010-09). 3rd Generation Partnership Project.

3GPP TS 25.308 (2004) *High speed downlink packet access (HSDPA)*. 3rd Generation Partnership Project.

3GPP TS 25.317 (2012) *High speed packet access (HSPA); requirements on user equipments (UEs) supporting a release-independent frequency band combination*. 3rd Generation Partnership Project.

3GPP TS 25.331 (2010) *Radio resource control* (RRC) Protocol Specification (Release 8) V8.12.0 (2010-09). 3rd Generation Partnership Project.

3GPP TS 34.108 (2010) *Common test environments for user equipment (UE) conformance testing* (Release 8) V8.10.0 (2010-03). 3rd Generation Partnership Project.

3GPP TS 34.121–1 (2010) *User equipment (UE) conformance specification; radio transmission and reception (FDD); Part 1: Conformance specification*. V9.2.0 (2010–09). 3rd Generation Partnership Project.

Korhonen, J. (2003) *Introduction to 3G Mobile Communications*, Artech House.

3

LTE Systems

3.1 LTE Cellular Systems

Chapter 2 discussed legacy cellular systems (2G and 3G). The performance of 3G did not meet the demand for driving future high-performance data voracious applications, such as full-motion video and wireless videoconferencing in terms of bit rate and latency requirements. Research also shows that the current UMTS standard has fundamental capacity limitations for high user loads. In particular, when the number of active users increases beyond a certain point the aggregate system capacity starts to decrease. In HSPA, terminal complexity for WCDMA or MC-CDMA systems is quite high, making equipment expensive, resulting in poor performing receivers. On top of that, as today's network is a plethora of different types of networks including WLAN, the 3G system cannot support seamless handover and mobility among these heterogeneous IP networks, including cellular networks and wireless local area networks (WLANs), which are the driving factors for 4G telecommunications networks and systems.

In March 2008, the International Telecommunications Union-Radio communications sector (ITU-R) came up with a set of requirements for 4G standards, known as the International Mobile Telecommunications Advanced (IMT-Advanced) specification. These are: (i) peak data rate: 100 Mbit/s for high mobility and 1 Gbit/s for low mobility; (ii) latency: round-trip time < 10 ms; (iii) network: should be based on all-IP packet switched optimized network; (iv) high level of mobility and security; (v) smooth handovers across heterogeneous networks; (vi) optimized terminal power efficiency; (vii) frequency flexibility and scalable channel bandwidths; (viii) higher system spectral efficiency indoors, 3 bit/s/Hz/cell in downlink and 2.25-bit/s/Hz/cell in uplink.

Mobile Terminal Receiver Design: LTE and LTE-Advanced, First Edition. Sajal Kumar Das.

Today, the 4G system is evolving mainly through the WiMAX and 3G LTE systems. Both are also part of the IMT-2000 family of standards. The IEEE 802.16e standard, also known as Wireless MAN, is commonly referred to as WiMAX (worldwide interoperability for microwave access). It is meant for supporting mobility using simpler Internet protocol (IP) based network architecture and OFDMA-based medium-access technology. On the other hand, 3GPP and 3GPP2 are developing their own version of beyond 3G systems based on the OFDMA technology. This is known as evolved universal terrestrial radio access (evolved UTRA) and is also widely referred to as LTE (Long-Term Evolution), whereas 3GPP2's version is called UMB (ultramobile broadband) (see Figure 2.6). LTE is designed to support only packet-switched services and aims to provide seamless IP connectivity between user equipment (UE) and the packet data network (PDN).

3.2 3GPP Long-Term Evolution (LTE) Overview

As shown in Chapter 2, Table 2.1, in December 2008, 3GPP released LTE specifications for long-term evolution of UMTS cellular technology. This was formally known as evolved UMTS terrestrial radio access (E-UTRA) and evolved UMTS terrestrial radio access network (E-UTRAN) but now it is more commonly referred to as LTE. It is designed to support only packet-switched services, in contrast to the circuit-switched model of the previous generation cellular systems. Based on downlink and uplink path duplexing, both frequency-division duplexing (FDD) and time-division duplexing (TDD) versions of LTE are defined. The LTE feasibility study and system objectives are captured in 3GPP Technical Report TS 25.912 and LTE requirements document TS 25.913.

3.2.1 LTE Design Goals

The design goals of the first released version of LTE (Rel-8) were:

- support scalable bandwidths of 1.4, 3, 5.0, 10.0 and 20.0 MHz;
- peak data rate DL: 100 mbps, UL: 50 mbps (for 20 MHz spectrum);
- support antenna configurations:
  - downlink: 4×2, 2×2, 1×2, 1×1;
  - uplink: 1×2, 1×1;
- mobility support: up to 500 kmph;
- latency:
  - C-plane: < 50–100 ms to establish U-plane;
  - U-plane: < 10 ms from UE to server.

As 3GPP LTE Rel-8 version does not satisfy all the ITU-R 4G requirements, so this version is loosely referred to as 3.9G.

3.3 3GPP LTE Specifications

The LTE specifications for the LTE E-UTRA and E-UTRAN are described in the 36 series and divided into the following categories:

- 36.100 series: covers radio specifications and evolved Node B (eNB) conformance testing;
- 36.200 series: covers layer 1 (physical layer) specifications;
- 36.300 series: covers layer 2 and 3 air-interface signaling specifications;
- 36.400 series: covers network signaling specifications;
- 36.500 series: covers user equipment conformance testing;
- 36.800 and 36.900 series: technical reports containing background information.

The SAE specifications are found in the 22 series, 23 series, 24 series, and 33 series of Release 8, with work being done in parallel in Release 9. The latest versions of the LTE and SAE documents can be found at http://www.3gpp.org/ftp/specs/latest/Rel-8/ (accessed April 30, 2016).

3.4 LTE Network Architecture

LTE has introduced a new OFDMA-based air interface, which is referred to as the Evolved UMTS Terrestrial Radio Access (EUTRA), and an evolved radio access network, which is referred to as evolved UTRAN (E-UTRAN) or eNode B (eNB). E-UTRA and E-UTRAN are commonly known as evolved RAN. The term "LTE" encompasses the evolution of the radio access through E-UTRAN. Similarly, nonradio related aspects have been evolved under the term "system architecture evolution (SAE)." Under the system architecture evolution (SAE) work item, 3GPP developed a new, flatter, all-IP, packet-only core network (CN) known as the evolved packet core (EPC). The complete packet system consisting of the EUTRAN and the EPC is called the evolved packet system (EPS). So, the terms LTE and EUTRAN refer to the evolved air-interface and radio access network based on OFDMA while the terms SAE and EPC are refer to the evolved flatter-IP core network. At a high level, the LTE network consists of two main components: the evolved core network (EPC) and the access network E-UTRAN. Figure 3.1 shows the LTE network architecture, various entities, and their interfaces (for more details please refer to TS 23.882 and 36.300).

- Core network (CN). In SAE, the core network is the EPC, which is mainly responsible for the overall control of the UE and establishment of the bearers. Apart from supporting LTE, the EPC also supports both legacy 3GPP (UTRAN, GERAN) and non-3GPP (cdma2000, 802.16, etc.) radio-access networks. The CN consists of many logical nodes. The main logical nodes of the EPC are mentioned briefly below and more details about these entities can be found in TS 23.401.
    - Serving gateway (S-GW). The main functions of S-GW are user protocol tunnel management and switching. It acts as a local mobility anchor and helps for forwarding

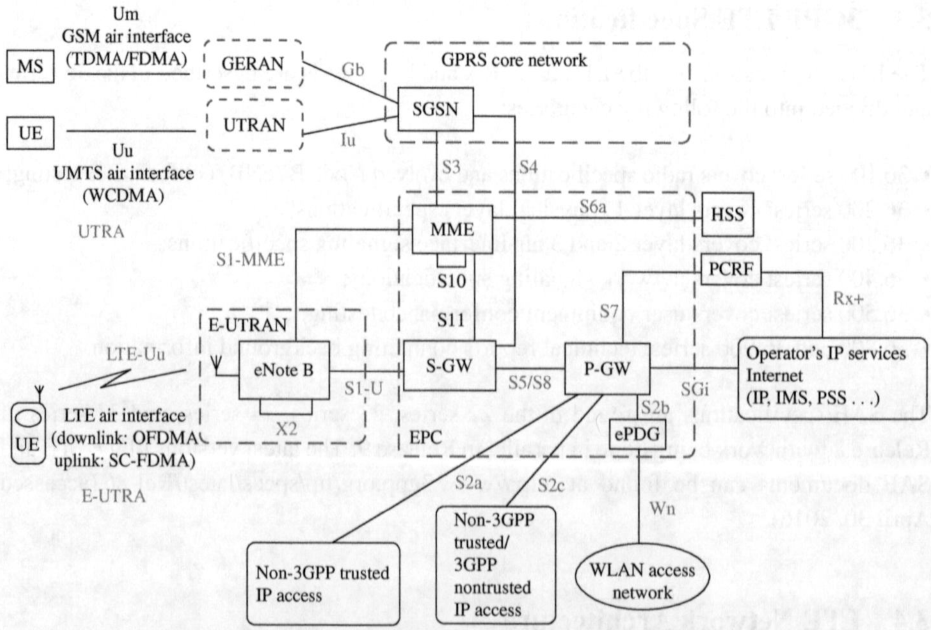

Figure 3.1 LTE network architecture and interfaces

and receiving the packets to and from the serving eNB to P-GW. All user IP packets are passed through the S-GW. It manages and stores UE contexts and also retains the information about the bearers when the UE is in the idle state. It also performs some administrative functions in the visited network.

- PDN gateway (P-GW). This is like a router between the EPS and external packet data networks. It interfaces with external packet data networks (PDNs) such as the Internet and the IMS. It is responsible for IP address allocation to the UE, as well as QoS enforcement and flow-based charging according to the rules from the PCRF. It also performs several IP functions such as address allocation, policy enforcement, packet filtering, and routing.
- Mobility management entity (MME). This is an important control node in the LTE network through which only signaling messages flow. User IP packets do not go through MME. Its main functions are NAS signaling, control and execution of paging retransmission, idle state mobility handling, roaming, tracking area list management, authorization, authentication, P-GW/S-GW selection, bearer management, NAS signaling, and so forth.
- Policy control and charging rules function (PCRF). This is responsible for policy control, decision making and controlling of flow-based charging functionalities.
- Home subscriber server (HSS). This is a central database that contains users' SAE subscription-related information. Its functionalities include mobility management, call and session establishment support, user authentication and access authorization.

It is based on pre-Rel-4 Home Location Register (HLR) and Authentication Center (AuC). The IP Multimedia Subsystem (IMS) is considered to be outside the EPS. The EPC NAS procedures are more or less similar to UMTS but here it allows concatenation of some procedures for faster establishment of the connection and the bearers. The NAS control procedures are specified in 3GPP TS 24.301.

- Access network (E-UTRAN). The access network is made up of the evolved Node B (eNB), which connects to UEs via an air interface. The RNC (as in the UMTS network) is eliminated from the data path and its functions are now distributed into multiple eNBs. The eNBs are responsible for all radio-related functions – radio resource management (RRM), IP header compression and encryption, signaling towards MME, selection of MME at UE attachment time, and routing of user plane data towards S-GW. The LTE system does not support soft handoff or macrodiversity. When the UE moves, the network transfers all information related to a UE (the UE context) together with any buffered data, from one eNodeB to another.
- User equipment (UE). It is used by an end user to communicate with the network. The UE has three main sections – radio, physical layer (PHY), protocol stack and applications. E-UTRA is designed to operate in the different operating bands. The RF operating bands and channel arrangement is given in 3GPP TS 36.101.

3.5 Interfaces

All the network interfaces are based on IP protocols. Each network elements is interconnected by means of well defined interfaces, which are standardized in order to allow multivendor interoperability, which allows network operators the possibility of sourcing different network elements from different vendors. The eNBs are interconnected by means of an X2 interface, which enables direct communication among eNBs and to the MME/GW entity by means of an S1 interface. The air interface between the eNB and UE is called as LTE-Uu interface. The other interfaces among different network entities are shown in Figure 3.1.

3.6 System Protocol Architecture

Figure 3.2 shows the user plane and control plane protocol layers existing among various network entities. In the control plane, the NAS protocol, which runs between the MME and the UE, is used for control purposes. All NAS messages are ciphered and integrity protected by the MME and UE.

- User plane. As shown in Figure 3.2, the user plane has several layers:
  - Packet data convergence protocol (PDCP, Ref TS 36.323). Its main functionalities in the user plane include decryption / encryption, compressing / decompressing the headers of user-plane IP packets using robust header compression (ROHC) to enable efficient use of air interface bandwidth, sequence numbering, and duplicate removal.

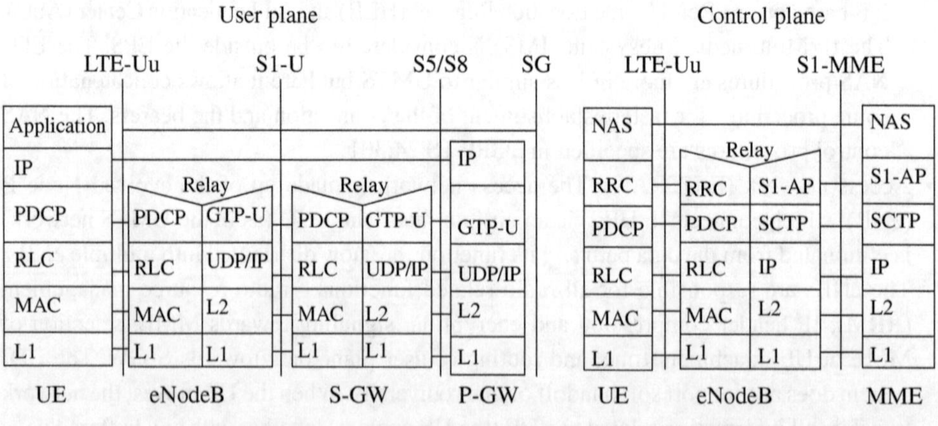

Figure 3.2 User and control plane protocol stack architecture over various entities in the LTE network

- ◦ Radio link control (RLC, Ref TS 36.322). Its main functionalities are error correction through ARQ, in-sequence delivery of service data units (SDUs) to the upper layers, prevention of duplicate SDUs from being delivered to the upper layers, segmentation according to the size of the transport block, concatenation of SDUs for the same radio bearer, and duplicate detection. Based on the reliability requirement, the RLC can be configured to the acknowledge mode (AM), unacknowledge mode (UM), or transparent mode (TM) for transfers. Generally, the UM mode is used for transport of real-time (RT) services, which are delay sensitive and cannot wait for retransmissions, whereas the AM mode is suitable for non-RT (NRT) services and the TM mode is used when the PDU sizes are known a priori, such as for broadcasting system information. AM, UM use uses RLC header, whereas the TM is used where effectively there is no header; it simply passes the message through it.
- ◦ Medium access control (MAC, Ref TS 36.321). Its main functionalities include multiplexing/demultiplexing of RLC PDUs, padding, error correction through HARQ, logical channel prioritization and scheduling information reporting, mapping between the logical and the transport channels, transport format selection. Uplink functions include random access channel scheduling, and transport format selection. There are two levels of retransmissions used for providing reliability – the hybrid automatic repeat request (HARQ) at the MAC layer and outer ARQ at the RLC layer. Any IP packet for UE is encapsulated by an EPC-specific protocol and tunneled between the P-GW and eNB for transmission to UE. For tunneling, different protocols are used across different interfaces. Generally, GPRS tunneling protocol (GTP) is used over the CN interfaces, S1 and S5/S8.

- Control plane. The control plane protocol stack layers between UE and MME are shown in Figure 3.2.

- Radio resource control (RRC, Ref TS 36.331) functionality is incorporated into UE and eNB on the network side. This is responsible for broadcast of system information, RRC connection control, initial security activation for ciphering and integrity protection, mobility control, and also for inter-RAT handovers, quality-of-service control, and measurement control. The RRC layer in eNB makes handover decisions based on neighbor cell measurements sent by the UE, and it controls UE measurement reporting such as the periodicity of channel quality information (CQI) reports, sends paging messages for the UEs, broadcasts system information, and allocates cell-level temporary identifiers to active UEs. It is also responsible for the setting up and maintenance of radio bearers. The NAS messages carried in RRC are effectively double ciphered and integrity protected, once at the MME and again at the eNB. The RLC and MAC sublayers in control plane perform the similar functions like the user plane. The PDCP layer performs decryption, integrity protection, sequence numbering, duplicate removal, and so forth.

3.6.1 User Plane Data Flow Diagram

As shown in Figure 3.3, a PDCP header is added to the IP packet, carrying information required for deciphering in the mobile terminal. Generally, IP header is 40 bytes (for IPv4) and application data has payload upto 1460 bytes. PDCP header is 2 to 3 bytes. The output from the PDCP is passed to the RLC. The RLC protocol performs concatenation and/or segmentation of the PDCP SDU and adds an RLC header, which is used for in-sequence delivery (per logical channel) in the terminal and for identification of RLC PDUs in the case of retransmissions. Typically RLC breaks the PDCP PDU to blocks of 40 bytes and add header of is 1 to 2 bytes. The RLC PDUs are forwarded to the MAC layer, which multiplexes a number of RLC PDUs (from different services) and attaches a MAC header to form a transport block. The transport-block size depends on the instantaneous data rate selected by the link adaptation mechanism. Thus, the link adaptation affects both the MAC and RLC processing. Finally, the physical layer attaches a CRC to the transport block for error-detection purposes, performs coding and modulation, and transmits the resulting signal, using single or multiple transmit antennas.

3.6.2 Protocol States

In the LTE system, two RRC states are defined – RRC IDLE and RRC CONNECTED (see Figure 3.4). In the RRC IDLE state, the UE is known in EPC and has an IP address but is not known in E-UTRAN/eNB. The UE can receive broadcast/multicast data, monitors a paging channel to detect incoming calls, performs neighbor cell measurements, does cell selection/reselection, and acquires system information. In the RRC IDLE state, to enable UE power savings, a UE-specific DRX (discontinuous reception) cycle may be configured by upper layers. LTE supports always connected experience by forcing the UE to continuously monitor control signals on PDCCH. In DRX, UE saves power by monitoring PDCCH less frequently, which allows UE to turn the modem in longer sleep state and activates UE

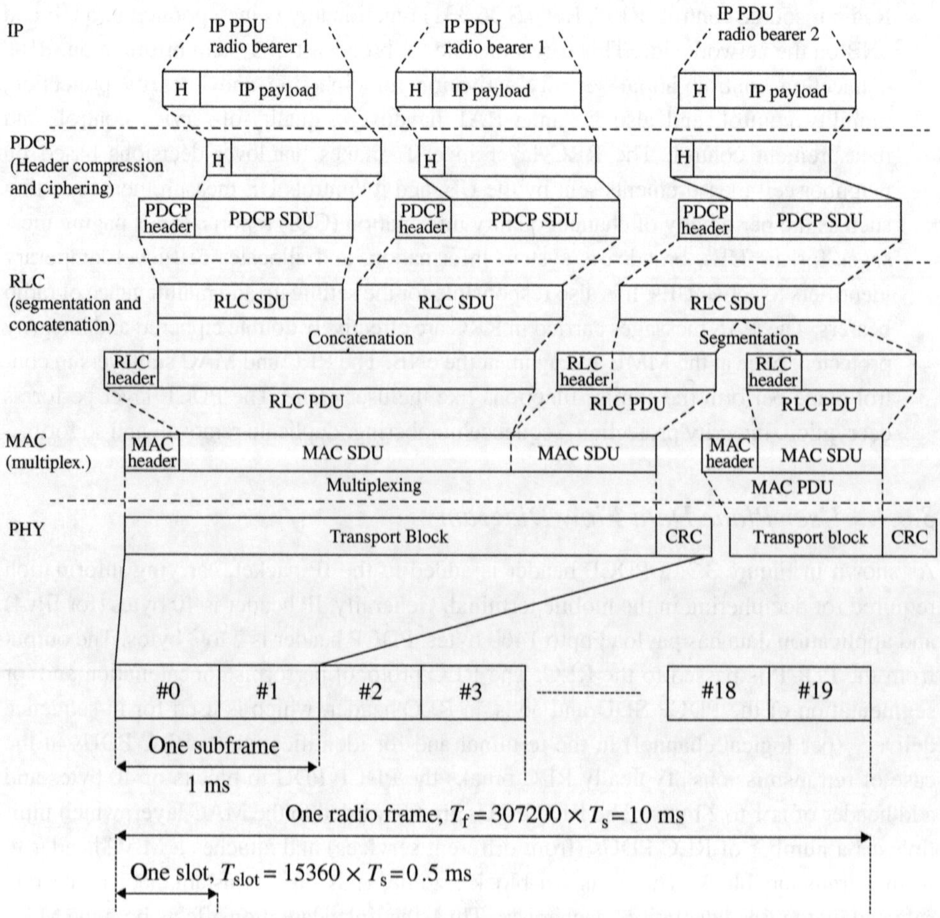

Figure 3.3 User plane dataflow

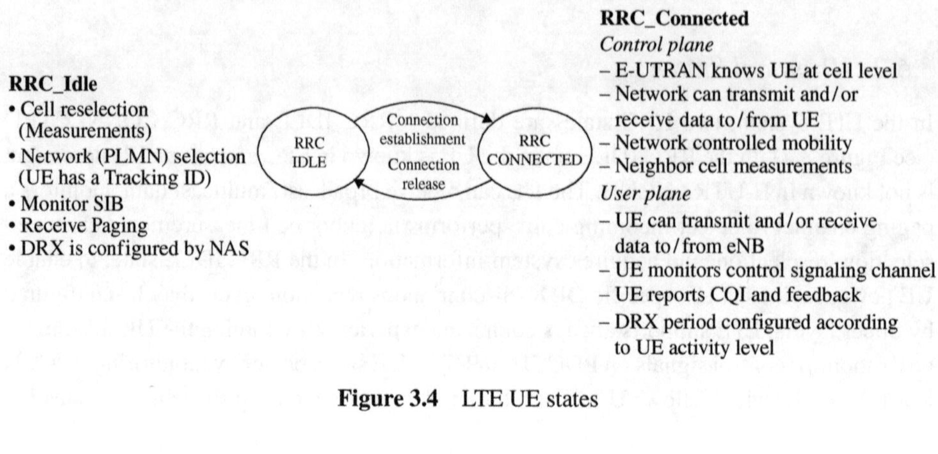

Figure 3.4 LTE UE states

only at well defined, suitable instants. After RRC camped on a LTE cell, UE activates the paging reception aligning with DRX period. Also, mobility is controlled by the UE in the RRC IDLE state.

In the RRC CONNECTED state, UE is known in EPC as well as in E-UTRAN/eNB and UE location is known at the cell level. In this state, transfer of unicast data to/from UE and the transfer of broadcast/multicast data to UE can take place and mobility is UE-assisted and network controlled. At lower layers, the UE can be configured with a UE-specific DRX/DTX. If the DRX is configured, the UE is allowed to monitor the PDCCH discontinuously according to the DRX period. The RRC controls the DRX operation by configuring the timers and DRX-Cycle. For more details please refer to 3GPP TS 3GPP TS36.133 and 3GPP TS36.321 section 5.7. The UE also monitors control channels associated with the shared data channel to determine if data is scheduled for it, provides channel quality feedback information, performs neighbor cell measurements and measurement reporting, and acquires system information. A UE moves from RRC IDLE state to RRC CONNECTED state when an RRC connection is successfully established, and the UE moves back from RRC CONNECTED to the RRC IDLE state by releasing the RRC connection.

3.6.3 Bearer Service Architecture

The EPS uses EPS bearers to route the IP traffic from PDN gateway to the UE. A bearer is an IP packet flow with a defined QoS. There could be one or more IP flows related to one or more services. The EPS sets up and release bearers as required by the applications. The bearers can be classified broadly into two categories depending on the nature of the QoS they provide – minimum guaranteed bit rate (GBR) bearers and non-GBR bearers. When UE is attached to the network, the UE is assigned an IP address by the P-GW and at least one bearer is established. This is called the default bearer, and it remains established throughout the lifetime of the PDN connection in order to provide low latency with always-on IP connectivity to that PDN.

3.7 LTE-Uu Downlink and Uplink Transmission Schemes and Air Interface

LTE introduces the new OFDMA-based air interface referred to as evolved UMTS terrestrial radio access (EUTRA). It is also known as LTE-Uu interface.

3.7.1 Downlink Transmission Scheme

Due to several disadvantages of WCDMA, and on the other hand several advantages of OFDM, a new access scheme was felt necessary in the LTE downlink. The LTE downlink is based on OFDMA with cyclic prefix (see [1] for more details about OFDMA).

3.7.1.1 OFDMA

In the air channel, as the data rate increases in a multipath environment, the channel fading goes from flat fading to frequency selective fading (the last reflected component arrives after the symbol period). The channel delay spread can cause heavy degradation of bit error rate in the signal transmission as a result of frequency selective fading and intersymbol interference (ISI). The most popular solutions to compensate for these problems are:

- Use of equalizers – adaptive compensation for time-variant channel distortion. But, as we move to higher data rates (i.e. > 1 mbps), the equalizer complexity grows to a level where the channel changes before you can compensate for it. So, there are practical difficulties in operating this equalization in real time at several Mb/s with compact, low-cost hardware.
- Adaptive array antenna – consider delayed waves as interference waves and eliminate them to avoid overlapping waves. This is a very complex and expensive solution.
- Multicarrier transmission, where a wideband frequency channel (carrier with large bandwidth) is broken up into several sub-bands (narrow band carriers) such that the fading over each subchannel becomes flat, thus helping to eliminate the ISI problem. Single carrier systems transfer data streams using a serial transmission in a single carrier, so the duration of a symbol becomes too small for higher data rates ($Ts = 1/r_b$), whereas a multicarrier system uses parallel transmission (divides the carrier bandwidth among several subcarriers and transmits data using each subcarrier), which results in a lower data-rate requirement in each of the parallel paths. This leads to larger symbol duration in each path. So, the BW requirement will be reduced (as now signal BW < coherence BW), which means this will be robust against multipath frequency selective fading and ISI (as shown in Figure 3.5). So, multicarrier transmission is the way forward for higher data rates.

Several typical implementation problems arise with the use of a large number of subcarriers. When we have large numbers of subcarriers, then we will have to assign the subcarriers' frequencies very close to each other. We know that receiver needs to synchronize itself to every subcarrier frequency in order to recover data related to that particular subcarrier. When spacing is very little, then the receiver synchronization components need to be very accurate, which is still not possible with low-cost RF hardware. So, bandwidth utilization will be very poor. And simply dividing the carrier BW among several carriers by using a conventional frequency Division multiplexing (FDM) approach will not help. The solution to this problem is to use orthogonal frequency carriers. This is known as OFDM. It is similar to FDM but much more spectrally efficient by spacing the subchannels much closer to each other. This is done by finding frequencies that are orthogonal, which means that they are perpendicular in a mathematical sense, allowing the spectrum of each subchannel to overlap with the other, without interfering with it. Mathematically, two signals are called orthogonal if the following condition is met:

$$\int_0^T S_1(t).S_2^*(t)dt = 0.$$

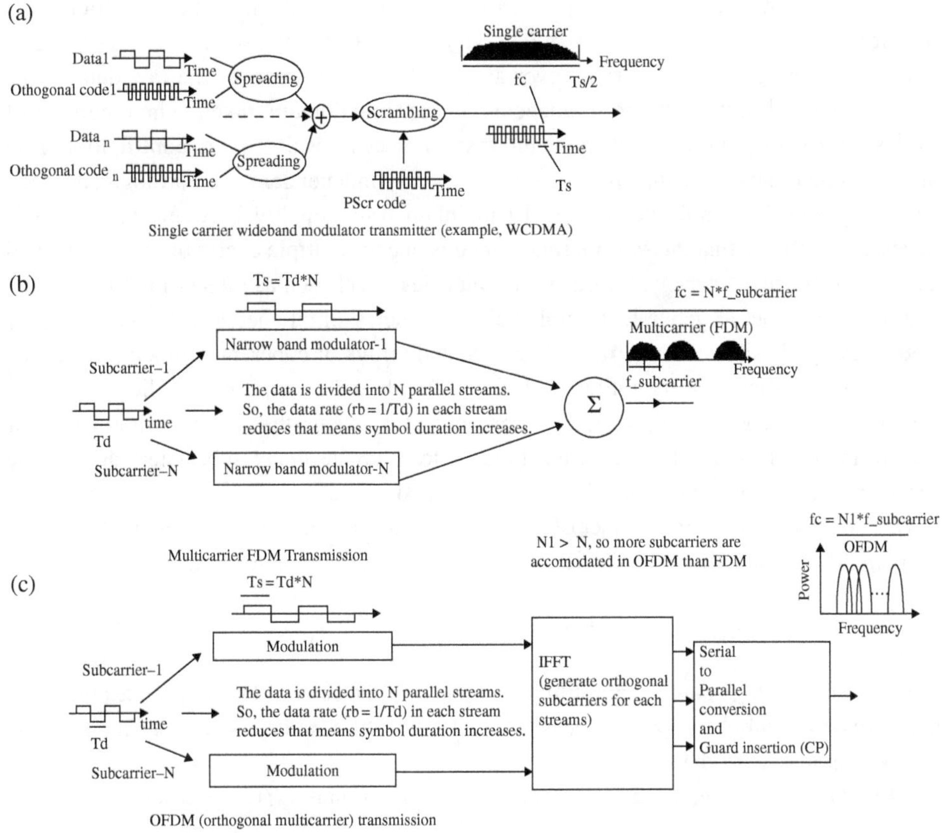

Figure 3.5 (a) Single carrier wideband modulator transmitter (WCDMA), (b) Multicarrier FDM transmission, (c) OFDM (orthogonal multicarrier) transmission

If we take a sine wave of frequency m and multiply with a sinusoid of frequency n, where both m and n are integers, then the integral over one period, which means the area will be: $E(t) = \sin m\omega t * \sin n\omega t = ½ (\cos(m-n)\omega t + \text{cost}(m+n)\omega t)$. As these two components are also sinusoids so, the integral or area under one period will also be zero.

$$\int_0^{2\pi} \frac{1}{2}\cos\left(m-n\right)\omega t - \int_0^{2\pi} \frac{1}{2}\cos\left(m+n\right)\omega t = 0-0.$$

We can conclude that when we multiply a sinusoid of frequency n by a sinusoid of frequency m (where m and n are integers), the area under the product is zero. In general, for all integer values of m and n, $\sin(nx)$, $\sin(mx)$, $\cos(nx)$, and $\cos(mx)$ are all orthogonal to each other. These frequencies are called harmonics. But, remember, when $m = n$, the above result is not zero – for example, the area is not zero. This principle is used in OFDM, where the orthogonality amongst the carriers allows the carriers to overlap and transmit simultaneously.

Now, we know OFDM will help us to transmit multicarrier in a bandwidth-efficient manner. The receiver acts as a bank of demodulators, translating each carrier down to DC, with the resulting signal integrated over a symbol period to recover the raw data. If the other carriers all beat down the frequencies (in the time domain, take a whole number of cycles in the symbol period T) then the integration process results in zero contribution from all these other carriers. Thus, the carriers are linearly independent (i.e., orthogonal) if the carrier spacing is a multiple of 1/T. To maintain orthogonality between carriers, it is necessary to ensure that the symbol time contains one or multiple cycles of each sinusoidal carrier waveform. Generally, the carrier frequencies are chosen as integer multiples of the symbol period, but to generate several such frequencies simultaneously we require many frequency synthesizers on the transmitter side and arrays of coherent demodulators at the receiver side. This made the OFDM solution difficult and expensive in earlier days. This problem is overcome using a digital approach. This means that inverse digital Fourier transform (IDFT) is used on the transmitter side to create many subcarriers and on the receiver side the inverse process (DFT) is performed.

OFDM transmits a large number of narrowband carriers, closely spaced in the frequency domain. Mathematically, each carrier can be described as a complex wave:

$$S(t) = A_c(t)\left[e^{j(2\pi f_c t + \varphi_c(t))}\right]$$

Both $A_c(t)$ and $\Phi_c(t)$, are the amplitude and phase of the carrier. The amplitude and phase can vary on a symbol-by-symbol basis. The values of the parameters are constant over the symbol duration period T.

OFDM consists of many carriers. Thus the complex signal $S_n(t)$ is represented by:

$$S_n(t) = \frac{1}{N}\sum_{n=0}^{N-1} A_n(t)\left[e^{j(2\pi f_n t + \varphi_n(t))}\right]$$

where, $\omega_n = \omega_0 + n.\Delta\omega$. This is, of course, a continuous signal. If we consider the waveforms of each component of the signal over one symbol period, then the variables $A_n(t)$ and $\Phi_n(t)$ take fixed values, which depend on the frequency of that particular carrier, and can be rewritten as $\Phi_n(t) = \Phi_n$ and $A_n(t) = A_n$. If the signal is sampled using a sampling frequency of $1/T$ then the resulting signal is represented by:

$$S_n(kT) = \frac{1}{N}\sum_{n=0}^{N-1} A_n(t)\left[e^{j(2\pi(f_0 + n\Delta f)kT + \phi_n)}\right]$$

At this point, we have restricted the time over which we analyze the signal to N samples. It is convenient to sample over the period of one data symbol. Thus we have a relationship: $\tau = \text{NT}$. Now, if we simplify the above equation, without a loss of generality by letting $f_o = 0$, then the signal becomes:

$$S_n(kT) = \frac{1}{N}\sum_{n=0}^{N-1} A_n(t)\left[e^{j2\pi.n.\Delta f.kT'}\right].e^{j\phi_n}$$

In above equation, the function $A_n e^{j\phi_n}$ is no more than a definition of the signal in the sampled frequency domain and $S(kT)$ is the time domain representation. The above equation can be compared with the general form of the inverse Fourier transform:

$$s(kT) = \frac{1}{N}\sum_{n=0}^{N-1} S\left(\frac{n}{NT}\right).e^{j2\pi nk/N}$$

The above two equations are equivalent if: $\Delta f = (\Delta\omega / 2\pi) = (1/NT) = 1/\tau$. This is the same condition that was required for orthogonality. Thus, one consequence of maintaining orthogonality is that the OFDM signal can be defined by using Fourier transform procedures. So, according to its mathematical distribution, on the transmitter side, inverse digital Fourier transform (IDFT) summarizes all sine and cosine waves of amplitudes stored in $S[k]$ array, forming a time domain signal (see Figure 3.6):

$$x[n] = \frac{1}{N}\sum_{k=0}^{N-1} X(k).e^{j2\pi nk/N} = \frac{1}{N}\sum_{k=0}^{N-1} X(k).\left(\cos\left(k.\frac{2\pi}{N}n\right) + j.\sin\left(k.\frac{2\pi}{N}n\right)\right)$$

Where, $n = 0,1,\ldots, N-1$. We can observe from the above equation that IDFT takes a series of complex exponential carriers, modulates each of them with a different symbol from the information array $S[k]$, and multiplexes all of this to generate N samples of a time domain signal. These carriers are orthogonal and frequency spaced with $\Delta f = 2\pi N$.

At the receiver side, the inverse process is performed. The time domain signal constitutes the input to a DFT block, which is implemented using the FFT algorithm. The FFT demodulator takes the N time domain transmitted samples and determines the amplitudes and phases of sine and cosine waves forming the received signal, according to the equation below:

$$X[k] = \frac{1}{N}\sum_{n=0}^{N-1} x(n).e^{-j2\pi nk/N} = \frac{1}{N}\sum_{n=0}^{N-1} x(n).\left(\cos\left(k.\frac{2\pi}{N}n\right) - j.\sin\left(k.\frac{2\pi}{N}n\right)\right)$$

Where $k = 0,1,\ldots\ldots,N-1$.

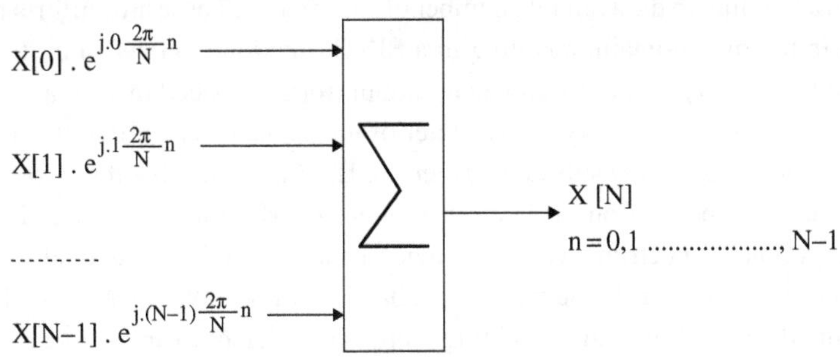

Figure 3.6 IDFT operation from different complex exponential carriers

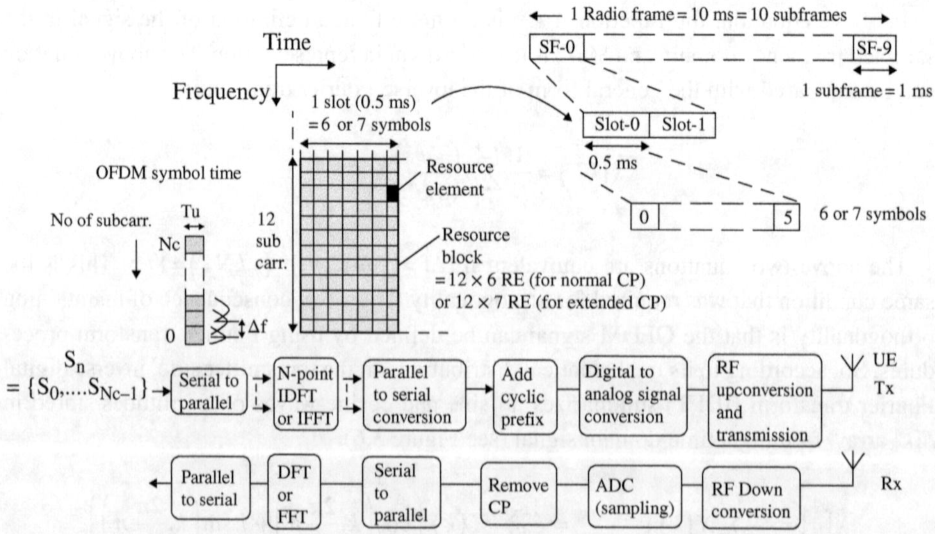

Figure 3.7 OFDM Transmitter receiver and LTE frame structure

OFDMA (OFDM access) is considered a multiple access technique because an individual carrier, or groups of carriers, can be assigned to different users. Users can each be assigned a predetermined number of carriers when they have information to send, or alternatively, users can be assigned a variable number of carriers based on the amount of information they have to send. The media access control (MAC) layer controls the assignments and schedules the resources based on user demand.

3.7.2 *LTE Downlink Frame Structure*

As shown in Figure 3.3, the data block (S_n) from a higher layer, referred to as transport block (codeword), is delivered to the PHY layer at every transmit time interval (TTI), where CRC addition, channel coding, rate matching, interleaving, and modulation bit mapping are performed to produce symbols. These symbols are divided into several parallel data streams according to the available number of subcarriers. These are uniform rectangular data pulses in the time domain, resulting in a SINC function (sin(x) / x) in the frequency domain (see Figure 3.7). If these frequencies are uniformly spaced in such a way that each one's zero crossings fall precisely at the center of the adjacent subcarrier, this results in no interference from neighboring subcarriers (zero ICI). This means that if the frequencies are orthogonal then the energy from one carrier will not interfere with the other. In an OFDM system, the available spectrum (BW_T) is divided into multiple orthogonal carriers (N_c), called subcarriers, where each one will have a bandwidth of $BW_S = BW_T / N_c$. The process of modulating data symbols and combining them is equivalent to an inverse Fourier transform (**IFFT**) operation. It converts N_c number of complex data symbols used as frequency

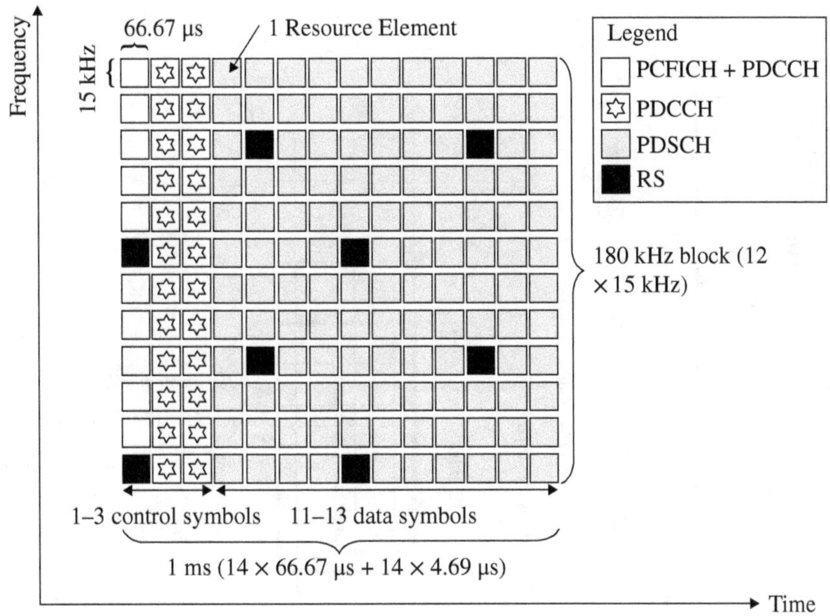

Figure 3.8 LTE subframe (=2 slots = 2 * 7 = 14 symbols) (for normal CP) and channel structure

domain bins into the time domain signal. Then the combined OFDM modulated and cyclic prefix (CP) added signal is up converted to an RF signal and transmitted. If there are S_n number of complex source symbols and N subcarriers available, then after serial to parallel conversion, each OFDM symbol will have duration T_u termed "useful symbol length," which is simply inversion of the carrier spacing ($1 / \Delta f$). In LTE, $\Delta f = BW_s = 15$ kHZ so Tu = 1/15 kHz = 66.7 μs. Again, to reduce intersymbol interference, in the time domain, a guard interval T_g is also added at the beginning of the OFDM symbol. The guard time interval, or CP is a duplication of a fraction of the symbol end. The total symbol length becomes $T_s = T_u + T_g$. The duration of the CP is determined by the highest anticipated degree of delay spread for the targeted application and cell size. The normal cyclic prefix is used in urban cells and high data rate applications while the extended cyclic prefix is used in special cases like multicell broadcast and in very large cells. Then the symbol of duration T_s is placed in the resource element (RE), which is the smallest element in the LTE time (T_s), frequency (BW_s) resource grid (see Figure 3.7).

As shown in Figures 3.8 and 3.9, the RE is the smallest unit made up of 1 symbol × 1 subcarrier. The resource element group (REG) is a group of four consecutive REs (excluding reference signal RE). The control channel element (CCE) is a group of nine consecutive REG. RB (Resource Block): This is a unit of 84 REs (12 subcarrier × 7 symbols for normal CP). The resource block group (RBG) consists of multiple RBs.

As shown in Figure 3.10, one radio frame (10 ms) is divided into ten subframes, each of 1 ms duration, and each subframe is again divided into two slots, each of 0.5 ms duration.

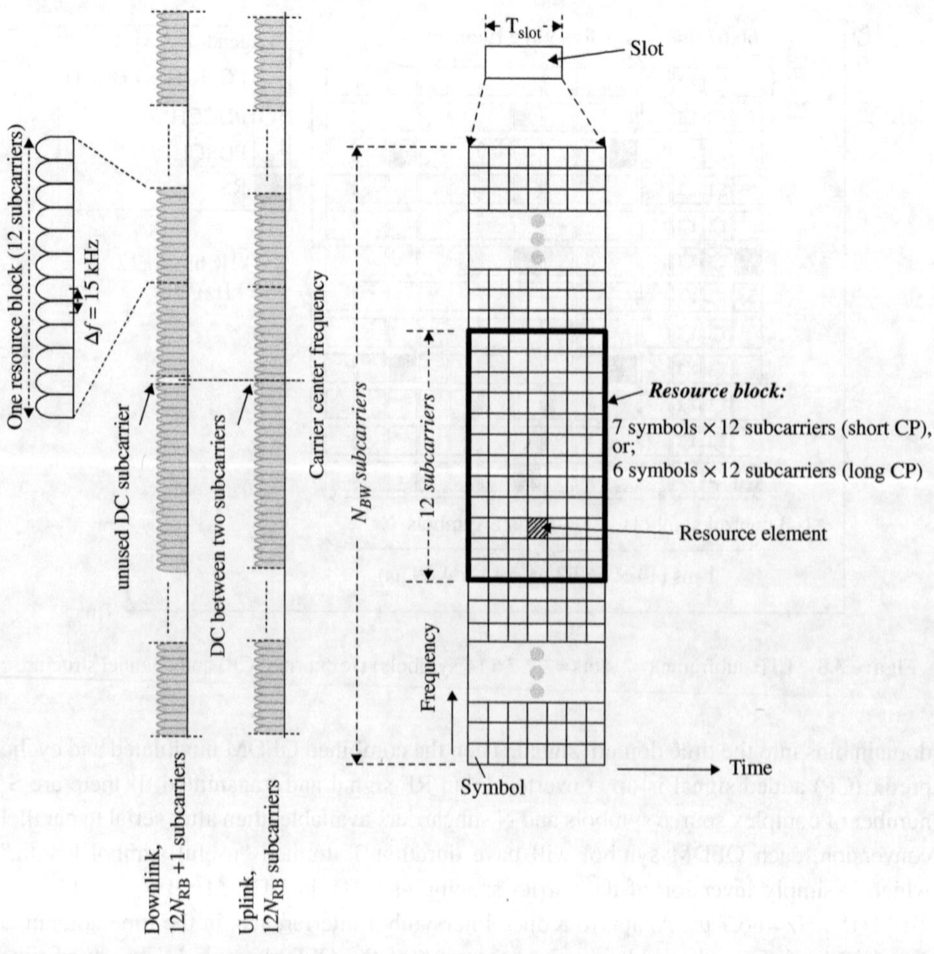

Figure 3.9 Resource grid (time on the *x* axis and frequency on the *y* axis) for uplink and downlink

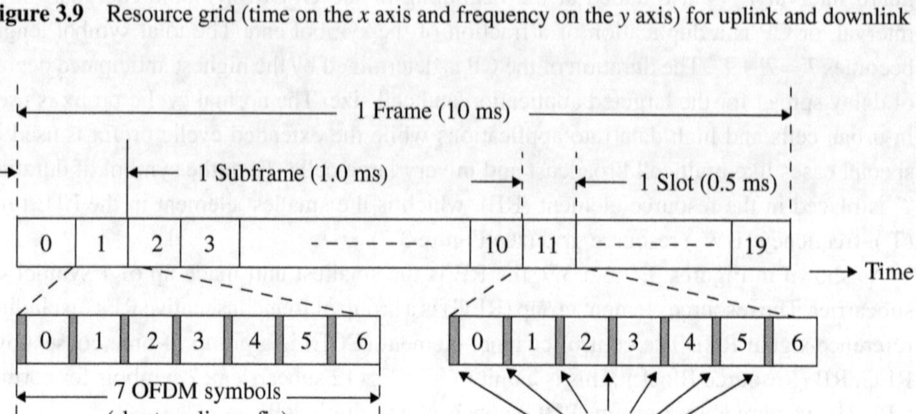

Figure 3.10 LTE frame structure (time domain) (1 slot = 15 360 * Ts = 0.5 ms, one radio frame = 307 200 * Ts = 10 ms)

Each slot in turn consists of a number of OFDM symbols, which can be either seven (for a normal CP) or six (for an extended CP). Users are allocated a specific number of subcarriers for predefined amount of time; these are referred to as physical resource block (RB). Each RB has a time (one slot=0.5 ms) and a frequency dimension (12 consecutive subcarriers, or 15×12=180 kHz) and each RB consists of 12×7=84 REs in case of normal cyclic prefix, and 12×6=72 REs for extended CP. Generally, the transmission is scheduled at the eNB by assigning multiple RBs. Physical resources are assigned on a basis of two resource blocks for one TTI (1 ms), for example one subframe such as one resource block pair. For the normal mode, the first symbol has a cyclic prefix of length $T_g = 160 \times T_S \approx 5.2$ μs. The remaining six symbols have a cyclic prefix of length $T_g = 144 \times T_S \approx 4.7$ μs. The reason for the different CP length of the first symbol is to make the overall slot length, in terms of time units, divisible by 15360. For the extended mode, the cyclic prefix is $T_{g\text{-}e} = 512 \times T_S \approx 16.7$ μs. The CP is longer than the typical delay spread of a few microseconds typically encountered in practice.

At the receiver side, RF signal is downconverted and sampled. If there are N_c subcarriers with spacing Δf, then minimum sampling rate requirement will be $f_s = 1/T_{sam} = N_c.\Delta f$, to guarantee that at least one sampling point will be there on each subcarrier. It is the nominal bandwidth of the OFDM signal. Generally, the sampling rate ($1/T_{sam} = N.\Delta f$) is chosen in such way that sampling theorem is sufficiently fulfilled ($N > N_c$). So, it will produce N samples per symbol period (T_s). Time-sampled OFDM signal is converted into the frequency domain by means of a fast Fourier transform (**FFT**). The resulting Fourier spectrum has discrete frequencies at: $k/N.T_{sam}$, $k = 0, 1, \ldots N-1$. Based on total number of subcarrier used over a symbol period the sampling rate varies, $f_s = N_c.\ \Delta f = N_{FFT}.\Delta f$, where N_{FFT} is the FFT period (number of subcarriers) and varies from 128 to 2048 depending on channel bandwidth. The largest FFT size in LTE is 2048, so the largest sampling rate will be 2048 * 15 kHz=30.720 MHz. This is a multiple or submultiple of the WCDMA chip rate of 3.84 Mcps. It sets the basic time unit $T_s = 1/30720000$ and radio frame length is defined as 10 ms ($T_{frame} = 307200 \times Ts$). Table 3.1 tabulates different system design parameters for LTE system for different BWs.

The radio is optimized for performance on the downlink, and for power consumption in uplink.

3.7.3 *Uplink Transmission Scheme and Frame Structure*

Although OFDMA is used in the downlink, due to its high peak-to-average power ratio (PAPR) (due to random constructive addition of subcarriers), which leads to more power consumption in UE, and high sensitivity to frequency offset problems, LTE uses single-carrier FDMA (SC-FDMA) with cyclic prefix on the uplink. As shown in Figure 3.11, OFDMA divides the whole information symbols among different subcarriers, whereas, SC-FDMA spreads the information of one symbol through all the available subcarriers. This is also known as DFT-spread OFDM (DFTS-OFDM), because extra DFT processing

Table 3.1 LTE system design parameters

System design parameters						
Frame duration	10 ms					
Subframe duration	1 ms					
Slot duration	0.5 ms					
Sub carrier spacing	15 kHz					
Transmission BW (MHz)	1.4	3	5	10	15	20
Sampling Frequency (MHz)	1.92	3.84	7.68	15.36	23.04	30.72
FFT size	128	256	512	1024	1536	2048
Number of occupied subcarriers (extra one for dc subcarrier)	72+1	180+1	300+1	600+1	900+1	1200+1

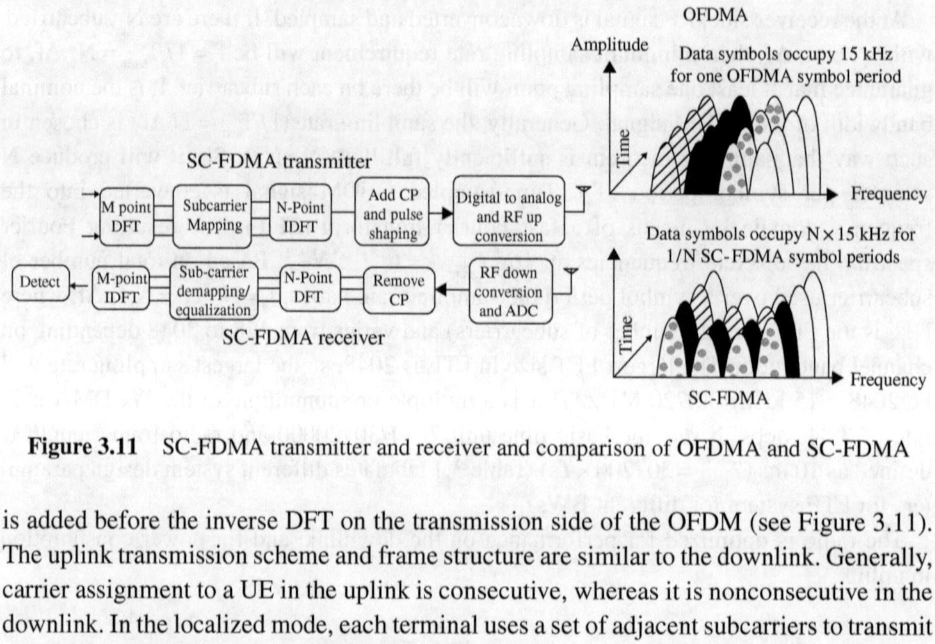

Figure 3.11 SC-FDMA transmitter and receiver and comparison of OFDMA and SC-FDMA

is added before the inverse DFT on the transmission side of the OFDM (see Figure 3.11). The uplink transmission scheme and frame structure are similar to the downlink. Generally, carrier assignment to a UE in the uplink is consecutive, whereas it is nonconsecutive in the downlink. In the localized mode, each terminal uses a set of adjacent subcarriers to transmit its symbols, whereas, in the distributed mode, the subcarriers used by a single terminal are distributed over the whole frequency band.

3.8 Channel Structure

Three types of channels are defined: logical, transport and physical channels. As shown in Figure 3.12, the SAP between the MAC and RLC sublayers provides the logical channels and the SAP between the PHY and MAC sublayers provide the transport channels, and the MAC performs multiplexing of logical channels on to the transport channels.

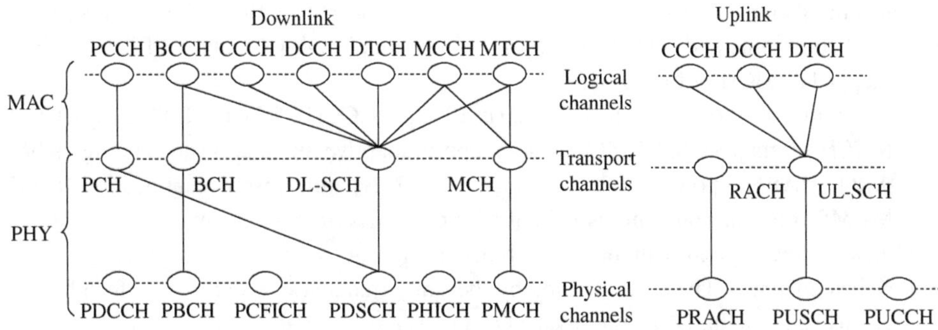

Figure 3.12 Downlink and uplink logical, transport and physical channels mapping

3.8.1 Downlink Channel Structure and Transmission Mechanism

3.8.1.1 Downlink Logical, Transport and Physical Channels

- Logical channels. These are characterized by the information that is transferred. The logical channels are divided into control channels and traffic channels. The control channels carry control-plane information and traffic channels carry user-plane information. Five control channels and two traffic channels are defined in downlink (see 3GPP TS 36.211).
  - Control channels. (i) Paging channel (PCCH): used for sending paging information. (ii) Broadcast control channel (BCCH): used for broadcasting system control information. (iii) Common control channel (CCCH): used for carrying control information between the network and UE and used by the UEs that have no RRC connection. (iv) Dedicated control channel (DCCH): this is a point-to-point bidirectional channel for exchanging control information, used by the UEs that have a RRC connection. (v) Multicast control channel (MCCH): this is a point-to-multipoint channel for transmitting MBMS control information and used by those UEs that are receiving MBMS.
  - Traffic Channels. (i) Dedicated traffic channel (DTCH): this is a point-to-point channel dedicated to a single UE for the transmission of user-specific information. (ii) Multicast traffic channel (MTCH): this is a point-to-multipoint channel used for the transmission of user MBMS data.
- Transport channels: characterized by how the data are transferred over the radio interface.
  - Paging channel (PCH): the PCCH is mapped to the PCH. For UE power saving, it supports discontinuous reception (DRX).
  - Broadcast channel (BCH): the BCCH logical channel is mapped either to a BCH or to a DLSCH depending on whether it carries a master information block (MIB) or a system information block (SIB). The BCH uses a fixed, predefined format as this is the first channel the UE receives after synchronizing with a cell. This is broadcast over the entire cell.

  - Multicast channel (MCH). This is required to broadcast over the entire cell. It supports SFN combining and semistatic resource allocation. The MCCH and MTCH are either mapped to MCH or to DL-SCH.
  - Downlink shared channel (DL-SCH): BCCH, CCCH, DCCH, DTCH, MCCCH, MTCH are mapped to DL-SCH. This supports adaptive modulation and coding, hybrid ARQ (HARQ), power control, semistatic and dynamic resource allocation, DRX, MBMS transmission. This is suitable for transmission over the entire cell coverage area or specific area with the use with beam forming.
- Physical channels. These are defined by the time, frequency resource used for physical transmission of data. Different physical channels are:
  - Physical broadcast channel (PBCH): PBCH is used for transmission of MIB (part of the BCH). It broadcasts the essential parameters for initial access of the cell, for example downlink system bandwidth, the physical hybrid ARQ indicator channel structure, and the most significant eight bits of the SFN. The PBCH transmission method is described in section 3.8.1.3.1.
  - Physical downlink shared channel (PDSCH). This is the main data bearing channel, which is allocated to users on a dynamic and opportunistic basis. The PCH and DL-SCH are mapped to this physical channel. It carries transport blocks (TB), which corresponds to one MAC PDU. These are passed from the MAC layer to the PHY layer once per TTI (1 ms). It is also used to transmit broadcast information not transmitted on the PBCH, which include system information blocks (SIB) and paging messages.
  - Physical multicast channel (PMCH) – the MCH is mapped to the PMCH, which is the multicell MBSFN transmission channel. It is defined to carry multimedia broadcast and multicast services (MBMS). It is transmitted in specific dedicated subframes where the PDSCH is not transmitted.
  - Control channels. The control channels occupy the first 1, 2, or 3 OFDM symbols in a subframe extending over the entire system bandwidth. In narrow band systems (less than 10 RBs), the control symbols can be increased to include the fourth OFDM symbol.
    - Physical downlink control channel (PDCCH). This is used to inform the UE about the resource allocation of PCH and DL-SCH and to indicate about the modulation, coding and hybrid-ARQ information related for DL-SCH. Generally, a maximum of three or four OFDM symbols can be used for PDCCH. The information carried on PDCCH is referred to as downlink control information (DCI). Depending on the purpose of the control message, different DCI formats are defined. Multiple PDCCHs can be transmitted in a subframe (see Figure 3.8) using control channel elements (CCEs). Once CCE consists of nine continuous resource element groups (REGs), where each REG consists of four resource elements (REs), as described in section 3.8.1.5 of this chapter in detail. QPSK modulation is used for the PDCCH and four QPSK symbols are mapped to each REG. Furthermore, 1, 2, 4, or 8 CCEs can be used for a UE, depending on channel conditions to ensure sufficient robustness.

- Enhanced physical downlink control channel (EPDCCH). This carries scheduling assignments and is transmitted using an aggregation of one or several consecutive enhanced control channel elements (ECCEs).
- The physical control format indicator channel (PCFICH). This carries the control frame indicator (CFI), which includes the number of OFDM symbols used for control channel transmission dynamically in each subframe (typically 1, 2, or 3). The unused OFDM symbols among the three or four PDCCH OFDM symbols can be used for data transmission. The 32-bit long CFI is mapped to 16 resource elements in the first OFDM symbol of each downlink frame using QPSK modulation.
- The physical hybrid ARQ indicator channel (PHICH). This is used to carry hybrid ARQ ACK/NACK for uplink transmissions (PUSCH). BPSK modulation is used with a repetition factor of three for robustness. The mapping of downlink and uplink logical, transport, and physical channels is shown in Figure 3.12.

Apart from these, there is one more channel called the relay physical downlink control channel (R-PDCCH), used for relaying.

3.8.1.2 Downlink Signals

In addition to these physical channels, some physical signals are also defined. These downlink physical signals correspond to a set of resource elements used by the physical layer only and do not carry any higher layer's information. The physical channels carry information bits, which are used by the upper layer, whereas physical signal does not carry any higher layer's information bits – rather these are mathematically designed signals used by the physical layer for synchronization and some other purposes. The downlink signals are broadly classified into two: (i) synchronization signals (primary synchronization and secondary synchronization signal); (ii) reference signals (CRS, UESRS).

Transmission of Synchronization Signals

Primary Synchronization Signal (PSS)

The PSS is constructed from a frequency-domain Zadoff–Chu (ZC) sequence of phase shifts with length 63, but the middle element (0) punctured to avoid transmitting on the DC subcarrier. The ZC sequences belong to a class of complex exponential sequences and are nonbinary unit-amplitude sequences, which satisfy a constant amplitude zero autocorrelation (CAZAC) property. The N_{ZC}-point DFT of ZC sequence also has constant amplitude, which limits the PAPR and generates bounded and time-flat interference for other users. This simplifies the terminal implementation as only phases need to be computed and stored, not the amplitudes. Any sequences of the same length have "ideal" cyclic autocorrelation, which means circularly shifted versions of itself is a delta function. This helps in detection of misaligned received signal with a reference sequence using correlation.The absolute value of the cyclic cross-correlation function between any two ZC sequences is also always constant.

In LTE the following PSS sequence {d(n), n ranges from 0 to 61} is used (see TS 36.211 subclause 6.11.1):

$$d_u(n) = \begin{cases} e^{-j\frac{\pi u n(n+1)}{63}} & n = 0,1,\ldots,30 \\ e^{-j\frac{\pi u(n+1)(n+2)}{63}} & n = 31,32,\ldots,61 \end{cases} \tag{3.1}$$

where the Zadoff–Chu root sequence index u has three possible values of 25, 29 and 34, based on cell physical layer identities $N_{ID}^{(2)} = 0, 1, 2$ respectively. So, three PSS sequences are derived and these represent three separate physical layer identities ($N_{ID}^{(2)}=0, 1, 2$), which are used during the cell search process. Each of the PSS sequences has a length of 62 (excluding 0) and these 62 symbols are mapped to 62 central subcarriers of the central six RBs (e.g. 6 * 12=72 subcarriers=1.08 MHz) in the frequency domain as shown in Figure 3.13. That means PSS (as well as SSS, see section 2.2) always occupies only 62 central subcarriers around the DC subcarrier (which is unused), and this does not

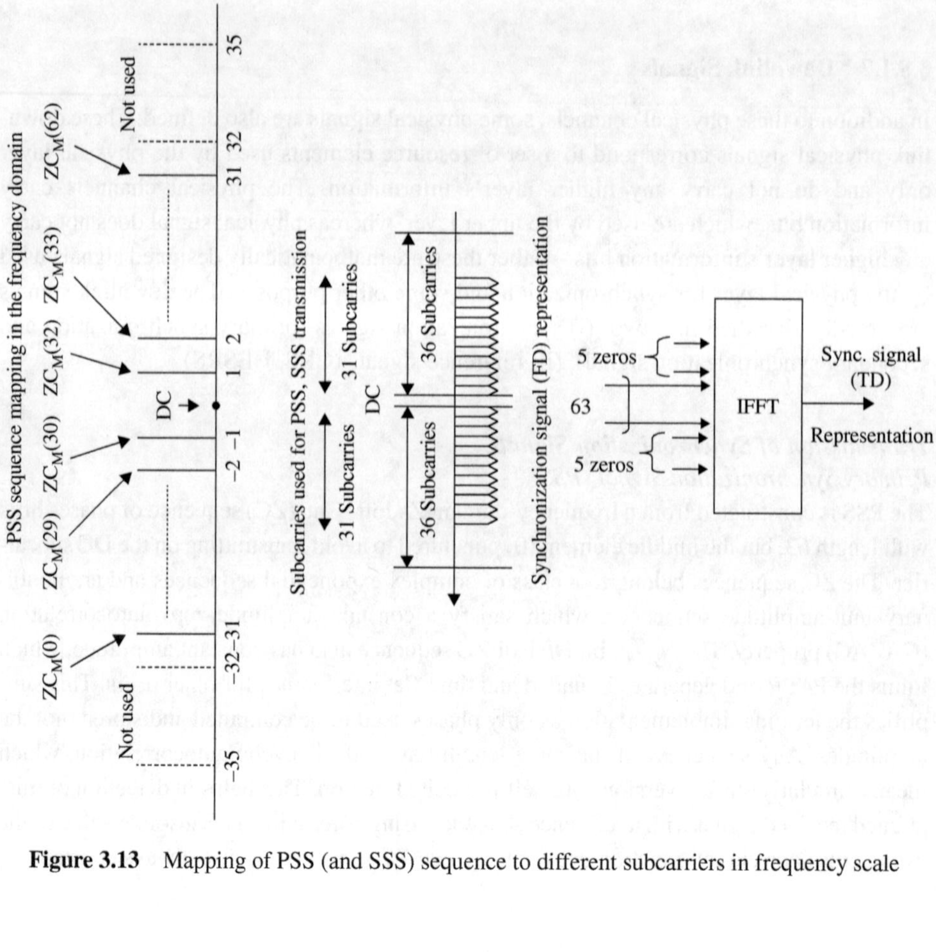

Figure 3.13 Mapping of PSS (and SSS) sequence to different subcarriers in frequency scale

change with respect to the used system bandwidth, which can vary from 6 to 110 RBs (1.4 to 20 MHz). So, during the initial sync-up procedure, this helps the UE to synchronize to the network without any a priori knowledge of the allocated bandwidth.

On the other hand, in the time scale, in the case of the LTE FDD cell, the PSS is always located in the last OFDM symbol of slot#0 (sub frame 0) and slot#10 (subframe 5) of each radio frame as shown in Figures 3.14(a) and (b). So, P-SCH is transmitted twice in each radio frame, especially to simplify the handover. Within one cell, the two PSSs within a frame are identical but, as discussed earlier, the PSS of a cell can take three different values depending on the physical layer cell identity of that cell. As shown in Figure 3.14(a), the five subcarriers (REs) at each side (each extremity) of the last OFDM symbol of the first (slot#0) and eleventh slots (slot#10) of each radio frame are unused and only the central 62 subcarriers are used, occupying a BW of 62 * 15 = 930 kHz. This helps the UE to detect the PSS (and SSS) using a size-64 FFT and a lower sampling rate correspondingly.

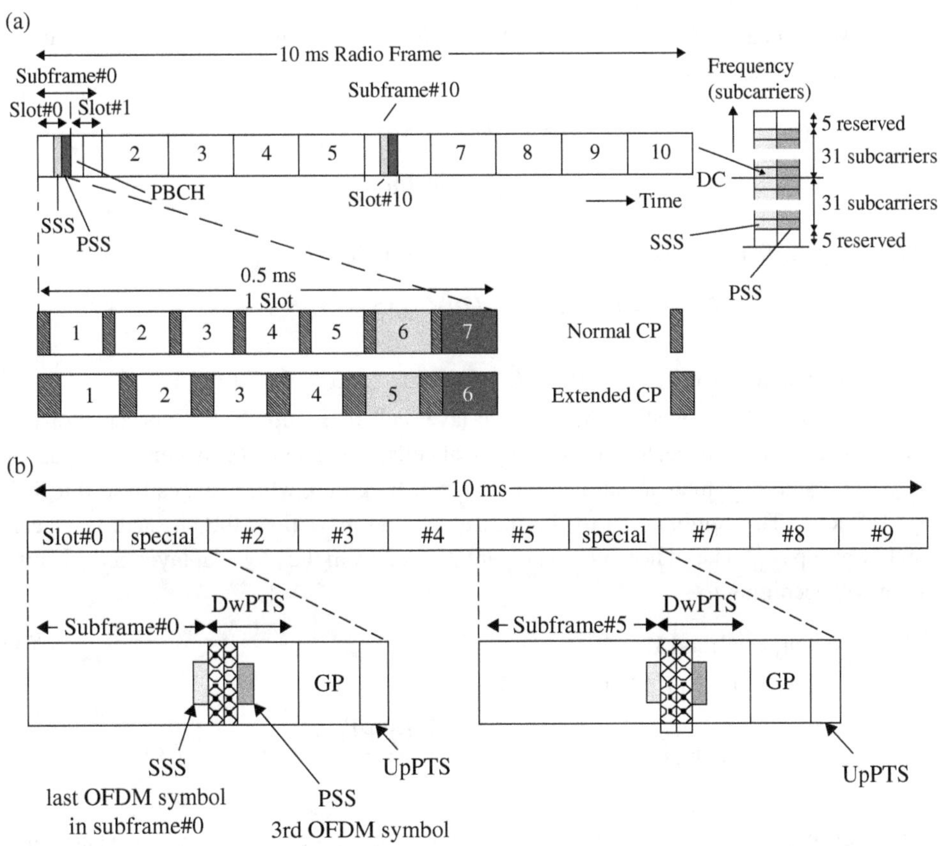

Figure 3.14 (a) PSS and SSS in FDD LTE system (type-1 frame structure). (b) Frame type 2 (TDD mode)

In TDD mode (Figure 3.14(b)), however, the PSS is placed in the third OFDM symbol positions of the downlink pilot time slot (DwPTS). The PSS sequences increase the distance between the SSS and the PSS, from being sent in adjacent symbols to being three symbols apart.

Transmission of Secondary Synchronization Signal (SSS)

In FDD mode, the SSS is placed adjacent to PSS – for example, the SSS is in the second-last OFDM symbol position in slot#0 and slot#10 as shown in Figure 3.14(a). In TDD, the cell PSS is sent in the third OFDM symbol of the third and thirteenth slots while SSS is transmitted three symbols earlier, as shown in Figure 3.14(b). This position difference allows for the detection of the duplexing scheme (FDD or TDD) used on a carrier, if this is not known in advance. The SSS sequences are based on maximum length sequences, known as M-sequences, which can be created by cycling through every possible state of a shift register of length n. The SSS signal consists of a frequency-domain sequence $d(0),\ldots,d(61)$ with the same length as the PSS (62), which is an interleaved concatenation of two length-31 binary sequences $s_0(n)$ and $s_1(n)$.

The two sequences $s_0^{(m_0)}(n)$ and $s_1^{(m_1)}(n)$ are defined as two different cyclic shifts of the m-sequence $\tilde{s}(n)$ according to

$$s_0^{(m_0)}(n) = \tilde{s}\left((n+m_0) \bmod 31\right)$$
$$s_1^{(m_1)}(n) = \tilde{s}\left((n+m_1) \bmod 31\right)$$

where, $0 \le n \le 30$, and $\tilde{s}(i) = 1 - 2x(i),\ 0 \le i \le 30$, is defined by

$$x(\bar{i}+5) = \left(x(\bar{i}+2) + x(\bar{i})\right) \bmod 2, \quad 0 \le \bar{i} \le 25$$

with initial conditions $x(0)=0,\quad x(1)=0,\quad x(2)=0,\quad x(3)=0,\quad x(4)=1.$

The SSS helps for indicating the physical layer cell ID group ($N_{\mathrm{ID}}^{(1)}$). In order to distinguish between different sector groups (physical cells), $s_0(n)$ and $s_1(n)$ depend on a pair of integers m_0 and m_1, which are unique for each cell ID group, which ranges from 0 to 167 e.g. 168 total. The combination of the indices m_0 and m_1 defines the physical layer cell identity group $N_{\mathrm{ID}}^{(1)}$. The indices m_0 and m_1 are derived from the physical layer cell-identity group $N_{\mathrm{ID}}^{(1)}$ according to

$$m_0 = m' \bmod 31$$
$$m_1 = (m_0 + m'/31 + 1) \bmod 31$$
$$m' = N_{\mathrm{ID}}^{(1)} + q(q+1)/2, \quad q = \frac{N_{\mathrm{ID}}^{(1)} + q'(q'+1)/2}{30}, \quad q' = N_{\mathrm{ID}}^{(1)}/30$$

Now, once the $s_0(n)$ and $s_1(n)$ (two length 31 binary sequences) are derived for a particular cell-ID group number, then these are scrambled. To randomize the interference from the neighboring cells, the concatenated sequence is scrambled with a scrambling sequence based on the PSS ($N_{\mathrm{ID}}^{(2)}$).

Scrambling Sequences

There are two types of scrambling sequences used: ($\tilde{c}(n)$, $\tilde{z}(n)$). These are derived as below.

For $\tilde{c}(n)$ sequences, there are two scrambling sequences $c_0(n)$ and $c_1(n)$, depending on the primary synchronization signal (PSS), and these are defined by two different cyclical shifts of the m-sequence $\tilde{c}(n)$ according to

$$c_0(n) = \tilde{c}\left(\left(n + N_{\mathrm{ID}}^{(2)}\right) \bmod 31\right)$$
$$c_1(n) = \tilde{c}\left(\left(n + N_{\mathrm{ID}}^{(2)} + 3\right) \bmod 31\right)$$

where, $N_{\mathrm{ID}}^{(2)} \in \{0,1,2\}$ is the physical layer identity within the physical layer cell identity group $N_{\mathrm{ID}}^{(1)}$ and $\tilde{c}(i) = 1 - 2x(i)$, $0 \le i \le 30$, is defined by

$$x(\bar{i}+5) = (x(\bar{i}+3) + x(\bar{i})) \bmod 2, \qquad 0 \le \bar{i} \le 25 \text{ with initial conditions}$$
$$x(0)=0,\quad x(1)=0,\quad x(2)=0,\quad x(3)=0,\quad x(4)=1.$$

$\tilde{z}(n)$ **Sequences:** the scrambling sequences $z_1^{(m_0)}(n)$ and $z_1^{(m_1)}(n)$ are defined by a cyclic shift of the m-sequence $\tilde{z}(n)$ according to

$$z_1^{(m_0)}(n) = \tilde{z}\left(\left(n + (m_0 \bmod 8)\right) \bmod 31\right)$$
$$z_1^{(m_1)}(n) = \tilde{z}\left(\left(n + (m_1 \bmod 8)\right) \bmod 31\right)$$

where, m_0 and m_1 are obtained from equations described in the section above (also please refer to Table 6.11.2.1-1 in 3GPP TS 36.211 version 10.0.0, release 10, for more details) and $\tilde{z}(i) = 1 - 2x(i)$, $0 \le i \le 30$, is defined by

$$x(\bar{i}+5) = \left(x(\bar{i}+4) + x(\bar{i}+2) + x(\bar{i}+1) + x(\bar{i})\right) \bmod 2, \qquad 0 \le \bar{i} \le 25$$

with initial conditions $x(0)=0,\quad x(1)=0,\quad x(2)=0,\quad x(3)=0,\quad x(4)=1.$

Like PPS, SSS is also transmitted twice per radio frame but here the transmitted sequences in subframe 0 (second-last symbol of slot 0 in the FDD) and transmitted sequences in subframe 5 (second-last symbol of slot 11) are different.

Generation of SSS in Subframe# 0

The even sequence (d(2n)) is derived by scrambling $s_0(n)$ with $c_0(n)$ and the odd sequence (d(2n + 1)) is derived by scrambling $s_1(n)$ with $c_1(n)$ and $z_1^{(m_0)}(n)$ as shown in Figure 3.15:

$$d(2n) = \begin{cases} s_0^{(m_0)}(n)\, c_0(n) & \text{in subframe 0} \end{cases}$$

$$d(2n+1) = \begin{cases} s_1^{(m_1)}(n)\, c_1(n)\, z_1^{(m_0)}(n) & \text{in subframe 0} \end{cases}$$

Generation of SSS in Subframe #5

The even sequence (d(2n)) is derived by scrambling $s_1(n)$ with $c_0(n)$ and the odd sequence (d(2n+1)) is derived by scrambling $s_0(n)$ with $c_1(n)$ and $z_1^{(m_1)}(n)$, similar to subframe 0, as shown in Figure 3.15, but $s_1(n)$ and $s_0(n)$ are interchanged:

$$d(2n)=\begin{cases} s_1^{(m_1)}(n)c_0(n) & \text{in subframe 5} \end{cases}$$

$$d(2n+1)=\begin{cases} s_0^{(m_0)}(n)c_1(n)z_1^{(m_1)}(n) & \text{in subframe 5} \end{cases}$$

Next, the 62 symbols ($d(0),\ldots,d(61)$) SSS signal is generated by concatenating above generated even and odd sequences for subframe 0 and 5.

In the time domain, in an FDD cell, the SSS is always located in the second-last OFDM symbol of the first and eleventh slots of each radio frame – for example, the SSS is located in the symbol immediately preceding the PSS.

In the case of the TDD, the SSS is transmitted in the last symbol of subframes 0 and 5 – that means three symbols ahead of the PSS.

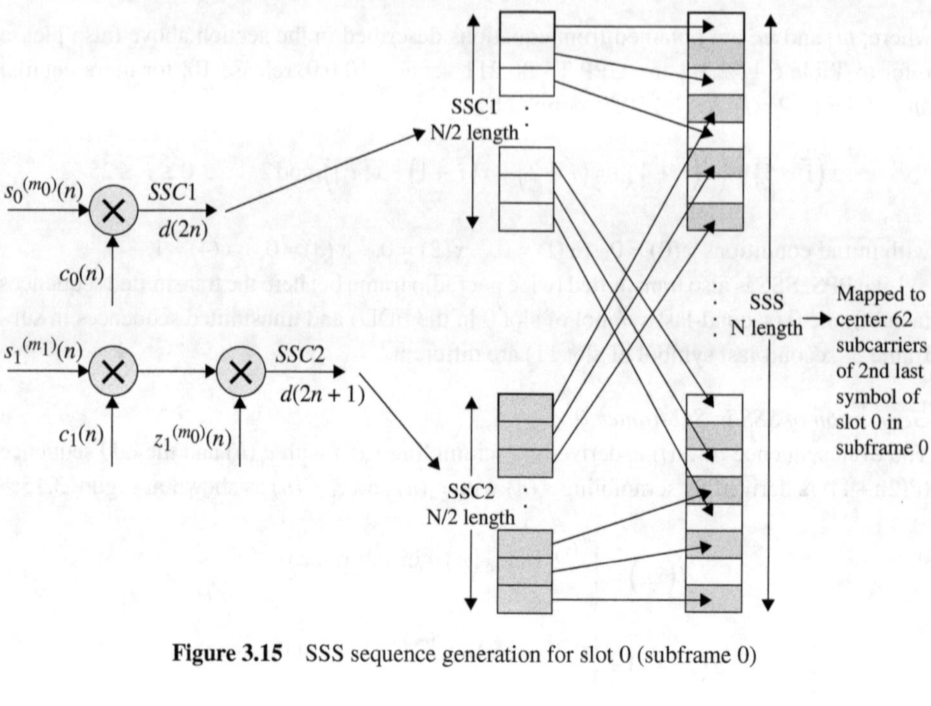

Figure 3.15 SSS sequence generation for slot 0 (subframe 0)

Reference Signals (RS)

The reference signal, as the name indicates, is a reference provided to the terminals whose reception quality is estimated by the terminal and compared against the reference for error estimate. The LTE downlink reference signals are:

- cell-specific reference signals (CRS);
- UE-specific reference signals (UESRS);
- MBSFN-specific RSs.

Transmission of Cell-Specific Reference Signal (CRS)

The CRSs are used for various downlink measurements as well as for demodulation of non-MBSFN transmissions. They can be used by the terminal for channel estimation for coherent demodulation of any downlink physical channel except for PMCH and for PDSCH in the case of transmission modes 7, 8, or 9. The measurements performed using cell-specific reference signals include channel quality estimation, MIMO rank calculation, MIMO precoding vector/matrix selection and measurements for handoff.

For CRS signal transmission, two types of sequences were considered. The first approach to reference signals sequence design is based on using orthogonal sequences for three cells (sectors) within an eNB. The orthogonal sequences are further scrambled by a PN sequence. A problem with orthogonal reference signal sequences is the loss of orthogonality in a frequency-selective channel. In the second approach, a simple cell-specific PN sequence is used as a reference signal sequence without any spreading using orthogonal sequences. There are 504 different reference-signal sequences defined for LTE, where each sequence corresponds to one of 504 different physical-layer cell identities. The CRS is cell specific and remains the same for the entire cell once configured.

In LTE, the CRSs are arranged in a specific manner in the time frequency two-dimensional lattice based on the system's requirements. It consists of reference symbols of predefined values inserted within the first and third last OFDM symbol of each slot in time dimension and with a frequency-domain spacing of six subcarriers. Furthermore, there is a frequency-domain staggering of three subcarriers for the reference symbols within the third last OFDM symbol. Within each resource-block pair (during one 1 ms subframe), consisting of 12 subcarriers, there are thus eight reference symbols.

The required spacing in time between the reference symbols can be obtained by considering the maximum Doppler spread (highest speed) to be supported, which for LTE corresponds to 500 km/h. The Doppler shift is $fd = (fc\ v/c)$ where fc is the carrier frequency, v is the UE speed in meters per second, and c is the speed of light ($3 \cdot 10^8$ m/s). Considering $fc = 2$ GHz and $v = 500$ km/h, then the Doppler shift is $fd = 950$ Hz. According to Nyquist's sampling theorem, the minimum sampling frequency needed in order to reconstruct the channel is therefore given by $Tc = 1/(2fd)$ 0.5 ms under the above assumptions. So, this implies that two reference symbols per slot (RB) are needed in the time domain in order to estimate the channel correctly.

The required spacing in frequency between the reference symbols can be obtained by considering the expected coherence bandwidth of the channel, which is in turn related to the channel delay spread. In particular the 90% and 50% coherence bandwidths are given, respectively, by $Bc_{,90\%} = 1/50\sigma_\tau$ and $Bc_{,50\%} = 1/5\sigma_\tau$ where σ_τ is the r.m.s delay spread. The maximum r.m.s channel delay spread considered is 991 ns, corresponding to $Bc_{,90\%}$ = 20 kHz and $Bc_{,50\%}$ = 200 kHz. In LTE, the spacing between two reference symbols in frequency, in one RB, is 45 kHz (15 kHz × 3, e.g. three subcarriers), thus allowing the expected frequency domain variations of the channel to be resolved. RS patterns are defined for multiple "antenna ports" at the eNodeB and, as shown in Figure 3.16, in the case of a single antenna the spacing between two CRS symbols is 3 × 2 = 6, whereas in the case of four antennas it is 3. The reference signal is repeated with a periodicity of 1 frame (10 ms).

In order to avoid reference signals collisions among the neighboring cells, a cell-specific frequency shift is applied to reference signal mapping to resource elements as:

$$v_{shift} = N_{cellID} \bmod 6.$$

This means that the CRS symbols are mapped to every sixth subcarrier and the start index of the subcarrier is determined by the physical layer cell ID.

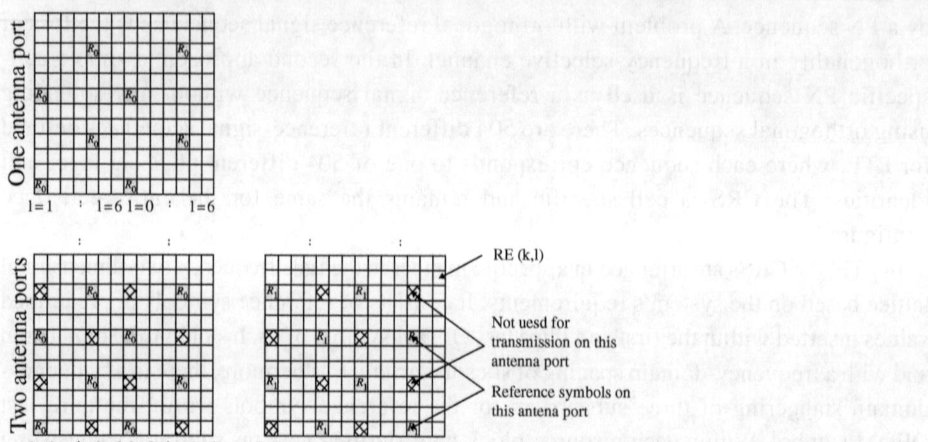

Figure 3.16 Mapping of downlink reference signals for different antenna ports (normal cyclic prefix)

The LTE system allows reference signal boosting where power on reference signal symbols can be different from the data symbols' power. But, in a synchronized system operation, if all cells in the system boost the reference signal power, the reference signals are going to experience higher interference, which can undermine the benefit of reference signal power boosting. So, as explained above, the reference signals of the two cells are mapped to different resource elements and hence collision probability reduces.

For cell-specific reference signals, up to a maximum of four antenna ports are supported.

In the LTE system, a maximum of four antenna ports defined and the CRS transmission in those ports are shown by reference signals [R0, R1, R2, R3] in Figure 3.16.

UE-Specific Reference Signal (UESRS)
UE-specific reference signals are transmitted in addition to the CRSs primarily for enabling beamforming of the data transmissions to specific UEs (see Figure 3.17). If the downlink data transmissions is configured (by a higher layer) for receiving UESRS, then UESRS will be transmitted in the RBs to which PDSCH is mapped for the UEs. If UESRSs are used, the UE uses them for channel estimation for demodulating the data in the corresponding PDSCH RBs. Thus the UERSs are considered as being transmitted using a distinct antenna port (antenna port 5), with its own channel response from the eNodeB to the UE.

MBSFN-Specific Reference Signal
These are transmitted only when multimedia broadcast single frequency network (MBSFN) operation is used. These signals are present only in subframes allocated for MBSFN. A pseudorandom sequence depending on MBSFN ID will be transmitted on antenna port 4.

3.8.1.3 System Information (SI) Transmission

In typical cellular systems, the basic system information (SI) is very essential for system configuration and operation. This is repeatedly broadcast by LTE eNB over the logical

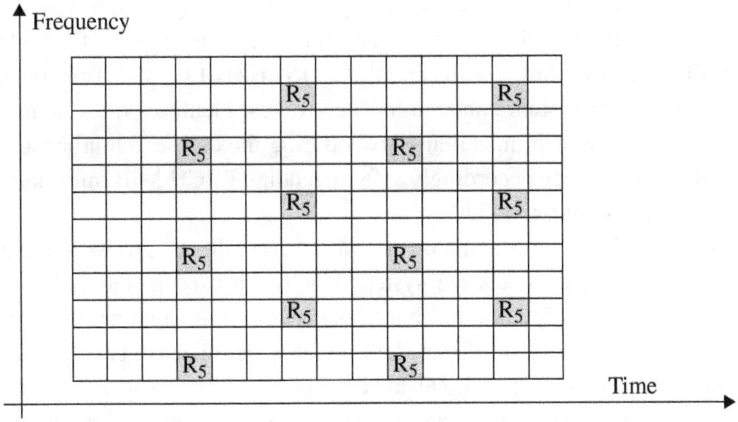

Figure 3.17 UE-specific RS arrangement with normal CP

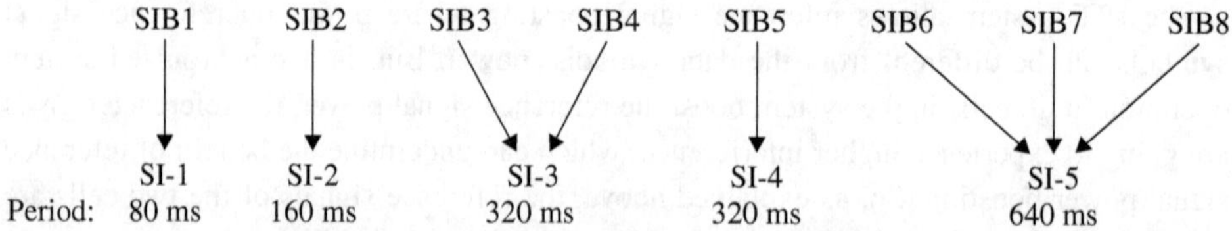

Figure 3.18 Different SIB transmission intervals

channel BCCH. There are two parts in SI – a static part and a dynamic part. The static part is known as master information block (MIB) and is transmitted using transport channel BCH. This is mapped to physical channel PBCH (see Figure 3.12) and MIB is transmitted once every 40 ms that means the BCH transmissions time interval (TTI) is 40 ms. Dynamic part is known as system information block (SIB) and is mapped on RRC SI messages (SI-1,2,3,4,5,6,7,8,9,10,11) which is mapped over DL-SCH channel and transmitted using PDSCH physical channel at periodic intervals (see Figure 3.18). The presence of system information on DL-SCH in a subframe is indicated by the transmission of a corresponding PDCCH marked with a special system-information RNTI (SI-RNTI). Similar to the PDCCH providing the scheduling assignment for "normal" DL-SCH transmission, this PDCCH also indicates the transport format and physical resource (set of resource blocks) used for the system-information transmission. System information blocks are grouped in SI containers. Each SI is composed of multiple SIBs. Each SI usually will have different transmission periodicity and will be sent in a single subframe as described in Table 3.2. Not all the SIBs need always be present.

PBCH Transmission

The master information block (MIB) consists of a limited number of the most frequently transmitted parameters, essential by UE for initial access to the cell. The MIB is transmitted via PBCH and the PBCH is always transmitted with a fixed bandwidth. The detectability of this bandwidth without the UE having prior knowledge of the system bandwidth is achieved by mapping the PBCH only to the central 72 subcarriers of the OFDM signal. That means PBCH is transmitted with central 72 * 15 kHz (=1.08 MHz) central bandwidth, regardless of the actual system bandwidth. The UE first identifies the system center-frequency (DC) from the synchronization signals during the cell search procedure and then reads the PBCH information accordingly. After decoding PBCH MIB information, UE get to know the actual system Bandwidth.

The size of the MIB is 14 bits and it is transmitted every 40 ms. The MIB contains (i) DL Bandwidth (3 bits) – it indicates the system bandwidth used in the cell (1.4 MHz … 20 MHz); (ii) PHICH Configuration (3 bits); (iii) system frame number (8 bits) – actually the SFN is 10 bits wide but out of that the two least significant bits of the SFN are not included in the MIB, which the terminal can detect indirectly; (iv) spare bits (10 bits).

The MIB also indirectly communicates one of the three antenna configurations (one, two or four) in use in the downlink (DL). The CRC of the transmitted MIB data undergoes

Table 3.2 LTE system information blocks

LTE system information blocks	Description
MIB	Transmitted on every 40 ms on PBCH. It contains channel bandwidth, PHICH configuration details, transmit power, number of antennas and SIB scheduling information etc. (Received in idle and connected mode).
SIB1	Transmitted on every 80 ms on DL-SCH. Indicates whether a terminal is allowed to camp on a cell or not. Carries cell ID, MCC, MNC, TAC, SIB mapping. Includes information about the allocation of subframes to uplink/downlink and configuration of the special subframe in case of TDD. (Received in idle and connected mode).
SIB2	Transmitted every 160 ms. Includes information that terminal needs in order to be able to access the cell. Carries information about common channel and shared channel, uplink power control, preamble power ramping, uplink cyclic prefix length, subframe hopping, uplink EARFCN, uplink cell bandwidth and random-access parameters. (Received in idle and connected mode).
SIB3	Transmitted every 320 ms. Carries cell reselection information and Intra frequency cell reselection information. (Received in idle mode.)
SIB4	Contains neighboring-cell-related information, including information related to neighboring cells on the same carrier, neighboring cells on different carriers, and neighboring non-LTE cells, such as WCDMA/HSPA, GSM, and CDMA2000 cells.
SIB5	Transmitted every 640 ms. Carries interfrequency neighbors (on different frequency); E-UTRA LTE frequencies, other neighbor cell frequencies from other RATs for cell reselection and handover. (Received in idle mode.)
SIB6	Transmitted every 640 ms. Carries WCDMA neighbors information (serving UTRA and neighbor cell frequencies) for cell reselection. (Received in idle mode if UE supports WCDMA.)
SIB7	Carries GSM neighbours information (GERAN and neighbor cell frequencies) for cell reselection as well as handover purpose. (Received in idle mode if UE supports GSM.)
SIB8	Carries CDMA-2000 EVDO frequencies, CDMA-2000 neighbor cell frequencies.
SIB9	Carries HNBID (Home eNodeB Identifier).
SIB10	Carries ETWS prim notification (public warning messages).
SIB11	Carries ETWS s notification.
SIB13	SIB13 contains information necessary for MBMS reception.

through a special mask to encode this information. On the UE receiver side, the UE tries several times iteratively to determine the mask applied, and from that it detects the DL antenna configuration.

For physical layer processing, the BCH data arrives to the encoding unit in the form of a maximum of one transport block every transmission time interval (TTI) of 40 ms. As discussed earlier, in PBCH there are 14 information bits and 10 spare bits, so a total of

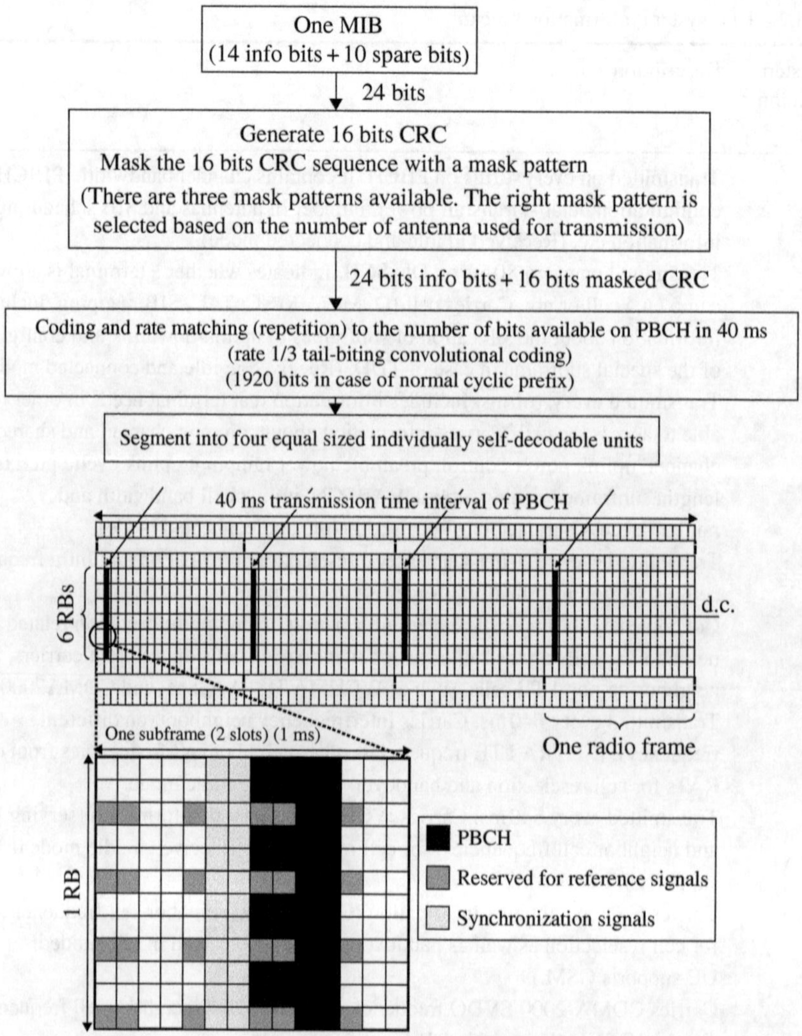

Figure 3.19 PBCH transmission steps

24 bits (see Figure 3.19). Then, from this information, the 16 CRC parity bits are computed. The eNB can use one, two or four antennas for transmission. The CRC bits are scrambled (masked), based on the one, two or four antennas used in the transmitter, and the scrambling is performed according to the eNode-B transmit antenna configuration with the sequence $x_{\text{ant},k}$ as indicated in Table 3.3 to form the sequence of bits c_k as below [1]:

$$c_k = \begin{cases} a_k & 0 \le k < A \\ \left(p_{k-A} + x_{\text{ant},k-A}\right) \bmod 2 & A \le k < (A+L). \end{cases}$$

where, $L = 16$.

Table 3.3 PBCH CRC masks for different antennas

Number of transmit antenna ports at eNB	PBCH CRC mask {Xant0, ... Xant15}
1	{0,0,0,0,0,0,0,0,0,0,0,0,0,0,0,0}
2	{1,1,1,1,1,1,1,1,1,1,1,1,1,1,1,1}
4	{0,1,0,1,0,1,0,1, 0,1,0,1,0,1,0,1}

As shown in Figure 3.19, using a 24-bit MIB data sequence, 16 bits parity is generated. Next, those 16 bits are appended to the MIB data sequence (24 bits) to form a 40-bit data sequence. Based on the multiantenna transmission scheme used in the eNB transmission, the CRC masking is performed according to Table 3.3. There are three sequences 0×0000, 0×5555 and $0\times ffff$ corresponding to 1, 2 or 4 eNB antennas. So, $14+10+16=40$ bits of data are generated after CRC addition. Once CRC is attached, the bit sequence c_k is coded using a rate 1/3 tail-biting convolutional coding. The reason for using convolutional coding for BCH, rather than the turbo code used for all other transport channels, is the small size of the BCH transport block. With such small blocks, tail-biting convolutional coding actually outperforms turbo coding. Next, the coded bits are rate matched using the circular buffer approach to obtain the rate matched sequence $b(0), b(1), \ldots, b(\text{Mbit}-1)$, where Mbit is the number of bits transmitted on the PBCH and it is dependent on the length of cyclic prefix. After processing, there will be a total of 1920 bits (for normal cyclic prefix prefix configuration) at the output of processing chain, before mapping to resources. This sequence $b(0), b(1), \ldots, b(\text{Mbit}-1)$ is scrambled with a cell-specific sequence and mapped to resource grid and modulated using QPSK modulation and transmitted. No channel interleaving is used on PBCH. The physical BCH channel is restricted to the 72 subcarriers around the DC in the resource grid. The PBCH is transmitted in the first four OFDMA symbols of the second slot of each radio frame as shown in Figure 3.19. In LTE, single antenna, two-antenna SFBC and four-antenna combined SFBC-FSTD transmit diversity schemes are supported on the physical broadcast channel.

3.8.1.4 PCFICH Transmission

As discussed earlier (see Figure 3.12), there are three types of downlink control channels: PCFICH, PHICH, and PDCCH. These are transmitted using resource element groups (REGs), where each REG contains four consecutive REs (or four REs separated by a CRS) on the same OFDM symbol and in the same resource block. Each subframe is divided into a control region followed by a data region. The size of the control region is expressed in OFDM symbols and can be dynamically varied to match the traffic situation. The number of control symbols per subframe can be two, three, or four for a system with bandwidth of 1.4 MHz and one, two, or three for other bandwidths, and can change from one subframe to the next. In the case of carrier aggregation, there is one control region per component carrier.

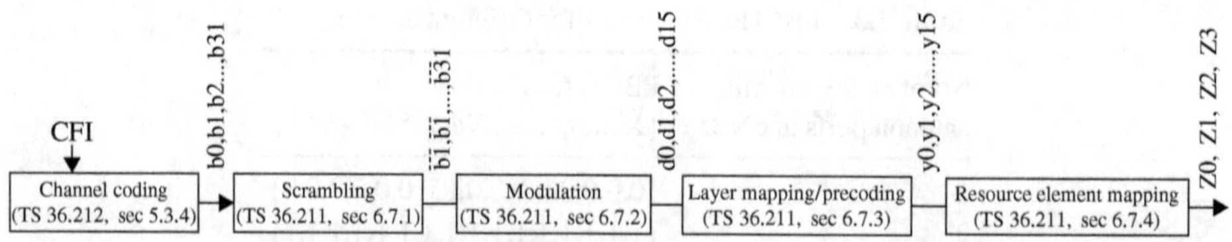

Figure 3.20 PCFICH processing

It is transmitted at the beginning of the subframe to allow terminals to decode downlink scheduling assignments as early as possible. In every subframe, eNB indicates the number of control symbols (used for PDCCH transmission) using the control format indicator (CFI) and transmits it on the PCFICH. It consists of two bits of information, corresponding to the three control-region sizes of 1, 2, or 3 OFDM symbols (or 2, 3, or 4 for a narrow bandwidth, $N_{RB}^{DL} \leq 10$), which are coded into a 32-bit codeword. This is scrambled with a cell- and subframe-specific scrambling code to randomize intercell interference, and QPSK modulated, then mapped to 16 resource elements of the first OFDM symbol of a subframe in groups of four resource elements (4 REGs). Based on the cell identity, the location of the four groups in the frequency domain is decided, which helps to avoid collisions between PCFICH transmissions in neighboring cells. The exact position of PCFICH can be measured from the cell ID and bandwidth using formulas given in 3GPP spec 36.211:

$$Z^{(p)}(0) \text{ is mapped to the REG represented by } k = k1$$

$$Z^{(p)}(1) \text{ is mapped to the REG represented by } k = k1 + \lfloor N_{RB}{}^{DL}/2 \rfloor . N_{sc}{}^{RB}/2.$$

$$Z^{(p)}(2) \text{ is mapped to the REG represented by } k = k1 + \lfloor 2.N_{RB}{}^{DL}/2 \rfloor . N_{sc}{}^{RB}/2.$$

$$Z^{(p)}(3) \text{ is mapped to the REG represented by } k = k1 + \lfloor 3.N_{RB}{}^{DL}/2 \rfloor . N_{sc}{}^{RB}/2.$$

where the additions are modulo $N_{RB}{}^{DL} N_{sc}{}^{RB}$, $k1 = (N_{sc}{}^{RB}/2).(N_{ID}{}^{cell} \bmod 2.N_{RB}{}^{DL})$.

where, $N^{RB}{}_{SC}$ = Number of frequency carriers per Resource block, $N^{DL}{}_{RB}$ = Number of resource blocks per bandwidth, $N^{cell}{}_{ID^{\circ}}$ = Physical Cell id.

Figure 3.20 shows the steps performed for PCFICH channel processing.

3.8.1.5 PDCCH Transmission

The PDCCH is used to carry downlink control information (DCI) and it contains downlink or uplink scheduling information as well as uplink power control commands for a UE or group of UEs and number of resource blocks, resource allocation type, modulation scheme, transport block, redundancy version, coding rate, and so forth. Multiple PDCCHs can be present in a subframe and a large number of possible PDCCH transmission formats is supported. DCI is therefore categorized into different DCI formats, where a format

Table 3.4 DCI formats

DCI Format	Use
Format 0	Uplink DCI formats. Used for transmission of resources to UE for sending their uplink data.
Format 1	Downlink DCI format. Used for downlink scheduling for 1 PDSCH codeword (SISO/SIMO modes).
Format 1A	Downlink DCI format. Compact version of format 1 scheduling for 1 PDSCH codeword or Dedicated preamble assignment to initiate random access.
Format 1B	Downlink DCI format. Used for transmission control information of multiple input multiple output (MIMO) rank 1 based compact resource assignment.
Format 1C	Downlink DCI format. It is used for very compact transmission of PDSCH assignment.
Format 1D	Downlink DCI format. Same as format1B with additional information about power offset (added attenuation or amplification over the base power).
Format 2	Downlink DCI format. Format 2 and Format 2A for transmission of DL-SCH allocation for closed and open loop MIMO operation, respectively.
Format 3	Uplink DCI formats. Format 3 and format 3A for transmission of TPC command for an uplink channel.

corresponds to a certain message size and usage. DCI formats and their uses are discussed in Table 3.4.

Resources are allocated in terms of CCE (control channel elements), where 1 CCE=9 REGs and 1 REG=4 REs. The number of CCEs used to transmit the control information will be variable based on the PCFICH value, bandwidth of the system from 1.4 MHz to 20 MHz and number of antenna ports present that in turn determines the presence of number of reference signals. The eNodeB divides CCEs into two parts, known as search space: (i) common search space – these CEs (maximum 16) are used for sending the control information, which is common for all the UEs; (ii) UE-specific search space: this is used for sending the control information for a particular UE only (can be decoded by a specific UE).

The PDCCH transmission chain is depicted in Figure 3.21. Multiple PDCCHs may be transmitted in a subframe (in case of dedicated control), one for each UE scheduled for uplink or downlink transmission.

The entire PDCCH payload is used to calculate the 16-bit-long CRC parity bits, which is used for error detection. The CRC parity bits generated are scrambled with the UE-RNTI (MAC ID) sequence (which is a UE-specific RNTI sequence and the same as MAC ID). Different UEs have different UE-RNTIs. So the UE-RNTI will be chosen depending on the UE to which the message will be sent. So only the intended UE can decode the DCI format and hence the corresponding PDSCH information. The uplink antenna selection information (when configured by higher layers and used with format 0 only) is also carried implicitly in the CRC. In this case, the CRC parity bits are scrambled with both the antenna selection mask and the RNTI. The CRC is attached to PDCCH information, then tail-biting

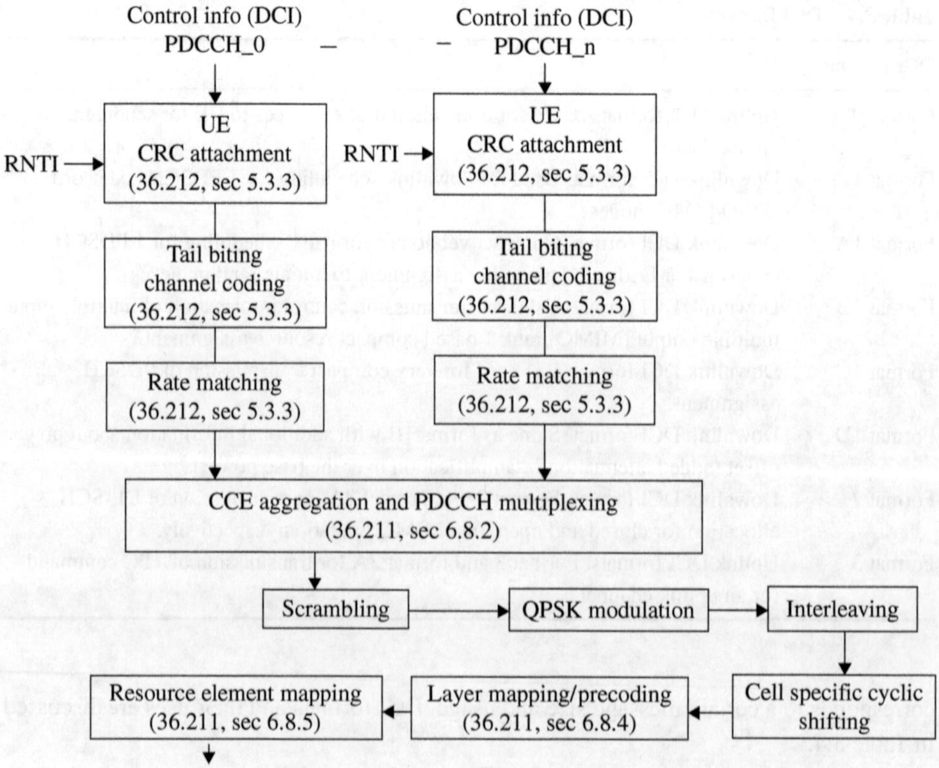

Figure 3.21 PDCCH transmission processing

convolutional coding and rate matching is performed separately on each PDCCH as shown in Figure 3.21. Next, the coded bits from multiple PDCCHs are multiplexed and then scrambled, modulated, transmit diversity precoded, and interleaved. After that, cell-specific permutation (to randomize interference to the neighboring cells) is applied and mapped.

3.8.1.6 Physical Hybrid-ARQ Indicator Channel (PHICH)

The physical hybrid-ARQ indicator channel in the downlink carries hybrid ARQ acknowledgements (ACK/NACK) of uplink data transfers. It is located in the first OFDM symbol of each subframe (for FDD, normal PHICH duration).

If PHICH is transmitted on a very narrow bandwidth (as information contains only 1 bit), it will create interference peaks in the neighboring cells. So, spreading is used, which increases bandwidth. Therefore the HIs for multiple UEs within a PHICH group are code multiplexed. A PHICH is carried by several resource element groups (REGs) and multiple PHICHs can share the same set of REGs and are differentiated by orthogonal

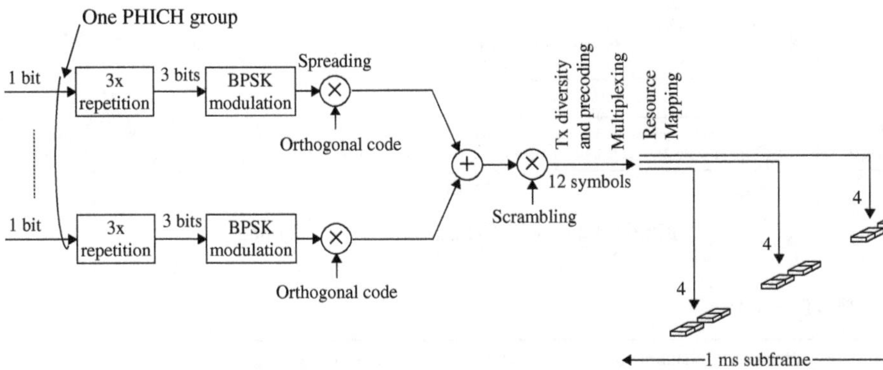

Figure 3.22 PHICH transmission processing

sequences. PHICHs that share the same resources are called a PHICH group. A PHICH group consists of eight PHICHs in the case of normal cyclic prefix. A specific PHICH is identified by two parameters: the PHICH group number, and the orthogonal sequence index within the group. The number of PHICH groups that can be supported in a system depends on the specific configuration. The actual number of PHICH groups can be derived from the downlink bandwidth and the parameter (Ng: 1, 2, 1/6, 1/2). These are broadcast in MIB (refer to 3GPP TS 36.211 section 6.9).

The PHICH transmission steps are described in Figure 3.22. As on PUSCH maximum, a single codeword or transport block can be transmitted, therefore 1 bit is enough for an ACK/NACK indication where "1" indicates a positive acknowledgment (ACK) and "0" indicates a negative acknowledgment (NACK). Then instead of using any sophisticated coding, simple repetition coding is used – for example, ACK is transmitted as "111" and NACK "000." These three modulation symbols are multiplied by the orthogonal sequences, which have the spreading factor of four for the normal cyclic prefix, resulting in a total of 12 symbols. Each REG contains four resource elements (REs) and each RE can carry one modulation symbol, so three REGs are needed for a single PHICH.

3.8.1.7 Physical Downlink Shared Channel (PDSCH)

The PDSCH carries user specific data (DL payload), known as transport blocks (TB), which correspond to a MAC PDU. These are passed from the MAC layer to the PHY layer once per transmission time interval (TTI), which is 1 ms. The PDSCH is also used to transmit system information blocks (SIB) and for paging and RRC signalling messages. The PDSCH transmission data processing is shown in Figure 3.23 and the details of each module are described in the next section.

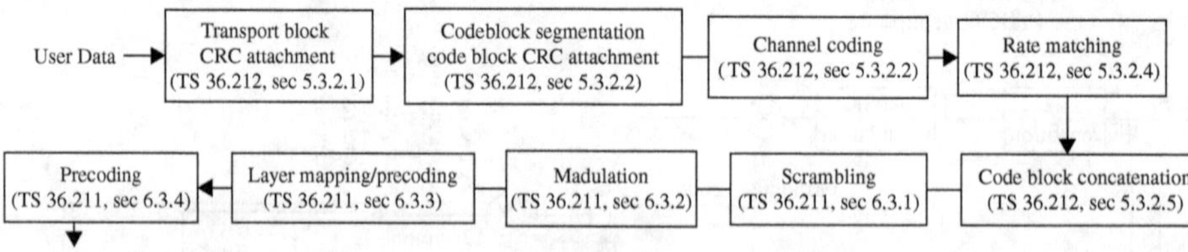

Figure 3.23 PDSCH transmission processing

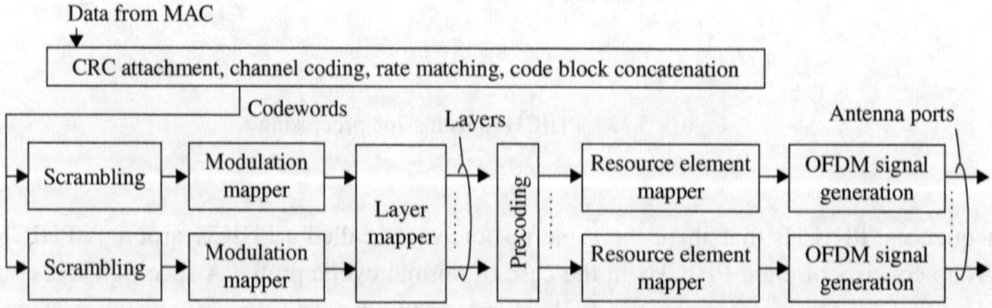

Figure 3.24 Overview of physical channel processing (DL)

3.8.2 *Downlink Physical Channel Processing*

As discussed in the previous sections, the data and control streams from/to the MAC layer are encoded/decoded to offer transport and control services over the radio link. As shown in Figure 3.24 and Figure 3.25, the physical layer processing scheme is a combination of error detection, error correction, rate matching, interleaving, and transport channel or control information mapping onto/splitting from various types of physical channels. As shown in Figure 3.25, the processing steps varies based on the channel type, which was discussed in detail in previous sections.

The following steps are involved (refer to 3GPP TS 36.212 section 5):

- CRC computation. At first the transport block is passed through a CRC encoder, it will generate 24 or 16 or 8 bit parity bits using cyclic generator polynomials based on different channel types. Then CRC bits are appended to the data bits.
- Code block segmentation and CRC attachment. If the input bit sequence is larger than the maximum code block size (6144), then segmentation of the input bit sequence is performed and an additional CRC sequence of length 24 bits is attached to each code block.
- Channel coding. Generally, tail-biting convolutional coding or turbo coding is applied to TrCHs. Different channels use different coding schemes as shown in Table 3.5.
- Interleaving and rate matching. The rate matching for turbo coded and convolutionally coded transport channels (and control information) is defined per coded block and consists of interleaving the three information bit streams $d_k^{(0)}$, $d_k^{(1)}$ and $d_k^{(2)}$, followed by the collection of bits and the generation of a circular buffer as depicted in Figure 3.26.

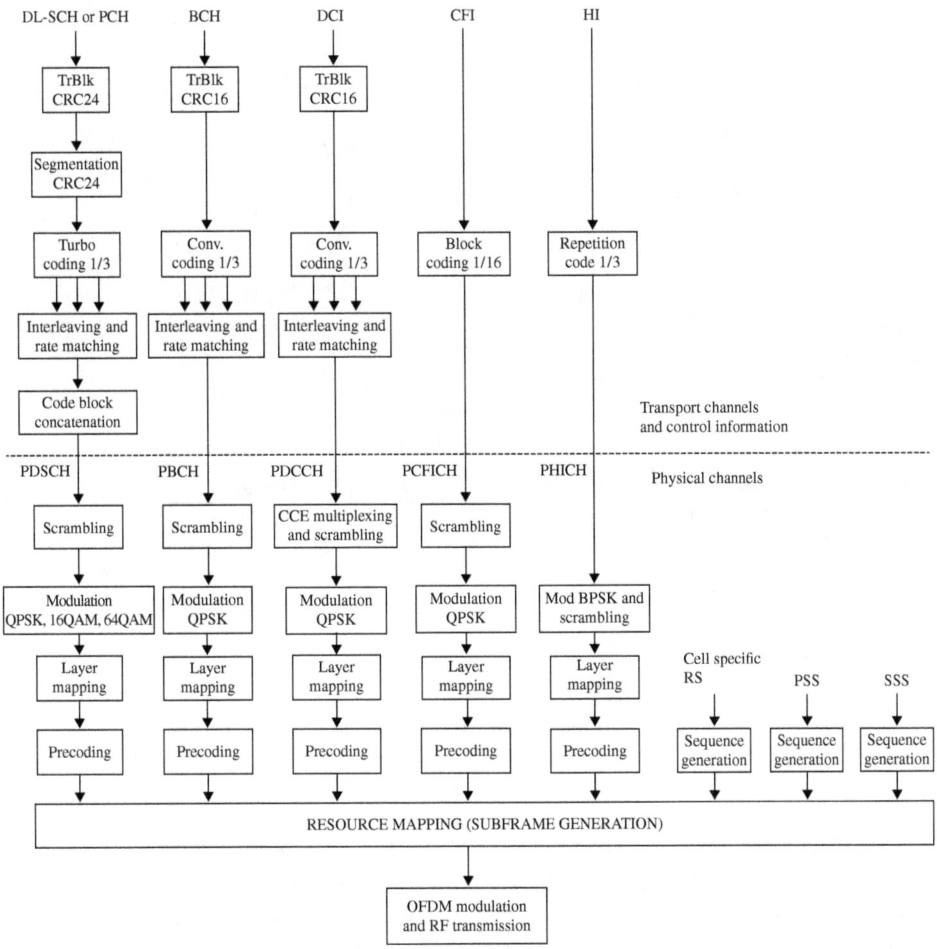

Figure 3.25 Processing blocks for different downlink transmission channels

Table 3.5 Channel coding scheme and coding rate

TrCH	Coding scheme	Coding rate	Control Information	Coding scheme	Coding rate
UL-SCH	Turbo coding	1/3	DCI	Tail-biting convolutional coding	1/3
DL-SCH			CFI	Block code	1/16
PCH			HI	Repetition code	1/3
MCH			UCI	Block code	variable
BCH	Tail-biting convolutional coding	1/3		Tail-biting convolutional coding	1/3

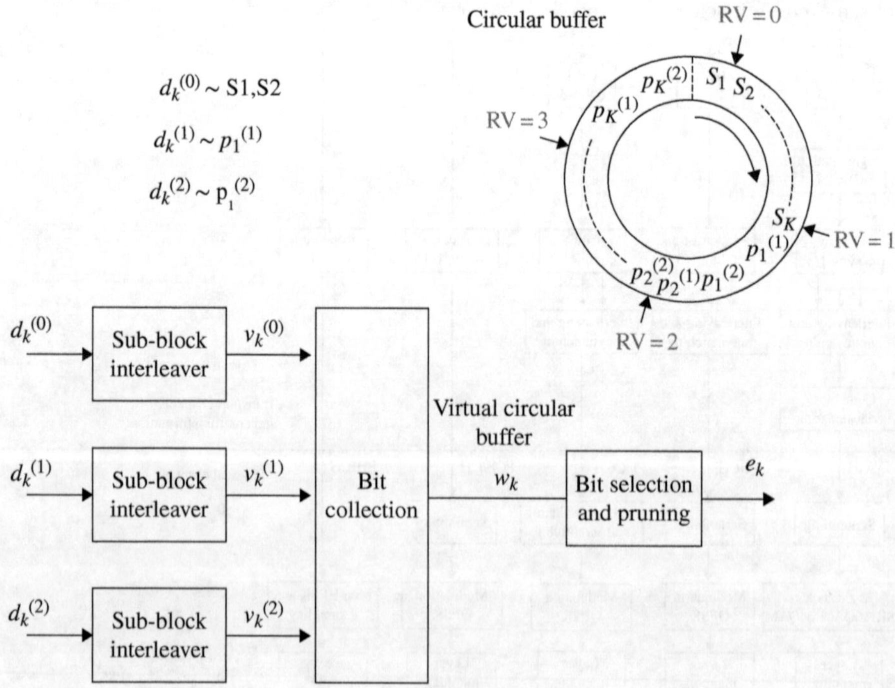

Figure 3.26 Rate matching

- Code block concatenation. The sequences e_{rk} are fed as input bit sequences to the code block concatenation block (where for $r = 0,\ldots,C-1$ and $k = 0,\ldots,E_r-1$). The output bit sequence from the code block concatenation block is the sequence f_k for $k = 0,\ldots,G-1$. Next, the resulting code blocks are reassembled into a single codeword.
- Scrambling. The block of code bits (delivered by the hybrid-ARQ functionality) is multiplied by a bit-level scrambling sequence. The downlink scrambling is applied to all transport channels as well as to the downlink L1/L2 control signaling. For all downlink transport-channel types (except MCH) and for the signaling, the scrambling sequences differ between neighboring cells (cell-specific scrambling). By applying different scrambling sequences for neighboring cells, the interfering signal(s) after descrambling is (are) randomized, ensuring full utilization of the processing gain provided by the channel code. This is achieved by having the scrambling sequences depend on the physical-layer cell identity. The scrambling sequence generator is reinitialized every subframe (except for the physical broadcast channel), based on the identity of the cell, the subframe number (within a radio frame), and the UE identity. The scrambling sequence in all cases uses an order-31 Gold code, which can provide 231 sequences that are not cyclic shifts of each other (Gold codes can be derived from the modulo two addition of two M-sequences).

Table 3.6 Different modulation schemes used for different channels in LTE

Downlink		Uplink	
Downlink channels	Modulation scheme	Physical channels	Modulation scheme
PBCH	QPSK	PUCCH	BPSK, QPSK
QPSK	QPSK	PUSCH	QPSK, 16QAM, 64QAM
PDSCH	QPSK, 16QAM, 64QAM	PRACH	uth root Zadoff–Chu
PMCH	QPSK, 16QAM, 64QAM	Physical signals	Modulation scheme
PCFICH	QPSK	Demodulation RS	Zadoff–Chu
PHICH	BPSK modulated on I and Q with the spreading factor 2 or 4 Walsh codes	Sounding RS	Based on Zadoff–Chu
Physical signals	**Modulation scheme**		
RS	Complex I+jQ pseudo random sequence (length-31 Gold sequence) derived from cell ID		
Primary synchronization	One of three Zadoff–Chu sequences		
Secondary synchronization	Two 31-bit BPSK M-sequence		

- Modulation. Then the digital modulation scheme is applied to transform the block of scrambled bits into a corresponding block of complex (I-Q) modulation symbols. Different modulation schemes used for different signals and various physical channels in downlink and uplink are shown in Table 3.6. (see TS 36.211).
- Antenna mapping. Antenna mapping is the combination of layer mapping and precoding, which processes the modulation symbols for one or two codewords to transmit them on different antenna ports depending on the specific transmission scheme being used. These antenna ports do not correspond to physical antennas but, rather, are logical entities distinguished by their reference signal sequences. Multiple antenna port signals can be transmitted on a single transmit antenna or, a single antenna port can be spread across multiple transmit antennas.
  - Layer mapping. In layer mapping, the modulation symbols for one or two codewords will be mapped onto one or more layers. In the case of a single transmit antenna (no diversity) the contents of the codeword are mapped to a single layer. But where there are two or more antennas, there are mainly two types of layer mapping: spatial multiplexing and transmit diversity. The layers in spatial multiplexing have the same meaning as "streams." They are used to transmit multiple data streams in parallel, so the number of layers here is often referred to as the transmission rank. In this case,

Table 3.7 Codeword mapping

Codewords	Layers	Mapping
1	1	Codeword mapped to single layer
1	2	Codeword symbols are split (even/odd) between two layers.
2	2	Each codeword is mapped to its own layer.
2	3	The first codeword is mapped to first layer, while the second codeword is split (even/odd) between the other two layers.
2	4	The first codeword is split (even/odd) between the first two layers, while the second codeword is split between second two layers.

there may be one or two codewords, but the number of layers cannot exceed the number of antenna ports. Here, one or two codewords may be distributed across one, two, three or four layers with mapping as shown in Table 3.7.

- The number of layers used in any particular transmission depends (at least in part) on the rank indication (RI) feedback from the UE, which identifies how many layers the UE can discern.
- In case of transmit diversity, there is only one codeword and the number of layers is equal to the number of antenna ports. The number of layers in this case is not related to the transmission rank because transmit-diversity schemes are always single-rank transmission schemes. For transmit diversity, it is almost as easy: the symbols from the codeword are distributed evenly across the two or four layers in a round-robin fashion. Transmit diversity for two antenna ports is based on space frequency block coding (SFBC), and transmit diversity for four antenna ports is based on a combination of SFBC and frequency shift transmit diversity (FSTD).
- Precoding. Precoding as a means of weighting the signals transmitted from different antennas in such a way that the signal-to-noise-plus-interference ratio (SNIR) at the receiver is maximized to support for beamforming. The layers are precoded according to the selected multiantenna transmission scheme.

• Resource element mapping. The resource-block mapping takes the symbols to be transmitted on each antenna port and then maps these to the resource elements of the set of resource blocks assigned by the MAC scheduler for the transmission.

3.8.3 Uplink Channel Structure and Transmission Mechanism

3.8.3.1 Uplink Logical, Transport and Physical Channels

• Logical channels. (i) Common control channel (CCCH): This is used for carrying information between the network and the UE. It is used for UEs that have no RRC connection. (ii) Dedicated control channel (DCCH): this is a point-to-point bidirectional channel for exchanging control information and is used by the UE, which has RRC connection. (iii) Dedicated traffic channel (DTCH). This is a point-to-point channel dedicated to a single UE for the transmission of user information.

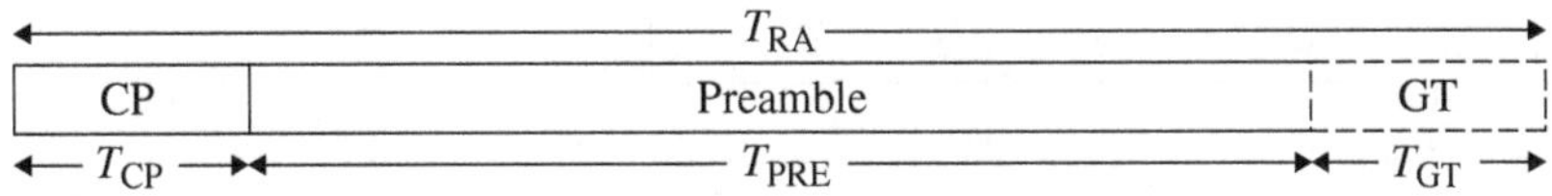

Figure 3.27 Random access preamble format

- Transport channels. (i) Uplink shared channel (UL-SCH): all the three uplink logical channels are mapped to this transport channel. It supports adaptive modulation/coding, HARQ, power control and semistatic/dynamic resource allocation. (ii) Random access channel (RACH): this channel supports transmission of limited control information and the possible risk of collision.
- Physical channels:
  - Physical random access channel (PRACH): the RACH transport channel is mapped to this channel. It carries the random access preamble, which the UE sends to access the network. It allows UE to synchronize timing with the eNodeB. It consists of 72 subcarriers in the frequency domain (six resource block, 1.08 MHz). The FDD LTE defines four different random access (RA) preamble formats with different preamble and cyclic prefix duration to accommodate different cell sizes. Figures 3.27 and 3.28 shows different preamble format.
  - Physical uplink shared channel (PUSCH): this carries user data, and UL-SCH transport channel is mapped to it. The uplink scheduling interval is 1 ms, similar to the downlink. In addition to carrying user data, the PUSCH carries any control information necessary to decode information such as transport format indicators and MIMO parameters. Prior to DFT spreading, control data is multiplexed with information data. It supports QPSK and 16 QAM modulation with 64QAM being optional.
  - Physical uplink control channel (PUCCH): this is a stand-alone uplink physical channel. It is used to carry downlink channel quality indication (CQI) reports, MIMO feedback (rank indicator, precoding matrix indicator), scheduling requests for uplink transmission and hybrid ARQ ACK/NACK for downlink transmissions. It is transmitted in a frequency region towards the edge of the system bandwidth. It consists of one RB per transmission at one end of the system bandwidth followed by a RB in the following slot at the opposite end of the channel spectrum thus making use of frequency diversity. A PUCCH control region comprises every two such RBs and PUCCH information is modulated using BPSK or QPSK.

In the preamble, a guard time (GT) needs to be introduced to avoid collisions with other transmissions as the random access (RA) mechanism is used by the UE when it is not yet synchronized on the uplink. The duration of the GT needs to account for the round-trip propagation time, which is dependent upon the supported cell size. With propagation speed of 1 km/3.33 μs, approximately 6.7 μs of guard time per km (2*3.33 μs) is required to accommodate the round-trip time. Now, to support cell size up to 100 km, as required for LTE, the guard time should be in the range of 670 μs. But, in case of small cell size this will be an overhead. So, to support that multiple random access, preamble formats with both small and

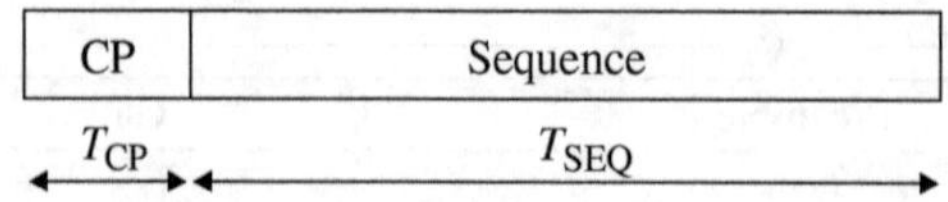

Preamble format	T_{CP}	T_{SEQ}
0	$3168-T_s$	$24576-T_s$
1	$21024-T_s$	$24576-T_s$
2	$6240-T_s$	$2.24576-T_s$
3	$21024-T_s$	$2.24576-T_s$
4	$448-T_s$	$4096-T_s$

Figure 3.28 FDD preamble format (0–3) and TDD preamble format (0–4)

large guard times are defined (Figure 3.28). In order to enable simple frequency-domain processing, the random access preamble also uses a cyclic prefix (CP). Its length accounts for both the propagations delays as well as the channel delay spread. There are various formats defined: (i) In format 0, both the CP and GT are equal to approximately 0.1 ms, which is sufficient to support cell sizes of up to approximately 15 km. Preamble sequence length is 0.8 ms. (ii) In format 1, the CP and GT are 0.68 s and 0.52 ms respectively, which are sufficient to support cell sizes of up to approximately 78 km. The preamble sequence length is 0.8 ms. (iii) Another aspect to consider is if preamble length of 0.8 ms provides enough energy for it to be successfully detected at the eNB. Therefore, to provide energy gain, in formats 2 and 3, preamble is repeated and the preamble sequence length is 1.6 ms. In format 2, both the CP and GT are equal to approximately 0.2 ms, which is sufficient to support a cell size of up to approximately 30 km. In format 3, CP and GT are approximately equal to 0.68 and 0.72 ms respectively, which is sufficient to support cell size of over 100 km. The preamble format to be used in a specific cell is informed to the UE using PRACH configuration index, which is broadcast in SIB-2. PRACH configuration index also indicates SFN and subframes, these indicate the exact position of random access preamble. The preamble format is defined in LTE (3GPP TS 36.211, section 5.7).

3.8.3.2 Uplink Physical Signals

Uplink physical signals are used in the physical layer and do not carry higher layer's information (Figure 3.29). Two types of UL physical signals are defined: reference signal and random access preamble.

- Uplink reference signal. There are two variants of the UL reference signal and both variants are based on Zadhoff–Chu sequences. (i) Demodulation reference signal (DM-RS) – this facilitates coherent demodulation and channel estimation in the eNodeB receiver. This is associated with transmission of PUSCH or PUCCH. It is transmitted in the fourth SC-FDMA symbol of the slot (for normal CP) and spans the same bandwidth

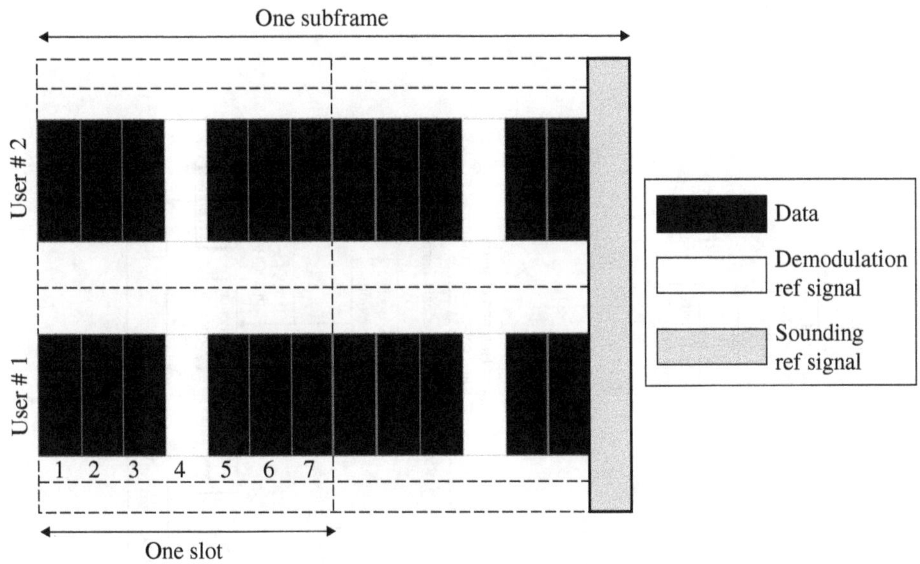

Figure 3.29 Uplink demodulation and sounding reference signals

as the allocated uplink data. (ii) Sounding reference signal (SRS). This is used to facilitate frequency-dependent scheduling decisions in the base station by estimating the uplink channel quality information, as the DM-RS cannot be used for this purposes as they are assigned over a particular bandwidth to a UE. The UE sends a sounding reference signal in different parts of the bandwidths where no uplink data transmission is available. The sounding reference signal is transmitted in the last symbol of the subframe. User data transmission is not allowed in this block, which results in about a 7% reduction in uplink capacity, so it is an optional feature. Users with different transmission bandwidth share this sounding channel in the frequency.

- Random access preamble. The random access procedure is used to request initial access, as part of handover, or to re-establish uplink synchronization. 3GPP defines a contention-based and a noncontention-based random access (Figure 4.8) procedure. Figure 3.29 shows the uplink data, demodulation and sounding reference signals.

3.8.4 Uplink Physical Channel Processing

The uplink channel processing steps are depicted in Figure 3.30 and these steps are similar to the downlink as explained in section 3.8.1.7:

- CRC calculation and addition. The CRC bits are computed and added to the transport block.
- Code block segmentation and CRC insertion. The code block is segmented if it is larger than the maximum code block size (6144 bits) and then CRC is computed and appended to it.

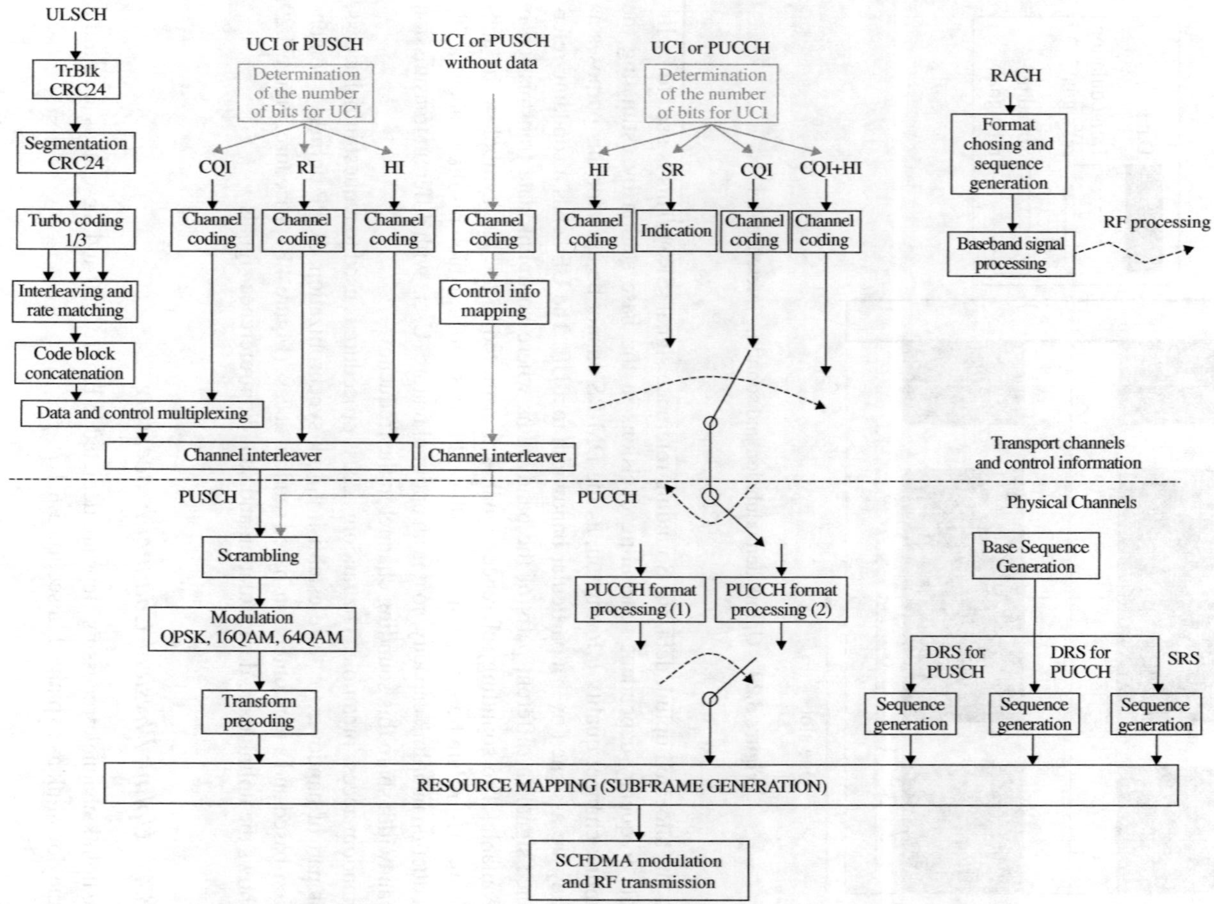

Figure 3.30 Processing blocks for different uplink transmission channels

- Channel coding. Rate-1/3 Turbo coding with QPP-based inner interleaving is used for the uplink.
- Rate matching and physical-layer hybrid-ARQ. This is similar to downlink. But the uplink hybrid-ARQ protocols, like, asynchronous versus synchronous operation is different.
- Bit-level scrambling. It randomizes the interference and similar to downlink.
- Digital modulation. Different modulation schemes are used for different uplink transport channels as given in Table 3.5.
- DFT precoding. The constellation mapper converts the incoming bit stream from higher layer to single-carrier symbols and then the serial / parallel converter formats time-domain SC symbols into blocks for input to the FFT engine and M-point DFT converts time domain SC symbol block into M discrete tones. The modulation symbols, in blocks of M symbols, are fed through a size-M DFT, where M corresponds to the number of subcarriers assigned for the transmission. The reason for the precoding is to reduce the cubic metric for the transmitted signal. SC-FDMA systems either use contiguous tones (localized) or uniformly spaced tones (distributed).
- Antenna mapping. The antenna mapping maps the output of the DFT precoder to antenna ports for subsequent mapping to the physical resource (the OFDM time–frequency grid).
- N-point IDFT. This converts mapped subcarriers back into time domain for transmission.
- Cyclic prefix and pulse shaping. Cyclic prefix is prepended to the composite SC-FDMA symbol to provide multipath immunity in the same manner as described for OFDM. As in the case of OFDM, pulse shaping is employed to prevent spectral regrowth.
- RFE. Converts digital signal to analog and upconvert to RF for transmission.

Processing blocks for different uplink channels are shown in Figure 3.30.

3.9 Multiple Input Multiple Output (MIMO)

MIMO techniques are about using multiple antennas at the transmitter and receiver to improve communication performance and data rate. This exploits the space dimension to improve wireless systems capacity, range and reliability. It offers significant increases in data throughput and link range without additional bandwidth or increased transmit power. It spreads the same total transmit power over different antennas to achieve an array gain that improves the spectral efficiency or achieves a diversity gain that improves the link reliability. The linear increase of channel capacity with the increase of number of antenna element in the MIMO system is shown in Figure 3.31(a), whereas the SIMO and MISO system shows logarithmic increasing of channel capacity.

There are different types of MIMO systems: (i) SISO (single input single output) – the transmitter and receiver have only one antenna; (ii) SIMO (single-input-multiple-output) – the receiver has multiple antennas whereas the transmitter has one antenna; (iii) MISO (multiple-input-single-output) – the transmitter has multiple antennas while the

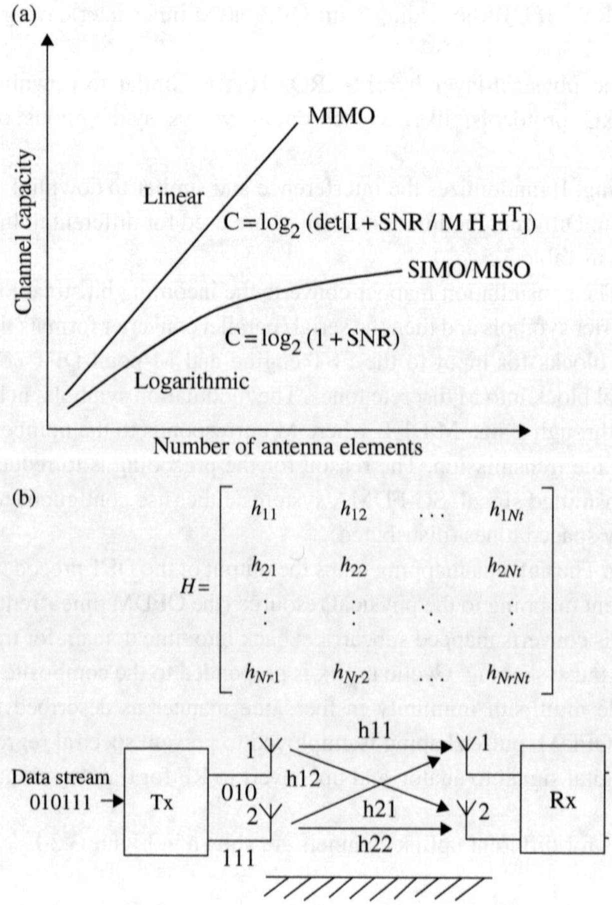

Figure 3.31 (a) Channel capacity versus number of antenna elements for MIMO and SIMO/MISO. (b) Spatial multiplexing (MIMO)

receiver has one antenna; (iv) MIMO (multiple-input-multiple-output) – the transmitter and receiver have multiple antennas. Different gains can be achieved depending on the MIMO mode used.

- Spatial multiplexing. This provides additional data capacity by using the different paths to carry additional traffic. It allows the transmission of different streams of data simultaneously on the same resource block(s) by exploiting the spatial dimension of the radio channel. These data streams can belong to one single user (single user MIMO/SU-MIMO) or to different users (multiuser MIMO/MU-MIMO). SU-MIMO helps to increase the data rate of one user, and MU-MIMO allows the overall capacity to be increased. As shown in Figure 3.31(b), if Nt is the number of transmit antennas, Nr is the number of receive antennas, the number of data streams that can be transmitted in parallel over the MIMO channel is given by min {Nt, Nr} and is limited by the rank of the matrix H. This form of MIMO is used to provide additional data capacity by utilizing the different paths to carry additional traffic – increasing the data throughput capability.

- Spatial diversity. This is often referred to as "transmit and receive diversity." These two methodologies are used to provide improvements in the signal-to-noise ratio and they improve the reliability of the system with respect to the various forms of fading. (i) Transmit diversity: instead of increasing data rate or capacity, MIMO can be used to exploit diversity and increase the robustness of data transmission. Typically an additional antenna-specific coding is applied to the signals before transmission to increase the diversity effect. Often, space-time coding is used according to Alamouti [2]. Switching between the two MIMO modes, like, transmit diversity and spatial multiplexing, is possible depending on channel conditions. (ii) Receive diversity – using two receive antennas at the receiver side this diversity gain is achieved.

3.9.1 MIMO in the LTE System

Different downlink MIMO modes are defined and used according to channel conditions, traffic requirements, and UE capability. There many transmission modes possible as below:

- single-antenna transmission, no MIMO;
- transmit diversity;
- open-loop spatial multiplexing with no UE feedback;
- closed-loop spatial multiplexing with UE feedback;
- multiuser MIMO;
- closed-loop precoding for rank = 1;
- beamforming.

Up to two codewords (one codeword represents an output from the channel coder) can be mapped onto different spatial layers in spatial multiplexing. The number of spatial layers available for transmission is equal to the rank of the matrix **H** (3GPP TS 36.211). To get the maximum capacity in spatial multiplexing, precoding is used on the transmitter side. This is achieved by multiplying the signal with a precoding matrix **W** before transmission. In the closed spatial multiplexing mode the UE estimates the radio channel and selects the optimum precoding matrix, and provides feedback to the eNodeB. An open-loop spatial multiplexing mode is also supported where no feedback is provided. The eNodeB will select the optimum MIMO mode and precoding configuration and conveyed the information to the UE as part of the downlink control information (DCI) on PDCCH. In case of transmit diversity, the same code is transmitted on different antennas, with different coding in different antennas or time switched (one antennas transits during that time other antenna is silent). One additional type of diversity, cyclic-delay diversity (CDD) is used in conjunction with spatial multiplexing in LTE.

- Uplink MIMO. For the uplink, MU-MIMO can be used. This means that multiple user terminals may transmit simultaneously on the same resource block. It is also referred to as spatial domain multiple access (SDMA). This requires only a single antenna at the UE side. If the UEs are sharing the same resource block then they have to apply mutually orthogonal pilot patterns. The antenna subset selection for two or more transmit antennas can also be used. The transmit antenna selection can be used too, in UE, to provide the best channel to the eNodeB using a switch (depending on the UE's capability). The decision is made according to feedback provided by the eNodeB. The CRC parity bits of the DCI format 0 are scrambled with an antenna selection mask indicating UE antenna port 0 or 1.
- UE feedback reporting. The UE reports the about the mobile radio channel to the eNB. A lot of different reporting modes and formats are available, which are selected according to MIMO mode of operation and network choice. The reporting may consists of
  - Channel quality indicator (CQI) – this indicates the downlink mobile radio channel quality as experienced by the UE. The "wideband CQI" is used for CQI measured over the complete system bandwidth and "sub-band CQI" is computed per sub-band of a certain number of resource blocks which is configured by the higher layers.
  - Precoding matrix indicator (PMI) – this indicates the optimum precoding matrix to be used in the eNB.
  - Rank indication (RI) – this is the number of useful transmission layers when spatial multiplexing is used. The reporting is periodic or aperiodic as configured by the network and UE sends the report on PUSCH.

3.9.2 Transmission Mode (TM)

As described earlier, in LTE, usually multiple Tx and Rx Antennas are used in the downlink and antennas can be used for diversity configuration or MIMO configuration for a better link performance or data rate. Apart from number of antennas, there are some other physical

Table 3.8 Different transmission schemes (PDSCH)

Transmission mode	Description
1	Single antenna port, no. of codeword = 1, no. of layer = 1, no. of antennas = 1
2	Transmit diversity, no. of codeword = 1, no. of layer = 2, no. of antennas = 2
3	Transmit diversity if the associated rank is 1, else large delay CDD. No. of codeword = 1 or 2, no. of layer = 2, no. of antennas = 2
4	Closed-loop spatial multiplexing, no. of codeword = 1 or 2, no. of layer = 2, no. of antennas = 2
5	MU-MIMO, no. of codeword = 1, no. of layer = 2, no. of antennas = 2
6	Closed-loop spatial multiplexing with a single transmission layer, no. of codeword = 1, no. of layer = 2, no. of antennas = 2
7	If the number of PBCH antenna ports is 1, otherwise transmit diversity , no. of codeword = 1, no of layer = 2 or 1, no. of antennas = 2

layer parameters like number of codewords, number of layers, precoding, code book index, and multiplexing. Based on these, the physical layer processing and hence the transmission techniques varies. In LTE, a special name is given for each of the ways of transmission and this is known as the transmission mode. Thus, SISO is known as TM1, transmit diversity is known as TM2, MIMO with no feedback is known as TM3, MIMO with feedback from UE (CQI, PMI, RI) is known as TM4, and so forth. Table 3.8 summarizes different transmission modes and the details can be found in 3GPP TS 36.213.

The TM designated for the UE is provided in an RRC message whenever the UE establishes an RRC connection (configuration of the transmission mode is optional).

3.10 Uplink Hybrid Automatic Repeat Request (ARQ)

The hybrid ARQ scheme used in LTE is a combination of physical layer coding (FEC) and the data link layer's repeat request mechanism. The eNB requests retransmissions of incorrectly received data packets. The ACK/NACK information is sent via PHICH in downlink. So, UE monitors corresponding PHICH once PUSCH is sent.

3.11 UE Categories

The LTE category defines the overall performance and the capabilities of the UE. Depending on the data rate and MIMO capabilities, different UE categories are defined. The LTE categories, or UE classes, are required to ensure that the eNB can communicate correctly with the user equipment, so the LTE UE category is indicated to eNB by the UE. These LTE categories define the standards to which a particular handset, dongle, or other equipment will operate. From consumer perspective the theoretical speed to be achieved using different categories of UEs are listed in Table 3.9.

There are some other UE categories also included in 3GPP Rel-12: **Cat-9**: peak data rate DL/UL : 450/50, **Cat-10**: peak data rate DL/UL : 450/100, **Cat-11**: peak data rate DL/UL : 600/50, **Cat-12**: peak data rate DL/UL : 600/100, **Cat-13**: peak data rate DL/UL : 390/150, **Cat-14**: peak data rate DL/UL : 3900/1500.

3.12 LTE UE Testing

To determine the compliance of UE (and other entities of the systems) with the LTE specifications, the 3GPP test spec for LTE contains a large number of different tests. The performance requirements for the various LTE physical channels under different configurations (LTE Transmitter and receiver tests) are specified in section 8 of 3GPP TS 36.521. UE transmitter measurements like TX power, frequency error, error vector magnitude (EVM), carrier Leakage, and Rx BER are given in 3GPP TS 36.521-1. Protocol confirmation testing can be found in 3GPP TS 36.523V9.3.0 and UE conformance testing can be found in 3GPP TS 36.523-3.

Table 3.9 LTE user equipment categories (3GPP TS 36.306)

LTE UE categories	CAT-1	CAT-2	CAT-3	CAT-4	CAT-5	CAT-6	CAT-7	CAT-8
LTE device introduced in 3GPP release	Rel-8	Rel-8	Rel-8	Rel-8	Rel-8	Rel-10	Rel-10	Rel-10
Peak 5ate DL/UL (mbps)	10 / 5	50 / 25	100 / 50	150 / 50	300 / 75	300 / 50	300 / 100	3000 / 1500
RF BW (MHz)	20	20	20	20	20	40	40	100
Modulation DL	64 QAM	64 QAM	64 QAM	64 QAM	64 QAM	64 QAM	64 QAM	64 QAM
Modulation UL	16 QAM	16 QAM	16 QAM	16 QAM	64 QAM	16 QAM	16 QAM	64 QAM
MIMO DL	Optional 2×2	2×2	2×2	2×2	4×4	2×2 or 4×4	2×2 or 4×4	8×8
MIMO UL	no	no	no	no	no	no	2×2	4×4

Note: Rx div is always used (e.g. Tx×Rx: 1×2).

References

[1] Das, Sajal Kumar. (2010) *Mobile Handset Design*, John Wiley & Sons, Ltd.

[2] Alamouti, S. M. (1998) A simple transmit diversity scheme for wireless communications. *IEEE Journal on Selected Areas in Communication* **16**(8), 1451–1458.

Further Reading

Chu, J. D. C. (1972) Polyphase codes with good periodic correlation properties. *IEEE Transactions on Information Theory* **18**, 531–532.

Fazel, K. and Kaiser, S. (2008) *Multi-Carrier and Spread Spectrum System*, John Wiley & Sons.

Frank, R., Zadoff, S., and Heimiller, R. (1962) Phase shift pulse codes with good periodic correlation properties. *IEEE Transactions on Information Theory* **8**, 381–382.

Halonen, T., Romero, J. and Melero, J. (2003) *GSM GPRS and EDGE Performance*, John Wiley & Sons, Ltd.

Holma, H. and Toskala, A. (2004) *WCDMA for UMTS*, John Wiley & Sons, Ltd.

Holma, H. and Toskala, A. (2009) *LTE for UMTS OFDMA and SC-FDMA Based Radio Access*, John Wiley & Sons, Ltd.

Khan, F. (2009) *LTE for 4G Mobile Broadband*, Cambridge University Press.

Korhonen, J. (2003) *Introduction to 3G Mobile Communications*, Artech House.

Seurre, E., Savelli, P. and Pietri, J.-P. (2003) *GPRS for Mobile Internet*, Artech House.

See also the GSM 3GPP Technical Specification series: TS 45.001, 45.002, 45.003, 45.004, 45.005, 45.008, the LTE 3GPP Technical Specification series: TS 36.201, 36.211, 36.212, 36.213, 36.133, 36.331, and the UMTS 3GPP Technical Specification series: TS 25.201, 25.211, 25.212, 25.213, 25.214, 25.215, 25.221, 25.331.

4

LTE UE Operations Procedures and Anatomy

4.1 UE Procedures

Once the UE is powered ON, it triggers the CPU reset and initially it performs the booting and system initialization related tasks including device self-test, and so forth, as described in Chapter 7. As discussed in Chapter 1, the mobile phone has an application unit as well as a modem unit and there are different CPUs used for execution of these. In the modem unit, once the system initialization is over and OS is loaded into the memory, then OS takes the control and starts executing different tasks/threads, where different modem sublayers and procedures are running as OS threads. The first procedure that runs at this stage is "network and cell selection," where the UE selects the public land mobile network (PLMN) to register (see Figure 4.1), selects a closed subscriber group (CSG) for registration using user permission, and selects and camps on the cell. The UE establishes the RRC connection and registers with the evolved packet core. After that, the mobile enters into the idle mode and executes the idle mode procedures, which include measurements for cell reselection, and paging reception using the DRX mechanism. When a paging message is received (an incoming call – for example a mobile terminated call) or when the UE wants to initiate a call (a mobile originated call) then the UE sends RACH for a channel request to the network. Once the channel is assigned, and the UE is engaged in communication, it is in dedicated mode. During that time, the UE performs measurements and reports to the network for handover initiation, handover execution (if a handover command is received), and so forth. Then the detach procedure helps to detach UE from the network during switch off.

Mobile Terminal Receiver Design: LTE and LTE-Advanced, First Edition. Sajal Kumar Das.

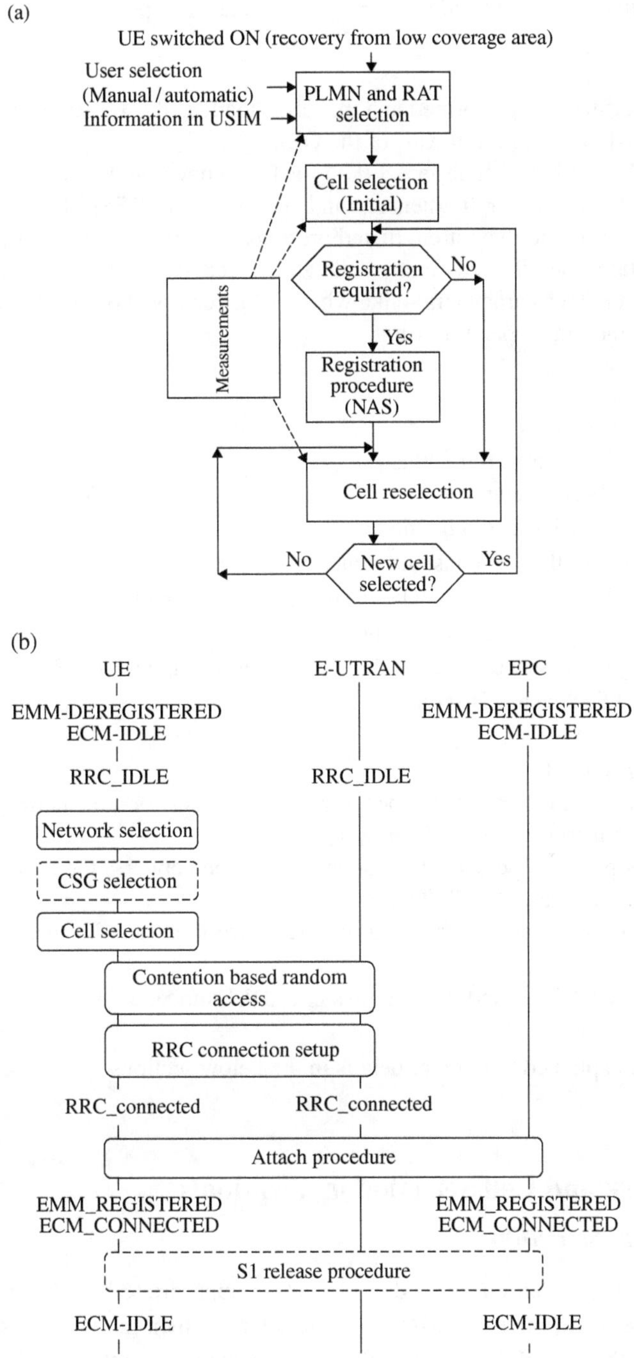

Figure 4.1 (a) PLMN, cell selection and registration. (b) Message flow for cell selection among different entities

Commonly, when UE is powered on, the following steps are performed:

1. UE is powered on.
2. Boot up, self-test are performed and the OS is loaded. The UE modem protocol layers are activated (as different threads of the OS).
3. Cell search. The NAS requests AS (L1) for frequency search. L1 programs the RF to receive different carrier frequencies and measures the RSSI. Then L1 ranks those carriers according to their strength and presents them to the higher layer.
4. Cell selection: the UE finds many cells and selects the best one based on the signal strength, and other parameters – like whether that cell is allowed or not, or whether it belongs to the same operator or not.
5. MIB decoding.
6. SIB decoding.
7. Initial RACH process.
8. Registration / authentication / attach.
9. Default EPS bearer setup.
10. At this stage, the UE is in idle mode.
11. Cell reselection: the UE constantly checks the presently camped cell's signal quality and if it diminishes then it switches to the best available one.
12. The UE keeps on listening to paging.
13. When a paging message comes or the user makes a call, the RACH process is invoked.
14. A dedicated EPS bearer is set up.
15. Data is received.
16. Data is transmitted.
17. If the UE's power is perceived to be too weak by the network, the network sends a TPC command to increase the UE Tx power.
18. If the UE's power is perceived as too strong by the network, the network sends TPC command to decrease the UE Tx power.
19. If the UE moves to another cell region, the network and the UE perform a handover procedure.
20. The user stops the call and the UE goes into IDLE mode.

These steps are explained with more details in the below sections.

4.2 Network and Cell Selection in Terminals

4.2.1 PLMN Selection

When the UE is powered on, it needs to select an appropriate PLMN and RAT to gain access to the network. First a PLMN has to be selected, then the UE will stay connected to the most suitable cell of that selected PLMN. The primary inputs to the PLMN and RAT selection algorithm are provided manually or fetched from the USIM. In practice, the IMSI in the USIM may be used to determine the home PLMN code and may serve as the

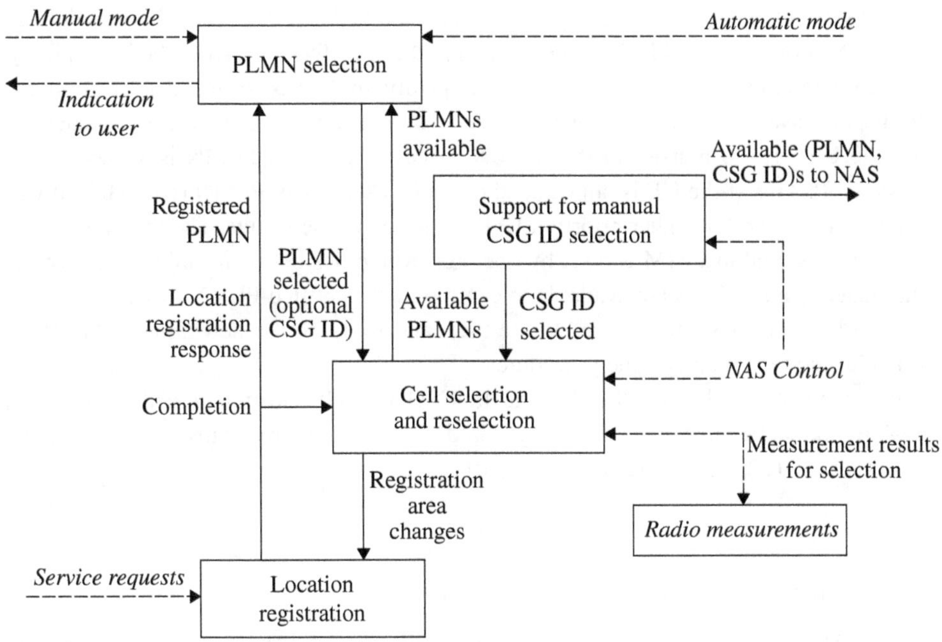

Figure 4.2 UE initial cell selecion procedures

basis for PLMN selection. Optionally, the USIM might also have an equivalent home PLMN (HPLMN) list, priorities among the PLMNs and also among the allowed RATs for each PLMN.

The NAS (part of the protocol layer – see Figure 4.2) is activated soon after all the various self-checks and USIM initialization phases have occurred. The LTE-compliant UE's NAS layer requests the AS layer to report for available PLMNs (all PLMNs or just the previously registered PLMN if that information was stored). After receiving that command, the AS layer in the UE scans all the RF channels within its supported RF bands (if there is no prior stored information exists in UE). On each carrier (RF channels), the UE searches for the strongest cell according to the cell search procedure and synchronize with it as described in next section and then read its system information (PLMN identity is broadcast within SIB-1) in order to find out which PLMN that the cell belongs to. PLMNs are reported to NAS as high quality, if RSRP $\geq -x$ dBm, ($x = 110$), and PLMNs not meeting high quality criteria are reported along with their RSRP value to NAS. (The UE can optimize this PLMN search procedure using stored information such as RF carriers and cell parameters. The NAS layer can stop the search at any instant – for example after finding the home PLMN.) After receiving the PLMN lists, the LTE NAS layer selects a PLMN from that list of reported PLMNs. The selection of the PLMN is done manually or automatically. In the case of automatic selection, the LTE UE selects the PLMN and RAT based on the availability of HPLMN or the highest priority EHPLMN; PLMN and RAT combinations defined within the user-controlled PLMN selector; PLMN and RAT combinations defined within operator-controlled PLMN selector; and other PLMNs reported to be high quality, selected in random order or decreasing signal quality. In the case of manual selection, the UE displays available PLMNs to the end user for selection. Once the PLMN is selected, then the UE tries to register in that PLMN, once the selected PLMN is known as the registered PLMN. If the UE is not camped on a highest priority PLMN (HPLMN), it will periodically search for higher priority PLMNs and report the results. The time interval of searching is stored in USIM and set by operator, which lies between 6 min to 8 h. On the other hand, if no PLMNs are available to camp on, or identified PLMNs are not allowed, then the UE indicates "no service" to the user, and waits until a new allowed PLMN is available, and then it repeats the procedure.

After the PLMN selection, the UE runs the CSG and cell selection procedures to find a suitable cell that belongs to the above registered PLMN. For more details refer to 3GPP TS 36.304, TS 23.122, TS 31.102, and TS 24.301.

4.2.2 *Closed Subscriber Group Selection*

The USIM contains closed subscriber groups (CSG) and identities of corresponding networks that the subscriber is allowed to use. If the USIM in the UE contains any CSG, then the UE has to run an additional procedure, known as CSG selection. This also operates in two modes: automatic and manual.

4.2.3 *Cell Selection Criteria*

As discussed above, the cell selection process finds and selects a suitable cell to camp on. Cell selection is triggered by PLMN selection or when UE leaves RRC_CONNECTED mode. The cell selection is executed without any stored information about the E-UTRAN frequencies (initial cell selection), or being assisted by stored information on carrier frequencies or when leaving RRC connected mode. During the cell-selection process, the UE detects a cell using the procedures mentioned above and then evaluates the suitability of the cell using cell selection criterion. A suitable cell is defined as a cell that will have following characteristics: (i) it should belong to the selected PLMN, registered PLMN, or an equivalent PLMN; (ii) should not be a barred cell; (iii) it should belong to at least one tracking area that is not forbidden; (iv) it should satisfy the following cell selection criteria (see 3GPP TS 36.304):

Srxlev > 0 and Squal > 0, where Srxlev = Qrxlevmeas – (Qrxlevmin + Qrxlevminofset) – Pcompensation and Srqual = Qqualmeas – (Qqualmin + Qqualminoffset)

where, Srxlev = cell selection Rx level value in dB, Squal = cell selection quality value (dB), Qrxlevmeas = measured cell Rx level value (RSRP), Qqualmeas = measured cell quality value

(RSRQ), Qrxlevmin = minimum required Rx level in the cell (dBm), Qqualmin = minimum required quality level in the cell (dBm), Qualmin = minimum required quality level in cell (dB): Qrxlevminoffset = offset to the signalled Qrxlevmin, Qqualminofset = offset to the signalled Qqualmin, Pcompensation = max (P_{EMAX} – PpowerClass, 0), P_{EMAX} = Max Tx power level, $P_{powerClass}$ = Max RF output power.

4.3 Cell Search and Acquisition

Once the UE is powered on, it needs to synchronize time and frequency with a cell. So, to facilitate that, the eNB broadcasts two types of downlink physical signals: (i) a primary synchronization signal (PSS); (ii) a secondary synchronization signal (SSS). The transmission procedures for these signals were discussed in Chapter 3. These two signals were especially designed to enable time and frequency synchronization; apart from that, these also indicate physical layer cell identity, cyclic prefix length, and frequency division duplex (FDD), or time division duplex (TDD) mode used in the cell.

4.3.1 Cell Search and Synchronization Procedure

There are two different types of cell search and cell synchronization procedures in LTE: (i) initial cell selection and synchronization – this occurs after the initial power on and when UE is not connected to LTE cell but now wants to access the LTE network; (ii) new cell identification – this type of synchronization procedure happens when the UE is already connected to the LTE cell and is searching for other new cells for cell reselection (idle mode) or handover (connected mode) purposes. In both cases, the UE uses PSS and SSS for time and frequency synchronization and to acquire some useful system parameters. Like WCDMA, LTE uses a hierarchical cell search scheme, in which 504 ($N_{ID}^{cell} \rightarrow 0$ to 503) physical layer cell identities (PCI) have been divided into 168 unique cells ($N_{ID}^{(1)} = 0$ to 167) identity groups (168 × 3 = 504, physical layer cell identity group), with each group containing three unique identities ($N_{ID}^{(2)} = 0$, 1, or 2, physical layer cell identities).

First, the UE tunes to different RF carriers in the selected/commanded PLMN and attempts to measure the wideband received power (received signal strength indicator – RSSI) for each carrier (EARFCN frequency channel number as commanded by the higher layer) over a set of supported frequency bands, one after another, as indicated by the higher layer. Next, after ranking these frequencies based on RSSI, the UE attempts the cell-search procedure using the downlink synchronization channels (PSS and SSS). The cell search procedure in LTE system is performed in three steps (see Figure 4.3 and 4.4).

Step1: Symbol Timing, Frequency Offset and Physical Layer ID Detection using PSS

In this stage, the symbol timing, frequency offset, and physical-layer ID are detected using PSS. As discussed above, the PSS occupies a bandwidth of 62 × 15 kHz around the

DC (0 Hz), so the low-pass filter can be used for extracting PSS signal from a larger spectrum. Then received PSS symbols are correlated with the reference locally generated PSS signal.

The received samples are fed to matched filters, which have three correlators per antenna. Generally, coherent detection involves detecting the sequence that maximizes the probability of transmitting the sequence using channel information. Coherent detection channel estimation needs to be done before the sequence detection can start. Where the channel estimation cannot be done before this, then noncoherent detection can be used. Most commonly, a noncoherent approach is used for the PSS. So, a noncoherent cross correlation can be performed between the received samples on both antennas with the three known types of P-SCHs in the time domain over a length-N FFT window.

The frequency domain autocorrelation properties of the Zadoff–Chu sequence are applicable to the time domain, so PSS has good autocorrelation properties in the time domain as well. In the frequency domain, a fairly large number of FFT operations needs to be performed on multiple timing hypotheses, which require complex implementation with increased power requirements, whereas, time-domain correlation has good autocorrelation properties and simple correlation can be performed with a time-domain PSS sequence, which provides reduced complexity and resources.

Then, for detection, a maximum likelihood (ML) approach can be implemented both for coherent and noncoherent cases. A maximum likelihood detector, finds the timing offset m^*_M that corresponds to the maximum correlation

$$m_M^* = \operatorname{argmax}_m \left| \sum_{i=0}^{N-1} Y[i+m] S_M^*[i] \right|^2 \tag{4.1}$$

where i is the time index, m is the timing offset, N is the PSS time domain signal length, $Y[i]$ is the received signal at time instant i and $SM[i]$ is the PSS with root M replica signal at time i as given in equation (1). So, the eNB sequence (PSS with root M replica signal) with the highest peak correlation is selected as a candidate.

So, once the terminal has detected and identified the PSS of the cell:

- The 5 ms timing boundary of the RB transmission in the cell, the symbol timing, the frequency offset, and the position of the SSS (which has a fixed offset from the PSS) are found or inferred. Once the timing error and frequency error are detected, these are compensated. This enables the UE to be synchronized on the subframe level. The PSS is also repeated in subframe 5, which means that the UE is synchronized on a 5 ms basis as each subframe is 1 ms.
- The cell identity (physical-layer ID) within the cell-identity group ($N_{\text{ID}}^{(2)}$) is detected.

Step-2: Radio Frame Timing and Cell Group ID Detection using SSS

Next, the radio-frame timing and cell group ID are detected using SSS in the frequency domain. As the SSS detection is generally performed in the frequency domain, FFT is applied to the sequence. For SSS detection the received signal is correlated with all possible sequences and, after applying the ML detector, the timing is obtained. As the channel can be known based on the PSS sequence so, either a coherent or noncoherent approach can be used for SSS.

The received SSS in frequency domain can be expressed as

$$R_{sss}[k] = d[k]H_{sss}[k] + W[k]$$

where $d[k]$ is SSS in the frequency domain, $H_{SSS}[k]$ is the channel frequency response (CFR) at SSS, and $W[k]$ is the additive white Gaussian noise. For coherent detection, the UE estimates the CFR by using the received PSS sequence. In the frequency domain, the channel compensated SSS can be written as

$$\hat{R}_{sss}[k] = R_{sss}[k]\hat{H}^{*}_{PSS}[k]$$

After the channel compensation, a deinterleaved and descrambled signal can be written as

$$a_{m_0}[l] = \hat{R}_{SSS}[2k]c_0[k]$$
$$a_{m_1}[l] = \hat{R}_{SSS}[2k+1]c_1[k]z_1^{(\hat{m}_0)}[k]$$

The correlation output of coherent detection is a cross-correlation between a descrambled signal and the cyclic shift of an SSS sequence, and can be represented as

$$\hat{m}_0 = \arg\max_i \left|\sum_{i=0}^{30} a_{m_0}[l]s^{(i)}[l]\right|^2$$
$$\hat{m}_1 = \arg\max_i \left|\sum_{i=0}^{30} a_{m_1}[l]s^{(i)}[l]\right|^2$$

where $m\hat{}_0$ and $m\hat{}_1$ are the i-th SSS sequence. Then, using the $m\hat{}_0$ and $m\hat{}_1$ detection device indicates the frame timing and cell ID group $N_{\mathrm{ID}}^{(1)}$.

Once the $N_{\mathrm{ID}}^{(2)}$ and $N_{\mathrm{ID}}^{(1)}$ are detected, then the physical layer cell identities will be computed as: $\mathrm{N_{ID}}^{\mathrm{cell}} = 3 * \mathrm{N}_{\mathrm{ID}}^{(1)} + \mathrm{N}_{\mathrm{ID}}^{(2)}$.

CP Detection

The LTE system supports normal and extended CP. The precise timing of the SSS changes depending on the CP type. Before SSS detection, the CP type is unknown to the UE, so it is blindly detected by checking for the SSS at the two possible positions.

Duplexing Mode Detection

Once the position of SSS and PSS are known in the frame structure, the terminal easily identifies the duplexing scheme (FDD or TDD) used on a carrier.

So, once the terminal has detected and identified the SSS of the cell:

1. Radio frame timing is detected.
2. The cell identity group (0 to 167) is detected.
3. The terminal notes whether normal CP or extended CP is used in the system that is detected.
4. The terminal notes whether FDD or TDD mode is used in the system that is detected .

4.3.1.1 Implementation of PSS and SSS Detection

As shown in Figure 4.3, the PSS sequence generator generates reference sync signals using the UE's clock circuit (local oscillator and digital circuits). It generates three reference P-SCH signals (l = 0, 1, 2) and these are supplied to a set of hierarchical matched filters (HMFs). These HMFs are used to perform noncoherent crosscorrelation in the time domain with these three generated P-SCH signals and the received PSCH signals. Then a pre-FFT detector detects the symbol timing and the index of the received P-SCH. A carrier frequency estimator and correction unit estimate and correct the frequency error. The ten surrounding subcarriers are removed by discarding the null carriers unit. A simple channel estimator could be used for S-SCH referenced to P-SCH and then, in some implementations, a coherent crosscorrelators bank could be used to coherently crosscorrelate with the 168 S-SCH signals. After that, maximal ratio combining (MRC) is performed over two receiver antennas. Next a post-FFT detector detects the index of SSCH and frame timing.

Step 3: Reception of Reference Signals

From steps 1 and 2 above, the UE obtains the physical layer identity and cell identity group number, then the UE determines the PCI for the cell (N_{ID}^{cell}). Once the UE knows the PCI of a cell, it also knows the location of the cell Reference signals in that cell's time-frequency grid structure (refer to Chapter 3) can be used now for channel estimation, cell selection/reselection, and handover procedures.

Next, based on the channel quality, the UE camps on a particular cell and proceed to the next step for CRS detection and then reads system information.

4.4 Cell-Specific Reference (CRS) Signal Detection

The CRS transmission procedure was described in Chapter 3; here the reception procedure will be discussed. As mentioned in the previous section, the terminal detects the physical layer identity of the cell, as well as the cell frame timing. Thus, the terminal knows the location of the first CRS position in the frequency domain (refer to equation $v_{shift} = N_{cellID}$ mod 6, as discussed in Chapter 3), as well as in time domain (symbol 0, 4, 7, 11 in a subframe in FDD).

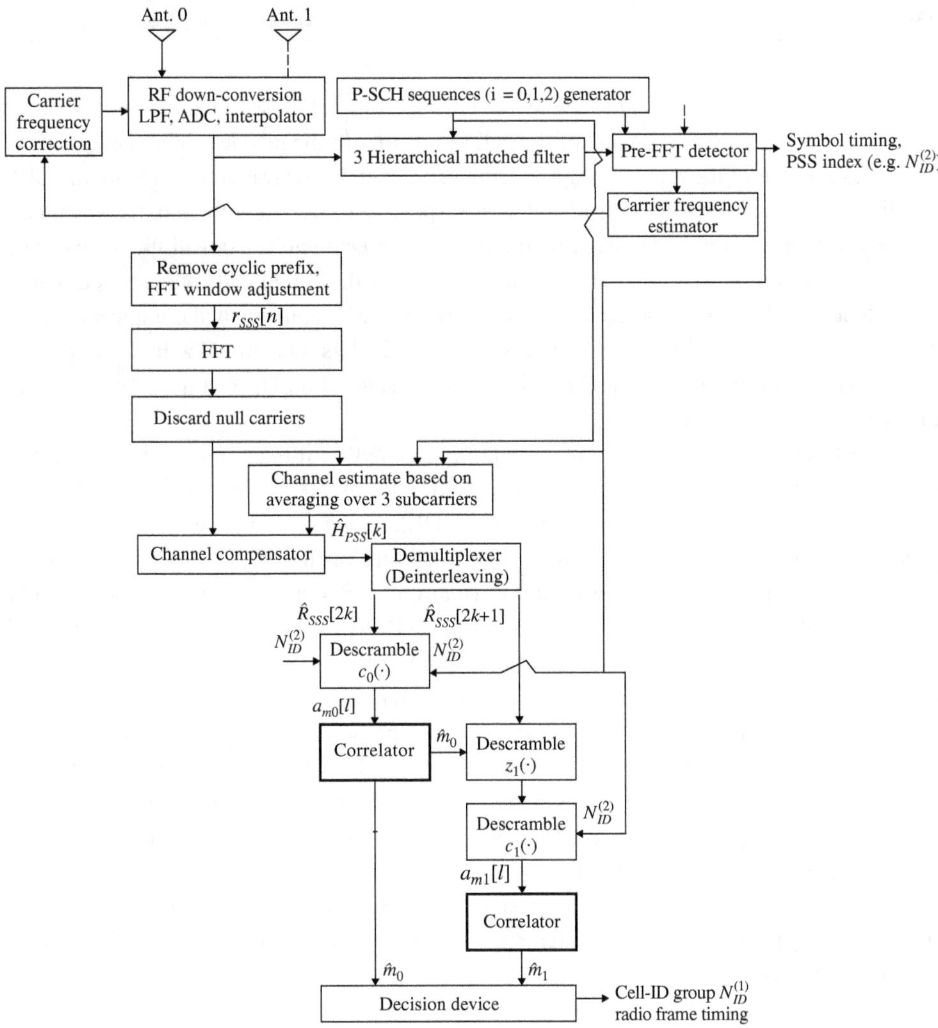

Figure 4.3 PSS and SSS detection blocks in UE receiver architecture

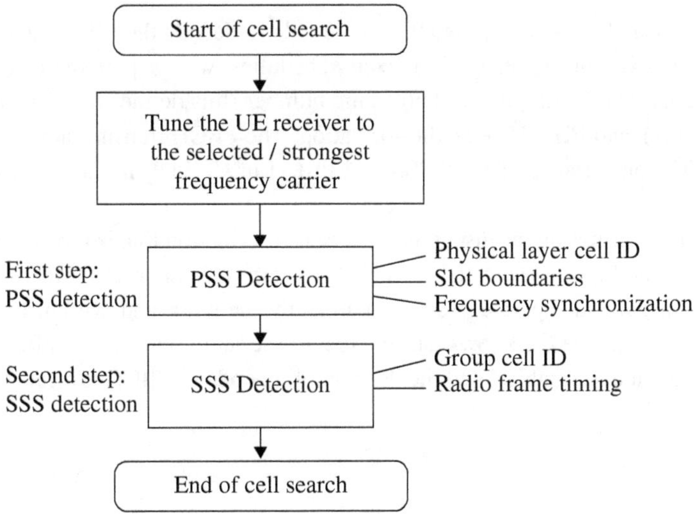

Figure 4.4 Cell selection steps

It also knows the reference signal sequence (the pseudo-random sequence used to generate the cell-specific reference signals) of the cell (according to the physical-layer cell identity) as well as the start of the reference signal sequence (given by the frame timing). So, the CRS locations in the time-frequency grid and their respective values are known to the terminal.

These reference signals are used by the terminal to estimate the downlink channel and perform equalization to remove the channel effect over the signal and coherent demodulation. Hence the UE will generate the local reference CRS sequence and compare with the received sequence to estimate the channel effect. The CRS is transmitted with some specific power, which is intimated to the UE using SIB messages, and UE can use that for calculating the multipath effect.

The behavior is slightly different depending on whether it is an initial cell search or a cell search for the purpose of neighboring cell measurements: In the case of an initial cell search – that is, the terminal state is in RRC_IDLE mode – the reference signal will be used for channel estimation and subsequent decoding of the BCH transport channel to obtain the most basic set of system information. Whereas, in the case of mobility measurements – that is, the terminal is in RRC_CONNECTED mode – the terminal will measure the received power of the reference signal. If the measurement fulfills a configurable condition, it will trigger sending of a *reference signal received power* (RSRP) measurement report to the network. Based on the measurement report, the network will conclude whether a handover should take place or not. The RSRP reports can also be used for component carrier management – for example decision making regarding whether an additional component carrier should be configured or if the primary component carrier should be reconfigured.

The UE then proceeds to the next step for PBCH reception to obtain the master information block (MIB) (see Chapter 3 for the transmission procedure for the channels mentioned below).

4.5 PBCH (MIB) Reception

After the cell search procedure, the UE will be able to decode the PBCH and read out the MIB. The MIB is transmitted with a fixed scheduling with a periodicity of 40 ms and contains the DL bandwidth, the system frame number (this defines the 8 most significant bits of the SFN), and PHICH-related information. The PBCH transmission is discussed in Chapter 3. The appearance of PSS, SSS and PBCH in LTE DL frame structure is shown in Figure 4.5.

The system information (as discussed above) is information that is repeatedly broadcast by the network and which needs to be acquired by terminals in order for them to be able to access and, in general, operate properly within the network and within a specific cell. On the receiver side, the UE knows the cell-specific scrambling code, so, after reception of the PBCH data, it descrambles, and the descrambling code of PBCH is a cell-specific code

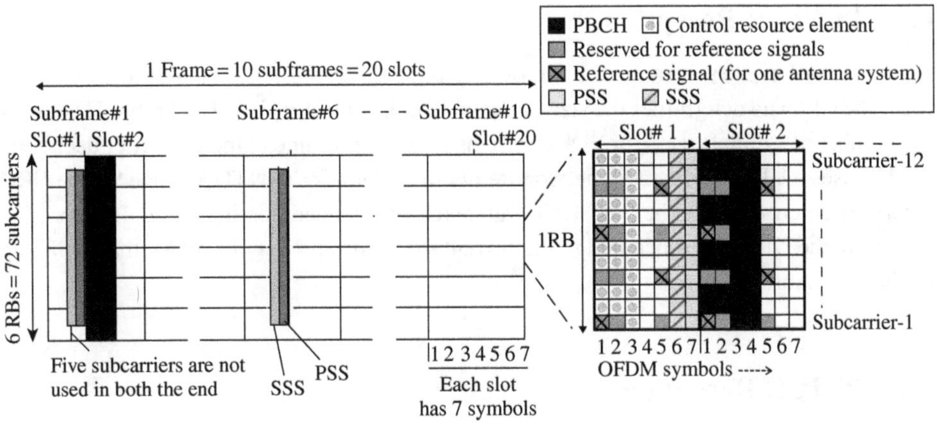

Figure 4.5 Appearance of PSS, SSS and PBCH in LTE DL frame structure

determined by the UE during the P-SCH/S-SCH. The TTI of the PBCH is 40 ms so, ideally, the UE has to decode PBCH in four consecutive radio frames to get the information transmitted in the MIB. But, at a high SNR this can be done earlier, for instance after the reception of two or three radio frames (as the PBCH contains a lot of redundancy). The information about the number of transmitting antenna used in the system is implied in the cyclic redundancy check (CRC) [1] mask of the PBCH. In the present method, the UE tries three times to discover the number of antenna ports used in the transmission. When receiving first time, the cell BCH, the terminal does not know to what set of four subframes a certain BCH transport block is mapped. Instead, a terminal must try to decode the BCH at four possible timing positions. Depending on which decoding is successful, indicated by a correct CRC check, the terminal can implicitly determine 40 ms timing or, equivalently, the two least significant bits of the SFN. This is the reason why these bits do not need to be included explicitly in the MIB.

The transmission of the PBCH is centered on the DC subcarrier because, when a UE accesses the system and tries to receive PBCH, it is unaware of the system bandwidth used. The total number of subcarriers used for the PBCH is 72 in the third and fourth OFDM symbols in the slot, which contain no reference signals. The PBCH does not use subcarriers reserved for reference signals of the four antenna ports irrespective of how many antennas are used for PBCH transmission. This is for reasons of simplicity because when a UE is receiving a PBCH it is unaware of the number of antennas used for transmission. The UE actually performs blind detection of the number of antennas used for PBCH with hypothesis of the single antenna, two antennas SFBC and four antennas.

The UE's next task is to start receiving the PCFICH channel to decode the control format indicator (CFI) to know the number of control symbols in the subframe.

4.6 PCFICH Reception

At the start of every subframe, the UE detects the REs that are occupied by the PCFICH, reads the CFI, and determines the size of the downlink control region. The PCFICH conveys the number of symbols used for PDCCH in the current subframe to the UE. After decoding, the UE uses a UE-specific radio network temporary identifier (RNTI) and checks the CRC errors to determine whether the DCI was actually intended for it or not. If errors are observed, the UE is not scheduled in the current subframe and it will wait for the beginning of the next subframe.

4.7 PHICH Reception

The PHICH carries HARQ ACK/NACKs for uplink data transfers. In the time domain, if the uplink transmission occurs in subframe N, the corresponding DL PHICH will be in subframe N + 4, because eNB processing time for PUSCH transmissions in the UL is equal to three subframes. In the frequency domain, it is indicated by uplink resource allocation with DCI format 0, where the specific PHICH (PHICH group number, orthogonal sequence index within the group) is derived from the lowest uplink PRB index in the first slot of the corresponding PUSCH transmission and the DMRS cyclic shift (refer to section 9.1.2 of 3GPP TS 36.213). To carry multiple ACK/NACKS on the same set of frequency resources, a code-multiplexing approach is used. The code multiplexing helps to exploit frequency diversity and to randomize intercell interference. The four-antenna-ports transmit diversity scheme used forACK/NACKs is different from that used for other downlink channels. The modified scheme avoids loss of code orthogonality.

Using the PHICH configuration, the UE can find out which of the remaining resource element groups are used by the PHICH and which are used by the PDCCH.

4.8 PDCCH Reception

Once the UE receives a DL subframe, decodes the PCFIH and PHICH, and finds out the number of symbols used for control information, then the UE calculates the number of REs used for the PDCCH using the equation: REs for PDCCH = total REs in the first N OFDM symbols − reference symbol RE's − PCFICH REs − PHICH REs. (Multiple PDCCHs can exist in the same subframe and each uses one, two, four, or eight control channel elements (CCEs) to contain the DCI.) Then UE will form the CCEs from the REs computed. It will use the reverse procedure, which was used by eNB PHY for arranging the CCEs onto REs. Then the UE will arrange the CCEs in a sequential manner. Next, the UE RRC will decide the RNTIs on which it needs to try decoding the CCEs. The UE will calculate the starting CCE index by employing the equation used by the eNB, using the RNTI, subframe number, number of CCEs, and the aggregation level. As explained in Chapter 3, the UE first searches on common search space in all aggregation levels on the indexes, calculated with the help of the RNTI provided by the RRC. Next, the UE will search in UE-specific search space on the CCE indexes, calculated with the RNTI provided.

4.8.1 Implementation of Control Channel Decoder

There could be one block for decoding control channels (PCFICH, PDCCH, PHICH, PBCH), termed the "control channel decoder." As specified in 3GPP TS 36.212, different physical channels carry different information. For example, the PCFICH carries the control format indicator (CFI), the PHICH carries the HARQ Indicator (HI), and the PDCCH carries DL control information (DCI).

PCFICH Decoder
The PCFICH is located in the first OFDM symbol of a DL subframe. Decoding of the PCFICH is typically executed immediately after soft bits for the first OFDM symbol are available. The control channel decoder assembles all PCFICH resource elements and performs maximum likelihood decoding of all four possible CFI values (see Chapter 3).

PDCCH Decoder
As the encoded DCI is mapped on the PDCCH in multiples of CCEs, and the number of CCEs determines the rate matching and hence the effective code rate, so it is unknown a priori to the UE which rate matching has been chosen per DCI. The CCEs are further decomposed into mini-CCEs and mapped onto the PDCCH in an interleaved manner. The control channel decoder reads all the soft bits for the PDCCH and assembles mini-CCEs and performs deinterleaving so that all CCEs are provided in original order and then tries for the blind decoding of all possible DCI candidates. The DCI candidates per subframe may have, at most, three possible payload sizes.

For PDCCH decoding, CDEC reads all the soft bits from the OFDM symbols with the PDCCH. It then assembles mini-CCEs and performs deinterleaving so that all CCEs are provided in their original order. The position of mini-CCEs within the OFDM symbols, as well as the interleaver specification, is configurable in the CDEC. There are two search spaces defined: UE-specific and common search space. All CCE combinations have to be evaluated, regardless of the actual search space. So, the control decoder performs a loop internally over all possible DCI candidates of one payload size. And each loop consists of rate matching, Viterbi decoding, and CRC checking with masking (distinguish multiple masking IDs (RNTIs)). For successful decoding (where a CRC check with masking results in an all-zero CRC), the decoder provides an index of masking ID (mapped to the corresponding RNTI), the DCI payload size, and decoded DCI data. For PDCCH decoding, up to three OFDM symbols for large system bandwidth have to be read. Thus, there are at most 1200 * 2 * 8 * 3 = 57600 bit = 57.6 kb of soft bits to be read. Execution of CDEC functionality is very time critical, so the reading must be done sufficiently quickly.

PHICH Decoder

The PHICH is located in the same OFDM symbols as the PDCCH. This is called control region. The length of the control region is derived from the CFI. The control decoder assembles all the PHICH resource elements and performs maximum likelihood decoding in order to obtain the ACK/NAK bit for the associated UL-DSCH. PHICH decoding is typically scheduled after PCFICH.

PBCH Decoder

As discussed in the earlier section, the number of antennas is initially not known from cell search, blind PBCH detection assuming three candidates (one, two, or four antennas) is required. For this purpose, soft bits for all three antenna configurations have to be generated and tested. The actual antenna configuration is indicated by eNB using configuration-dependent CRC scrambling of PBCH, so the UE is able to verify the correct antenna configuration.

There are 480 soft bits per frame. After the reception of PBCH related data in subframe n, the four hypotheses have to be evaluated for each supported antenna configuration in order to decode PBCH.

A typical implementation of Control and PBCH channel decoder is shown in Figure 4.6.

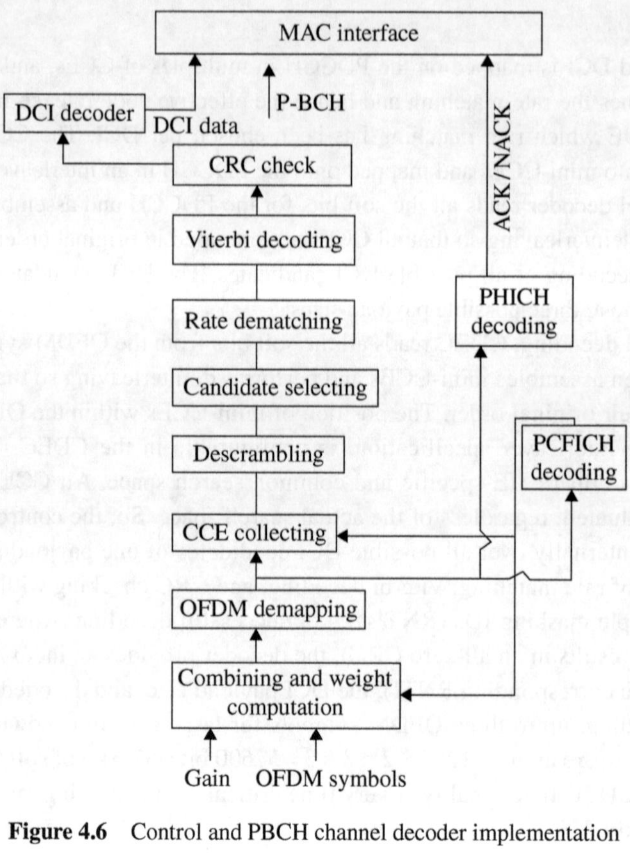

Figure 4.6 Control and PBCH channel decoder implementation

4.9 PDSCH Reception

The downlink subframes are mainly occupied by the PDSCH, which contains the individual user data. Based on channel feedback and the scheduling algorithm, the eNB will allocate data for each user using physical resource block (PRB). The eNB uses the PDCCH channel's DCI to inform the UE about where the data is located and the modulation and coding scheme (MCS) is used for that. The DCI may also contain updates on TPC, hybrid automatic repeat request (HARQ), uplink grants, and so forth.

The PDSCH follows directly after the PDCCH, and therefore the UE is forced to buffer the PDSCH across the whole channel bandwidth because it does not know if and where it is scheduled when the channel data starts to arrive. This reception and buffering is a waste of energy if the UE, after decoding the DCIs, learns that it is not scheduled. So, it is better to decode the PCFICH and PDCCH as fast as possible and then stop buffering the PDSCH and power down the receiver if the data is not intended for the UE.

The reception and decoding PDSCH is similar to the control channels as discussed above, and the steps involved were already explained in Figures 3.23 and Figure 3.25 of Chapter 3 (in reverse order).

4.10 SIB Reception

After receiving the MIB through the BCCH channel, the UE configures the PDSCH (BCCH-DL-SCH) channel and maps it on the PDSCH to receive the SI block type1 (SIB1), which is transmitted with a periodicity of 80 ms and contains the information relevant to cell access. SIB2 contains radio resource configuration information that is common for all UE. After SIB2's reception, UE configures the random access channel and common shared channels. Later, as required, it starts UL synchronization using random access procedure.

4.11 Paging Reception

Once UE enters into RRC idle mode, it starts listening to the incoming paging message to know about incoming calls and do the periodic measurements for cell reselection.

The LTE paging procedure is used (i) to initiate mobile terminated (MT) PS calls; (ii) to initiate mobile terminated CS fallback calls; (iii) to trigger LTE UE to reacquire system information; (iv) to provide an earthquake and tsunami warning system (ETWS) indication.

As shown in Figure 3.12 of Chapter 3, the downlink paging message is transmitted on the PCH transport channel. The PCH transport channel is again mapped to the PDSCH physical channel, and the resource blocks corresponding to it are indicated in the PDCCH physical channel. The paging indication on the PDCCH is a single fixed indicator known as the paging RNTI (P-RNTI). The P-RNTI is derived from the IMSI of the subscriber to be paged and is constructed by the eNB (as an idle mode terminal it does not have an allocated C-RNTI). The UE need not have to monitor the PDCCH continuously (at every 1 ms) for P-RNTI to listen to paging message as this will drain the battery, so, instead of that, the UE is allowed to monitor the P-RNTI in the PDCCH, only at a predetermined period (say 60 ms, 100 ms etc.) known as a paging occasion. Every paging occasion, the UE receives and checks for a paging indication and if it is not found then it will enter into sleep mode until the next paging occasion, when it has to wake up again for paging reception. This process of monitoring the paging message discontinuously in RRC_Idle is known as discontinuous reception (DRX). Different groups of UEs (grouped according to their IMSI) monitor different subframes for their paging messages.

4.11.1 Calculation of Paging Frame Number

The UE computes the paging frame number using the following relations:

$$\text{Paging frame number} = \text{SFN mod T} = (\text{T/N}) \times (\text{UE\_ID mod N})$$

where, T = DRX cycle length in radio frames. The DRX cycle is broadcast in SIB2 and can have values of 32, 64, 128 or 256 radio frames. These correspond to time intervals of 320, 640, 1280 and 2560 ms. The UE can also propose its own DRX cycle length within ATTACH REQUEST and TRACKING AREA UPDATE REQUEST messages). N = Min(T,nB), nB is broadcast in SIB2 and can have values of {4T,2T,T,T/2,T/4,T/8,T/16, T/32}, N can have values of {T,T/2,T/4,T/8,T/16,T/32}, UE_ID = IMSI mod 1024 (refer to 3GPP TS 36.304).

4.11.2 Paging Procedure

The LTE paging procedure is applicable to the UE in the ECM IDLE state. The UE in this state is in RRC IDLE mode and does not have S1 connectivity with the MME. The MME is responsible for the initiation of the LTE paging procedure and it forwards S1AP paging message to one or more eNBs. (As the location of the terminal is typically not known on a cell level, the paging message is typically transmitted across multiple cells in the so-called tracking area.) The MME starts the timer T3413 after sending a S1AP paging message for PS data call, and LTE UE is addressed by S-TMSI instead of IMSI. The eNB receives the UE_ID from the MME in PAGING S1AP message as “UE Identity Index Value” for MME initiated paging message and constructs the RRC paging message. The UE wakes up at every paging occasion and then receives and searches for the P-RNTI within the PDCCH of subframe belonging to the paging occasion. If the P-RNTI has value of FFFE, it indicates that the UE may have a paging message on PDSCH. The UE finds the P-RNTU in the PDCCH and decodes resource allocation information. This information directs the UE to the PDSCH RB where the paging message has been sent. So, the UE decodes RRC message from PDSCH RBs and checks UE identity in all the records. If the UE does not find its identity in the paging record then it will return to check the PDCCH for the P-RNTI at each paging occasion. If the UE find its identity then it indicates that a paging message is present for the UE, so it will trigger random access procedure to establish RRC connection as described in the next section.

The UE sends a RRC connection request message and the eNodeB responds with RRC connection setup message. The UE includes a service request NAS message or extended service request NAS message within RRC connection setup complete message based on whether it is the paging procedure for PS data call or the paging procedure for a CS fallback call, respectively.

Once the eNodeB forwards the NAS message to the MME then it will stop T3413 and will proceed to establish a connection with the UE. A paging retransmission will be triggered if T3413 expires prior to the MME receiving a NAS message from the UE. The UE also checks for an RRC paging message for SI modification flag and ETWS flag and if the former is present then the UE reacquires BCCH SI. If the latter is present, the UE reads ETWS notifications in SIB10 and/or SIB11. The paging procedure is shown in Figure 4.7, and for more details refer to 3GPP TS 36.304, TS36.331, and TS24.301.

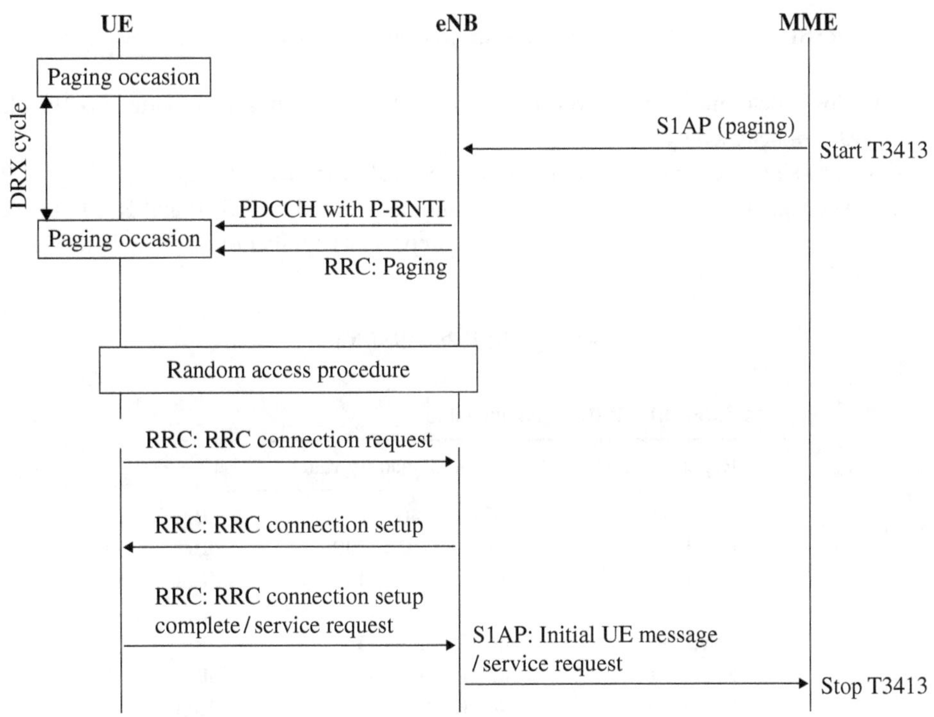

Figure 4.7 Paging procedure (ECM idle and RRC idle mode)

4.12 UE Measurement Parameters

In LTE, UE needs to measure the signal strength and quality of its own and neighboring cells periodically, for cell reselection (selecting the best available permitted cell) during idle, and in connected mode measure, and it reports this to the network for effective handovers. The UE measures RSRP and RSRQ as mentioned below.

- Reference Symbol Received Power (RSRP). This is the average power received from a single reference signal resource element and is calculated using the useful part of an OFDMA symbol, without a cyclic prefix. The average is taken in linear units. It is measured over the resource elements that carry cell-specific reference signals within the considered measurement frequency bandwidth (not the wideband power). The antenna connector of the UE is considered as reference point. It is used in cell selection, cell reselection and handover. The RSRP measurements are mapped to the integer number value ranging from 0 to 97 and then included in the RRC messages for reporting in connected mode. The reporting range of RSRP is defined from −140 dBm to −44 dBm with 1 dB resolution as shown in Table 4.1.
- Received signal strength indicator (RSSI). It represents the total received power over wideband by the UE. It is measured only in all symbols containing reference signals. It includes power from serving cell as well as co-channel interference and noise.

$$RSSI = \text{wideband power} = \text{noise} + \text{serving cell power} + \text{interference power}.$$

  - It helps in determining interference and noise information. It is not reported to eNodeB by UE, rather used by itself.
- Reference signal received quality (RSRQ). The RSRQ indicates the quality of received reference signal. Its measurement and calculation is based on RSRP and RSSI values. The RSRQ formula is shown below, where N represents number of resource blocks over which RSSI is measured:

$$RSRQ = RSRP/\ (RSSI/N)$$

Table 4.1 RSRP mapping table

Reported value	Measured quantity value	Unit
RSRP_00	RSRP < −140	dBm
RSRP_01	−140 ≤ RSRP < −139	dBm
RSRP_02	−139 ≤ RSRP < −138	dBm
…	…	…
RSRP_95	−46 ≤ RSRP < −45	dBm
RSRP_96	−45 ≤ RSRP < −44	dBm
RSRP_97	−44 ≤ RSRP	dBm

Table 4.2 RSRQ mapped values (see 3GPP TS 36.133)

Reported value	Measured quantity value	Unit
RSRQ_00	RSRQ<−19.5	dB
RSRQ_01	−19.5≤RSRQ<−19	dB
RSRQ_02	−19≤RSRQ<−18.5	dB
…	…	…
RSRQ_32	−4≤RSRQ<−3.5	dB
RSRQ_33	−3.5≤RSRQ<−3	dB
RSRQ_34	−3≤RSRQ	dB

- RSRQ takes the antenna connector of the UE as its reference point. The RSRQ values are mapped from dBm to integer numbers ranging from 0 to 34 (as shown in Table 4.2) and then included in RRC messages for reporting. The reporting range of RSRQ is defined from −19.5 dB to −3 with 0.5 dB resolution.

4.13 Random Access Procedure (RACH Transmission)

After initial synchronization, it can be considered that UE is time synchronized with eNB in the DL but may not be synchronized in the UL, so a random access procedure is required to be performed (for acquisition of UL timing). At this stage, the mobile terminal requests for a connection setup (when the UE wants to connect to the network or to make a call) using the RACH procedure. It is used for several purposes, such as (i) for initial access when establishing a radio link – when the UE moves from the RRC_IDLE state to the RRC_CONNECTED state; (ii) DL data arrival during RRC_CONNECTED when UL is "nonsynchronized," UL data arrival during RRC_CONNECTED when UL is "non-synchronized"; (iii) to re-establish a radio link after radio-link failure; (iv) for a handover when uplink synchronization needs to be established to the new cell; (v) as a scheduling request if no dedicated scheduling-request resources have been configured on PUCCH.

For the random-access transmission procedure there are two possibilities: (i) the contention-based random access procedure – here there are many UEs in the same area/cell sending the same request in which there is possibility of collision among the requests coming from various other UEs (this is used during RRC idle to RRC connected, UL data transfer, and RRC connection re-establishment); (ii) the contention-free or non-contention-based random access procedure – the network can inform the UE to use a unique identity to prevent its request from colliding with requests coming from other UEs (this is used during Intra-system handover and DL data arrival, for example if synchronization is lost).

Contention Based Random Access Procedure
This has the following steps – see Figure 4.8(a):

1. The UE listens to a DL broadcast signal (SIB2) to obtain the transmission timing. It is also informed of the available signatures, frequency bands, and time slots for a random access. There are 64 preambles in a cell, which are grouped to indicate the length of the needed resource. Some are also reserved for contention-free access. The UE selects one of the 64 available RACH preambles and sends it along with the UE identity, which is called random access radio network temporary identity (RA-RNTI) and is determined from the time slot number in which the preamble is sent.
2. Next, eNodeB sends a "random access response" to the UE on the DL-SCH, addressed to RA-RNTI and calculated from the timeslot in which the preamble was sent. Random access response contains (i) cell radio network temporary identity (C-RNTI) – another identity given by the eNB to the UE for further communication; (ii) timing advance value – the eNB causes the UE to change its timing so that it can compensate for the round-trip delay caused by the UE's distance from the eNB; (iii) uplink grant resources – the network assigns initial resources to the UE, which it can use for UL-SCH.
3. Then using UL-SCH, the UE sends an "RRC connection request message" to the eNodeB. The RRC connection request message contains: UE identity (TMSI or a random value) and a connection establishment cause. The UE is identified by a temporary C-RNTI.
4. Once the message from the UE is successfully received by the eNB, it responds with a contention resolution message. This message is addressed towards the TMSI value or a random number but contains the new C-RNTI, which will be used for further communication.

Contention-Free Random Access
During handover, a temporary valid preamble will be issued dedicated to the UE. Here, no contention resolution is needed – see Figure 4.8(b).

4.13.1 *Preamble Transmission by UE*

The preamble is transmitted on the time–frequency resource known as the physical random-access channel (PRACH) as discussed in Chapter 3. The network broadcasts about the PRACH resources in SIB-2 (see Table 3.2 in Chapter 3.2), which includes information that the terminal needs in order to be able to access the cell. As part of the first step of the random-access procedure, the terminal selects one preamble to transmit on the PRACH. The PRACH resource has a BW corresponding to six resource blocks (1.08 MHz). The basic random-access resource is 1 ms in duration but it is also possible to configure longer preambles.

The preamble sequences are generated from cyclic shifts of root Zadoff–Chu sequences. From the Zadoff–Chu sequence, cyclic shifted sequences are obtained by cyclic shifts as shown in Figure 4.9(a). The reception of the random-access preamble is shown in Figure 4.9(b), where the principle is based on correlation of the received signal with the root Zadoff–Chu sequences. The received time domain samples are collected and converted into a frequency domain representation using an FFT. The output of the FFT is multiplied

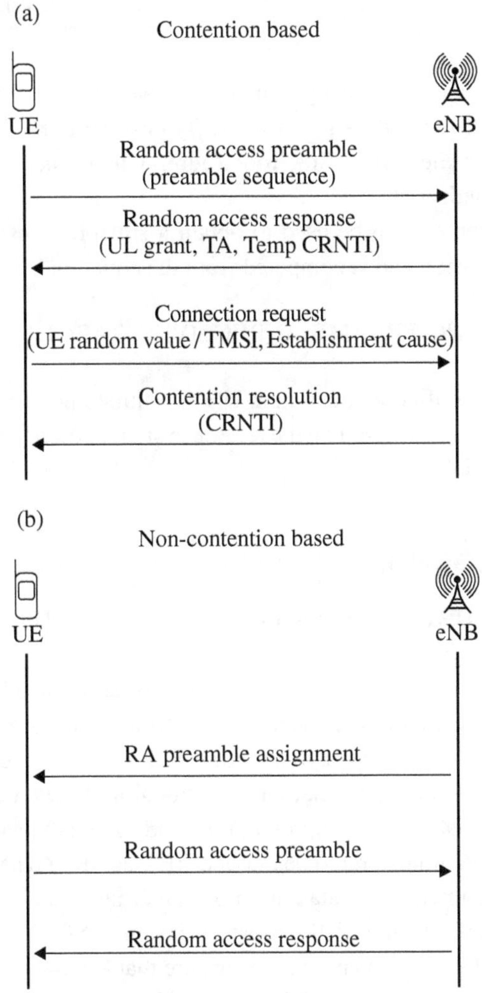

Figure 4.8 Contention based and contention-free random access

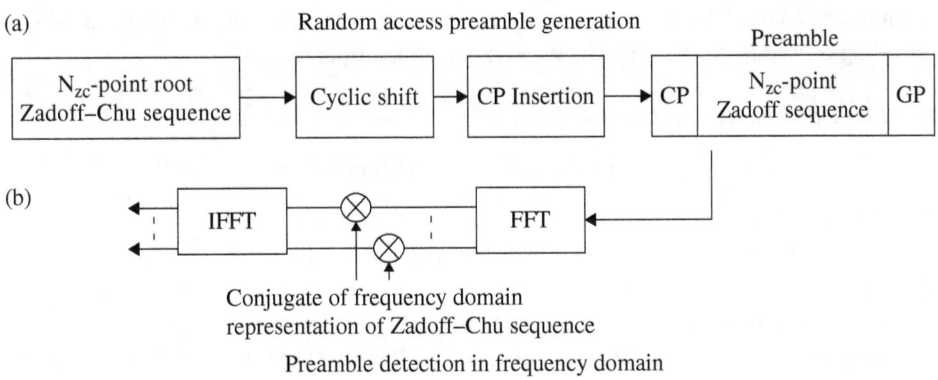

Figure 4.9 (a) Random access preamble generation and transmission in UE. (b) Detection in eNB

by the complex-conjugate frequency-domain representation of the root Zadoff–Chu sequence and the result is fed through an IFFT. By observing the IFFT outputs, it is possible to detect which of the shifts of the root Zadoff–Chu sequence has been transmitted and the delay corresponds to it.

Open-loop power control can be used to obtain a suitable transmission power for the initial PRACH transmission and is computed from the equation below (see TS 36.213):

PPRACH = min {PCMAX , PREAMBLE_RECEIVED_TARGET_POWER + PL} [dBm].

where, PCMAX is the configured maximum UE transmitted power, PL is the DL pathloss estimate calculated in the UE, based on the reference signal power signaled in SIB2 and the measured RSRP at the UE.

4.14 Data Transmission

Figure 4.10 shows the UE data-transmission procedure using PUSCH, and corresponding reception in eNB.

As an example, let us consider application data that is passed to a UE modem via a TCP/IP interface as shown in Figure 4.11 and Ethernet frames are used, where the Ethernet frames have roughly 1500 bytes of payload. For a TCP/IP transfer, out of total 1500 bytes of packet size, 1460 bytes are payload, and additional 40 bytes are header information. Next, the PDCP layer enables the compression of headers and reduces the header size to 3 or 5 bytes. Then, for signaling, a radio bearer and RRC control signal overhead are added. Next, the RLC layer segments the data unit into smaller units and adds headers to each of these segments (protocol data unit (PDU) = service data unit (SDU) + header). The sequence number for each RLC PDU can be used to make sure that a correct reassembly takes place at the receiver end. In this case, the RLC header could be assumed to be slightly more than 2 bytes long. Then, during MAC layer processing, the MAC layer adds a protocol overhead greater than 2 bytes. On top of that, it also adds 16 bits for alignment purpose. During PHY processing, the CRC is added in layer 1 (the physical layer), which is a multiple of 8 bits (commonly, 24 bits CRC are used). Next, for encoding purposes, most commonly a 3/4 rate encoding is used to protect the data over channel impairments.

Baud Rate at the Physical Layer

1 radio frame = 10 subframes

1 subframe (1 ms) = 2 time slots

1 time slot = 7 modulation symbols when normal CP length is used or 6 when extended CP is used

1 modulation symbols = 6 bits if 64 QAM is used as modulation scheme.

For CAT-1, 2, 3, 4, 5: total RBs available in 20 MHz ~ 100 RBs (see 3GPP TS 36.306)

1 RB = 1 sub-frame * 12 subcarriers = 2 slots * 12 subcarriers

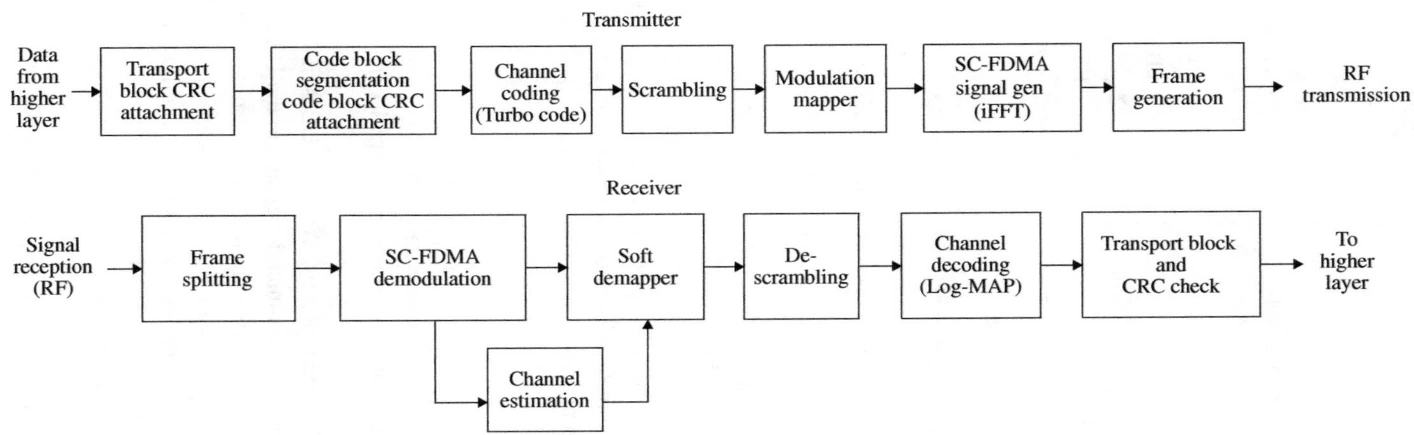

Figure 4.10 PUSCH transmission (in UE transmitter) and reception (in eNB)

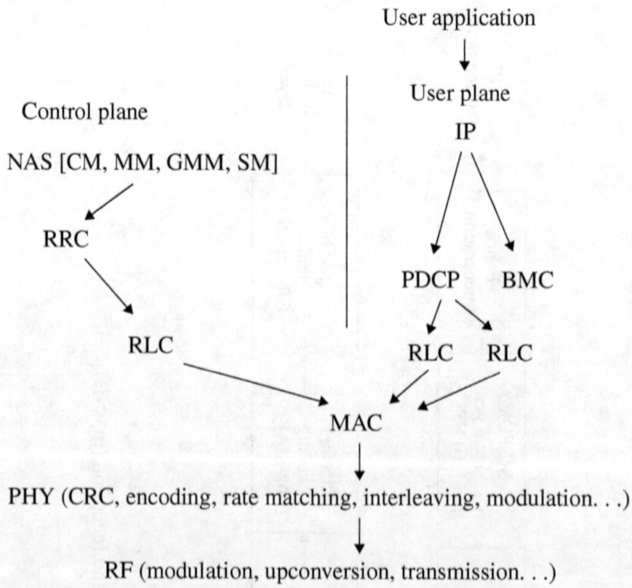

Figure 4.11 User packet flow inside the UE protocol stack

Therefore, the number of bits in a subframe = 100 RBs × 12 subcarriers × 2 slots × 7 modulation symbols × 6 bits = 100800 bits.

Again, 1 subframe = 1 ms. Hence, the data rate = 100800 bits/1 ms = 100.8 mbps. For a device that uses 4 × 4 MIMO (Cat-6) the peak data rate will be = 4 × 100.

8 mbps = 403 mbps. In the physical layer, data encoding reduces the rate. If 3/4 coding is used to protect the data, then the rate will be (¾) × 403 mbps = 302 mbps.

4.15 Handover

The handovers in the RRC Connected state are network controlled and UE assisted. The following stages are performed for handovers (see Figure 4.12):

1. Measurement configuration and reporting. When the UE is in the RRC Connected state, the eNB sends the measurement command message to the UE for configuring the different measurement types. The information in the message includes the measurement object (which cells needs to be measured), reporting configuration (whether the event is triggered or periodic), the measurement ID, quantity configuration, and measurement gap. Reporting criteria for the E-UTRA report include events A1, A2, A3, A4 and A5 (refer to 3GPP TS 36.331) whereas those for the inter-RAT measurement report include events B1 and B2.
2. Handover decision. When the signal quality on target cell (eNB2) becomes better than the signal quality on serving/source cell (eNB1), the handover decision is taken by the network (as computed from the measurement report).

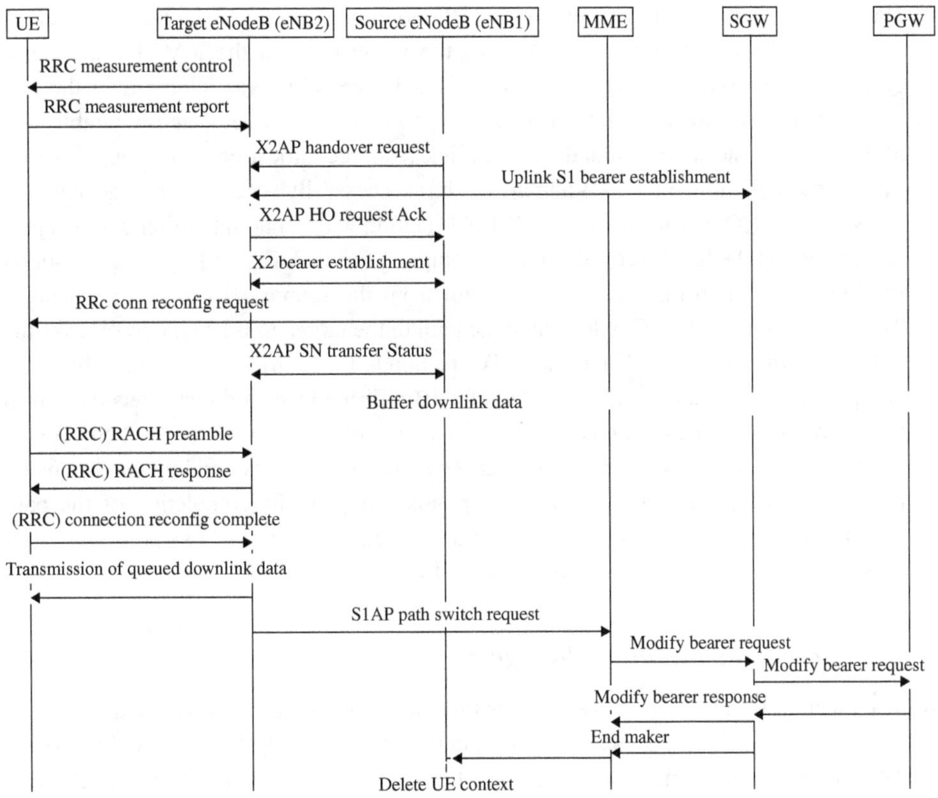

Figure 4.12 Message flow during handover

3. Handover preparation. As a preparation for handover, the source eNB1 sends the coupling information and the UE context to the target eNB2 (handover request message) on the X2 interface, which contains information about active E-RABs and security keys. The GTP connect for the uplink side is established between the target eNB2 and the serving SGW. The target eNB2 accepts the session if resources are available and allocates radio resources for the UE, allocates a RACH preamble to it (optional), and reserves a C-RNTI (which provides a unique UE identification at the cell level identifying the RRC connection). The target eNB2 may perform admission control depending on the received EPS bearer QoS information. The Target eNB2 responds to the source eNB with a handover request acknowledge message. This message carries the handover command message (RRC connection reconfiguration request) in a transparent container. The source eNB1 commands the UE (HO command) to change the radio bearer to target eNB2. The UE receives the HO command with the necessary parameters (i.e. new C-RNTI, optionally dedicated RACH preamble, possible expiry time of the dedicated RACH preamble, etc.) and is commanded by the source eNB1 to perform the HO.

4. Handover execution. After receiving the HO command from eNB1, the UE synchronizes with the target eNB2 and accesses the target cell via the RACH following a contention-free procedure (if a dedicated RACH preamble was allocated in the HO command) or following a contention-based procedure (if no dedicated preamble was allocated). The network responds with uplink resource allocation and timing advance (TA) to be applied by the UE. When the UE has successfully accessed the target cell, the UE sends the HO confirm message (C-RNTI) along with an uplink buffer status report indicating that the handover procedure is complete for the UE. The target eNB requests the MME to switch the message flow path from the source eNB to the target eNB. The MME requests the SGW to switch the path to the target eNB and the SGW asks the PGW to switch the path. Then the SGW responds to the MME, signaling about the completion of the path switch. The target eNB will buffer data directly received from the SGW until all the data received via the source eNB has been transmitted. This is needed to maintain the transmission order. Then the SGW will be sending the data using the target eNodeB TEID. The MME responds to signal the completion of the path switch. The end marker has been received at the target eNodeB. At this point the target asks the source eNodeB to release resources for the UE.

4.15.1 *Idle State Mobility Management*

A location area (LA) is a set of base stations that are grouped together and identified as LA identity (LAI). The routing area (RA) is used in the packet-switched domain and identified as RAI. Generally, it is smaller than LA. The tracking area (TA) is introduced in LTE and similar to the location area and routing area. A tracking area is a set of LTE cells. When UE is in an RRC idle state for mobility management purposes, the concept of a tracking area (TA) is introduced. Generally, a tracking area covers multiple eNBs. The tracking area identity (TAI) information, indicating in which TA an eNB belongs, is broadcast in a system information message. The UE also knows its current TA information (stored) and if it detects the change of tracking area. The UE updates the MME with its new TA information (TA update message) as it moves across TAs. Whenever P-GW receives data for any UE, it buffers the packets and queries the MME for the UE's location and gets the response about UE's current TA. (A UE can be registered in multiple TAs simultaneously.) This helps to reduce the constant update of UE location with the MME and resulted in UE power saving. (Refer to 3GPP TS 36.300.)

4.15.2 *Interoperability with Legacy Systems (I-RAT)*

E-UTRAN interoperates with GERAN, UTRAN, and other systems, and interoperability with other radio systems is possible at the IP level. Today, different interoperability mechanisms have been standardized to cater for different deployment options: (i) an inter-RAT handover framework allows optimized handovers of packet data sessions from E-UTRAN to UTRAN/GERAN; (ii) a network assisted cell change (NACC) framework allows the

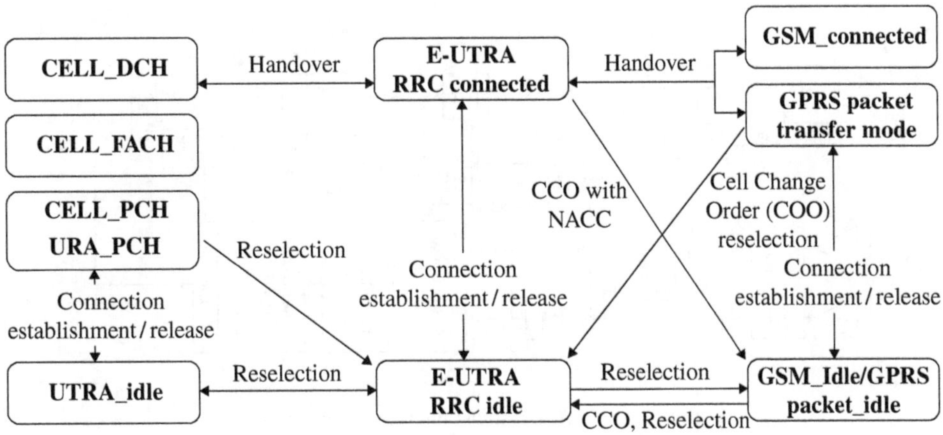

Figure 4.13 E-UTRA and inter-RAT mobility

handover of packet sessions from E-UTRAN to GERAN; (iii) single-radio voice-call continuity (SR-VCC) – handover supported from E-UTRAN to UTRAN/GERAN.

As the initial deployment of LTE has low coverage, it will mainly be used in hotspots. So, if the user makes a voice-over LTE (VoLTE) call and goes out of the LTE coverage area, then the call will be dropped. In order to avoid this, methods have been defined for transferring the call to legacy (2G or 3G) networks. These are SRVCC and packet-switched handovers (for voice-call continuity in HSXPA). When UE returns to a good LTE coverage area again the UE needs to be handed over to the LTE network from the legacy network for better data service. For this reverse handover case, R-SRVCC is being standardized, whereas the packet switched handover (PSHO) mechanism is applicable for both directions. (Please refer to 3GPP TS 43 239, version 6.13.0, Release 6, for more details about PSHO.)

The network-sharing feature in LTE architecture enables service providers to reduce the cost of owning and operating the network by allowing the service providers to have separate CN (MME, SGW, PDN GW) while the E-UTRAN eNBs are jointly shared by them. The inter-RAT mobility is shown in Figure 4.13.

4.16 Anatomy of an LTE UE

The anatomy of a typical UE is shown in Figure 4.14. Its different blocks are discussed below.

- RF. The details about the front-end block are discussed in Chapter 6. The RF block receives the LTE carrier frequencies as tuned by the baseband and passes the I,Q samples to the digital front-end (DFE) unit. On the transmission side it receives the I,Q signal from the baseband, converts it to a high-frequency analog signal, amplifies it using a power amplifier, and transmits.

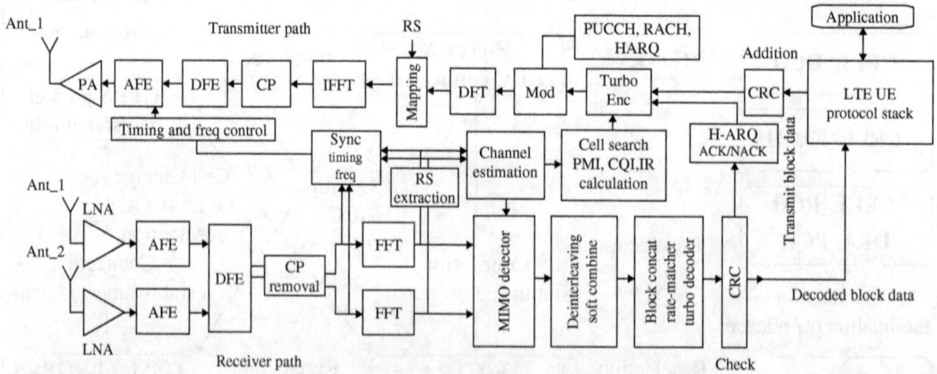

Figure 4.14 LTE UE transmitter and receiver blocks

- DFE. The function of the DFE is usually gain control, sample rate conversion, pulse shaping, matched filtering and sometimes phase adjustment.
- Baseband. The baseband receiver receives the digitized signal as complex I,Q samples from the analog to digital converters (ADCs) via the radio frequency and baseband module (RF-BB) (like the Digital RF interface standard, as explained in Chapter 5).
  - OFDM demodulator. The OFDM decoder unit extracts time samples from an OFDM symbol and transforms them into the frequency domain by an FFT in order to retrieve the modulation symbols for further processing in the receiver chain. The demodulator consists of three main blocks: (i) cyclic prefix removal: this reads the time domain input samples from the RF and then removes the CP; (ii) FFT: this can perform an energy preserving 128-2048 bit FFT or scaled FFT; (iii) writes the output frequency domain (FD) samples back to memory.
  - Inner receiver. Channel estimation is performed using reference signals extracted from the RB. The estimated channel impulse response is used to equalize the received symbol and soft bits are generated. These are discussed in detail in the next section.
  - Outer receiver: Next, the hard bits (frame data) are detected from the soft bits using the deinterleaving and decoding process.
  - Protocol processing. The received data is passed to the higher layer for processing. Signaling data goes to the signaling plane and is used by the protocol layer whereas the user data passes via the user plane to the user application.
  - Transmitter. In the transmit path, data from higher layers is encoded, interleaved, and passed to the RF unit for modulation, RF upconversion, amplification, and transmission.

4.17 Channel Estimation

To mitigate or combat the channel effect on the received signal, the receiver employs a method called channel estimation; its task is to estimate the channel parameters (channel impulse response) when the signal is propagating through the channel medium.

Channel estimation plays an important role in a communication receiver to reduce the bit error rate and to improve the system performance. Channel estimators can be categorized as non-data aided or data aided.

- Non-data-aided or blind channel estimators do not use any pilot or training sequence bits. Rather, they exploit certain underlying mathematical information regarding the type of data being transmitted and estimate channel response by statistics about the received signals. These methods are bandwidth efficient but still have high bit-error rate.
- Data-aided channel estimators require known reference (training or pilot) signals to be transmitted along with unknown user data. The channel response can be estimated by comparing the received reference signals (which are actually the known transmitted reference signals impaired by the channel) with the respective known transmitted reference signals. The estimation accuracy increases with the number of training symbols, but throughput or system efficiency decreases, so the optimum number of such reference signals must be inserted according to the degree of channel variation, namely the coherence time and coherence bandwidth of the channel under estimation.

In general, the fading channel of OFDM systems can be viewed as a two-dimensional (2D) signal (time and frequency). The optimal channel estimator in terms of mean-square error is based on 2D Wiener filter interpolation but such a 2D estimator structure is too complex for practical implementation. So, one-dimensional (1D) channel estimations are also adopted in OFDM systems. The two basic 1D channel estimations are block-type pilot channel estimation and comb-type pilot channel estimation, in which the pilots are inserted in the frequency direction and in the time direction respectively.

- Block-type pilot channel estimation. The task here is to estimate the channel conditions (specified by H_k) given the pilot signals (specified by X_k) and received signals (specified by Z_k), with or without using certain knowledge of the channel statistics. The estimation can be based on least-square (LS), minimum mean-square error (MMSE), and modified MMSE.
- MMSE estimator. Let us assume equivalent time-domain channel impulse response, g, is a random vector with a Gaussian distribution and is uncorrelated with the noise (v). Assume that v has a covariance matrix $\sigma_v^2 I_N$, where I_N is an N×N identity matrix. So, the minimum mean-squared error (MMSE), which minimizes $E\{(\hat{g}-g)^H(\hat{g}-g)\}$ can be written as: $\hat{\mathbf{g}}_{\text{MMSE}} = \mathbf{R}_{\mathbf{gz}}\mathbf{R}_{\mathbf{zz}}^{-1}\mathbf{z}$, where, $\mathbf{R}_{\mathbf{gz}} = E\{\mathbf{gz}^H\} = \mathbf{R}_{\mathbf{gg}}\mathbf{F}^H\mathbf{X}^H$ is the covariance matrix between g and z.

Then the auto covariance matrix of z will be: $\mathbf{R}_{\mathbf{zz}} = E\{\mathbf{zz}^H\} = \mathbf{XFR}_{\mathbf{gg}}\mathbf{F}^H\mathbf{X}^H + \sigma_v^2\mathbf{I}_N$, where R_{gg} is the auto covariance matrix of g and is considered to be known. So, the MMSE frequency domain channel response will be:

$$\hat{\mathbf{h}}_{\text{MMSE}} = \left[\hat{H}_{-N/2} \cdots \hat{H}_{N/2-2}\ \hat{H}_{N/2-1}\right]^T = \mathbf{F}\hat{\mathbf{g}}_{\text{MMSE}}.$$

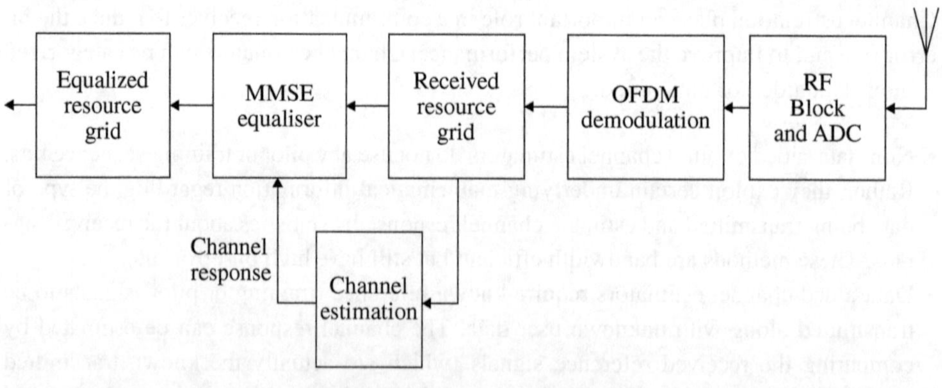

Figure 4.15 Channel estimation

- Least-squares estimator (LSE). This is a maximum likelihood (ML) estimator and assumes that the time domain channel impulse response is deterministic and tries to find $\hat{g}_{LS}$, which minimizes $(z-XFg)^H(z-XFg)$, where, X and F are invertible square matrices. So, the LS solution will be:

$$\hat{\mathbf{h}}_{LS} = \mathbf{F}\left(\mathbf{F}^H\mathbf{X}^H\mathbf{X}\mathbf{F}\right)^{-1}\mathbf{F}^H\mathbf{X}^H\mathbf{z} = \mathbf{F}\left(\mathbf{F}^{-1}\mathbf{X}^{-1}\left(\mathbf{X}^H\right)^{-1}\left(\mathbf{F}^H\right)^{-1}\right)\mathbf{F}^H\mathbf{X}^H\mathbf{z} = \mathbf{X}^{-1}\mathbf{z}.$$

With the extra statistical information, the MMSE estimator can outperform the LS estimator. In high-SNR scenarios, the MMSE estimator becomes equivalent to the LS estimator. The low-rank approximation loses statistical information contained in the noise subspace and performs less well than the original MMSE estimator.

In LTE, as in many OFDM systems, known symbols, called pilots, are inserted at specific locations in the time frequency grid in order to facilitate channel estimation (see Figure 3.8). Channel estimates can first be obtained at the pilot positions using simple least-squares demodulation (see Figure 4.15). The remaining channel coefficients can then be calculated using interpolation techniques in both time and frequency directions.

As for MIMO-OFDM, when antenna port 0 is transmitting its pilot symbols, the other antenna is silent. This implies that pilot transmissions from the two antenna ports are completely orthogonal – that MIMO channel estimation is a straightforward extension of SISO channel estimation techniques.

4.18 Equalization

Once channel estimates at data subcarriers are derived, the receiver performs equalization to compensate for signal distortion by using the estimated channel impulse response. A receiver can obtain estimation for channel gains for all subcarriers and then apply estimated channel gains to equalize the data symbols using various equalization techniques.

- One-tap equalizer. OFDM systems are favored over single-carrier modulations in that the simple one-tap frequency-domain equalizer (FDE) can equalize OFDM signals that go through frequency selective fading channels. In channels whose impulse responses remain constant within one OFDM symbol period, the received signal at each subcarrier takes the form of

$$Z_{i,k} = H_{i,k} X_{i,k} + V_{i,k}.$$

One-tap equalizers restore the transmitted signal by:

$$\hat{X}_{i,k} = H_{i,k}^{-1} Z_{i,k},$$

Regardless of noise, the zero-forcing equalizer simply uses the inverse of the channel response and forces the frequency-selective-faded signals back to flat faded ones. However, it may result in noise enhancement in subcarriers that suffer deep fading.
- MMSE equalizer. A linear MMSE equalizer appears to be the first choice for separating the layers in the spatial multiplexing mode. The MMSE equalizer tries to minimize $\left\{\mathrm{E}\left|\hat{\mathrm{X}}_{\mathrm{i,k}} - \mathrm{X}_{\mathrm{i,k}}\right|^2\right\}$ taking the noise component into account and equalizes the signal by:

$$G_{i,k} = \frac{H_{i,k}^*}{\left|H_{i,k}\right|^2 + \dfrac{1}{SNR}}.$$

where $G_{i,k}$ is the equalizer coefficient at the k-th subcarrier during the i-th symbol. This equalizer has the advantage that the noise enhancement problem in low-SNR cases is gone. When SNR is high enough, it is also clear that the MMSE equalizer approaches the zero-forcing equalizer. In channel estimation problems, adaptive algorithms can be used to adjust the equalizer coefficients to minimize $\left\{\mathrm{E}\left|\hat{\mathrm{X}}_{\mathrm{i,k}} - \mathrm{X}_{\mathrm{i,k}}\right|^2\right\}$ without prior channel information. The equalized signal is compared with the reference signal to obtain error signal ($e_{i,k}$). Then, the equalizer coefficients are adjusted according to the error signal:

$$G_{(i+1),k} = G_{i,k} - g_{i,k} e_{i,k},$$

where, $g_{i,k}$ is gain factor.
- Multiple-tap equalizer. In fast-fading channels, channel response not only changes from the previous symbol to the current symbol but also varies within one symbol period. The fast channel variation within one symbol period brings about intercarrier interference, which causes the system performance to deteriorate further. One-tap equalizers cannot cope with such situations, so multiple-tap equalizers which may cancel intercarrier interference from adjacent subcarriers, are required.

4.19 Detection

The number of layers transmitted in parallel depends on the UE category. Linear equalization cannot achieve a diversity order equal to the number of receiver antennas. So, to improve the throughput, MIMO is used with time-invariant channels and low frequency-selectivity over the codeword length. But the diversity order of 2 is only achieved with maximum likelihood (ML) decoding. For MIMO systems, a major challenge is the separation and detection of the transmitted symbols at the receiver. Due to the amount of operations involved, such as matrix inversion, it cannot be handled by the baseband processor itself. Among different detection algorithm, maximum likelihood (ML) detection is optimal. To approach ML performance efficiently, tree-search schemes known from sequential decoding, sphere decoding, or the M-algorithm could be used. For complexity reduction, an important prerequisite is defining a tree that reduces the number of visited tree nodes. It turns out that the mean square error is a more appropriate metric than the Euclidean metric, with only a small performance penalty. Sorting the layers based on a sorted QR decomposition or a permuted Cholesky decomposition is the second key ingredient. The choice of the complex baseband representation or its equivalent real-valued representation as the underlying signal model also makes a difference due to additional degrees of freedom for sorting in the real valued model. An alternative advanced MIMO

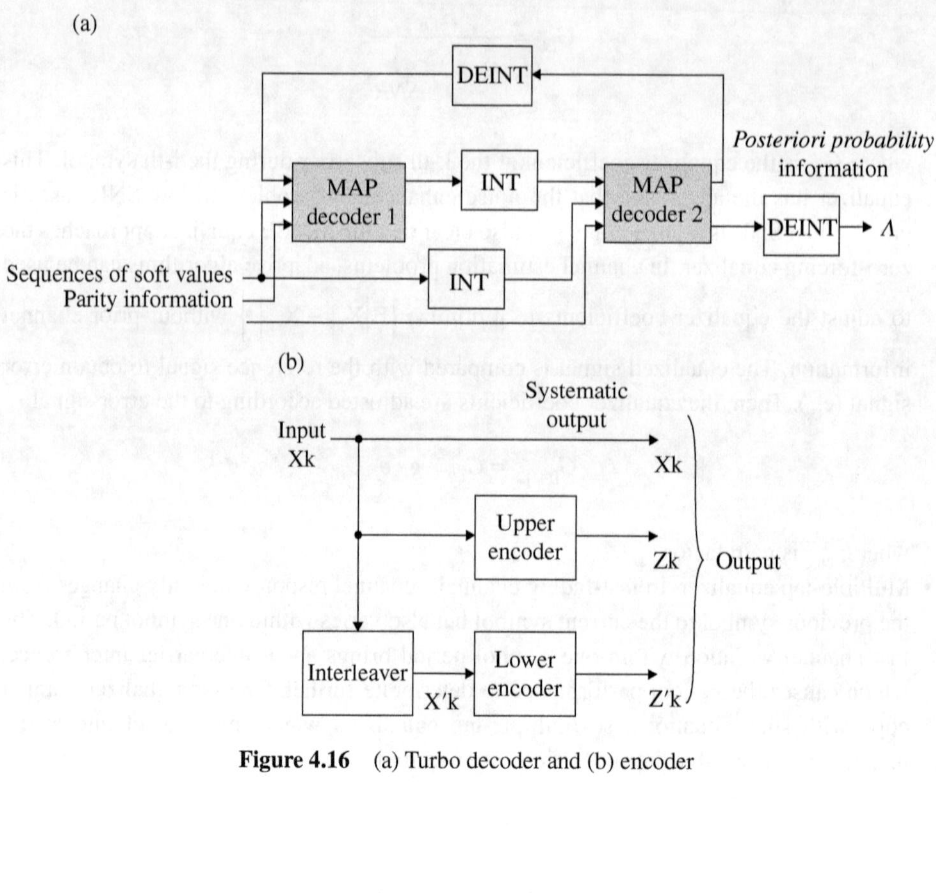

Figure 4.16 (a) Turbo decoder and (b) encoder

receiver is serial interference cancellation (SIC), for example in the case of a 2×2 system where two codewords are transmitted in parallel. Here, the re-encoding of the data stream detected first introduces additional latency. In all equalizer modes, generation of LLR is necessary to provide an input to the turbo channel decoder.

4.20 Decoder

The commonly used turbo code (TC) decoder architectures split a received data block into so-called windows for parallel decoding (as shown in Figure 4.16). An acquisition process estimates the initial values at the window borders. For TC decoding, a complete data block is iteratively processed in a loop that comprises two component decoders using the Log-MAP algorithm, where these two decoders exchange so-called a posteriori probability information. The iterative exchange continues until a stopping criterion (comparing with a threshold) is fulfilled. The codeblock-specific CRC that is attached in addition to the CRC over the entire payload block can serve as an early stopping criterion for the iterative turbo decoding process, which helps to reduce UE power consumption. Quadratic permutation polynomial (QPP) turbo code interleaver is standardized for the LTE (see TS 36.212).

Reference

[1] Das, Sajal Kumar. (2010) *Mobile Handset Design*, John Wiley & Sons, Ltd.

Further Reading

3GPP (2016) *3GPP Specification Series*, http://www.3gpp.org/DynaReport/36-series.htm (accessed May 4, 2016).

3GPP TR 25.814 (2005) *TSG RAN, Physical layer aspects for Evolved UTRA*, 3GPP.

3GPPTS 36.212 (2007) *Evolved Universal Terrestrial Radio Access (E-UTRA); Multiplexing and channel coding* (Release 10), 3GPP.

Benvenuto, N. and Cherubini, G. (2002) *Algorithms for Communications Systems and their Applications*, John Wiley & Sons Ltd.

Ikuno, J. C., Pendl, S.,Šimko, M., and Rupp, M. (2012) *Accurate SINR Estimation Model for System Level Simulation of LTE Networks*. Proceedings of 2012 IEEE International Conference on Communications, Institute of Communications, Vienna University of Technology.

Sesia, S., Toufik, I, and Baker, M. (eds.) (2011) *LTE – The UMTS Long Term Evolution, From the Theory to Practice*, John Wiley & Sons, Ltd.

Van de Beek, J. J., Edfors, O., Sandell, M., *et al.* (1995) On Channel Estimation in OFDM Systems. Proceedings of the IEEE 45th Vehicular Technology Conference (VTC 1995), vol. 2, pp. 815819, 1995, 10.1109/VETEC.1995.504981.

Van Nee, R. and Prasad, R. (2000) *OFDM for Wireless Multimedia Communications*, Artech House.

Wang, Z. and Giannakis, G. B. (2000) Wireless multicarrier communications. *IEEE Signal Processing Magazine* **17**(3), 29–48.

5

Smartphone Hardware and System Design

5.1 Introduction to Smartphone Hardware

A smartphone contains several components including processors. Figure 5.1 shows the different modules of a smartphone and Figure 5.2 lists different hardware components. These components are discussed in more details in this chapter and the radio frequency (RF) part is discussed in Chapter 6. The baseband part is responsible for a cellular system's modem bit detection, bit transmission, protocol processing, sleep and power management, audio processing, and so forth, and the application part is responsible for running different applications. In the UE's modem the physical layer receiver and transmitter signal processing modules are either implemented in customized HW IP blocks in "System on a Chip" (SoC) or software running in a digital signal processor (DSP). Generally, the protocols from layer 1 up to layer 3, as well as higher layers and applications are implemented in software running in a general-purpose processor (MCU). Digital signal processors and MCUs communicate by employing shared memory interfaces or some other IPC/RPC (Inter Processor Communication / Remote Processor Communication) mechanisms.

Figure 5.3 shows an internal view of the upper and lower surfaces of a reference smartphone device (Blackberry Bold). The different internal blocks of this smartphone are shown in Figure 5.4.

5.2 Smartphone Processors

As discussed in Chapter 1, a smartphone incorporates the functionalities of a handheld computer to run different user applications along with the functionalities of a cellphone modem. In most common designs, a smartphone is equipped with several processors. To access a

Mobile Terminal Receiver Design: LTE and LTE-Advanced, First Edition. Sajal Kumar Das.

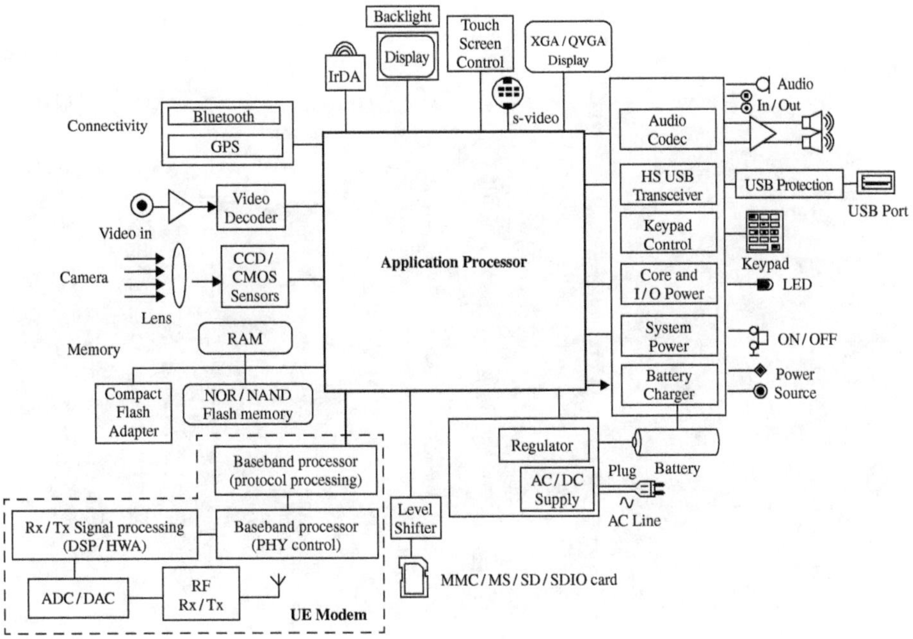

Figure 5.1 Internal blocks of a mobile terminal (smartphone)

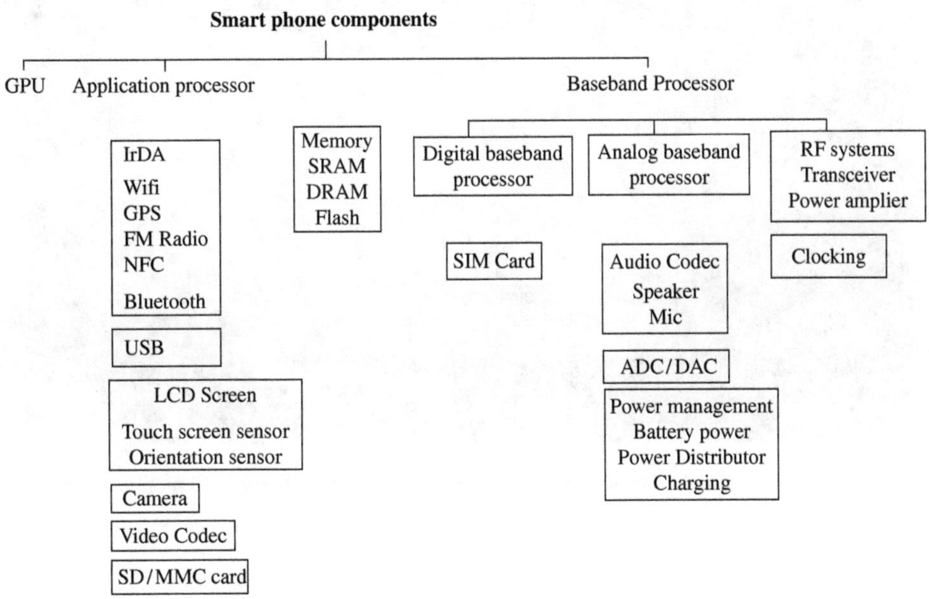

Figure 5.2 Different components of smartphone

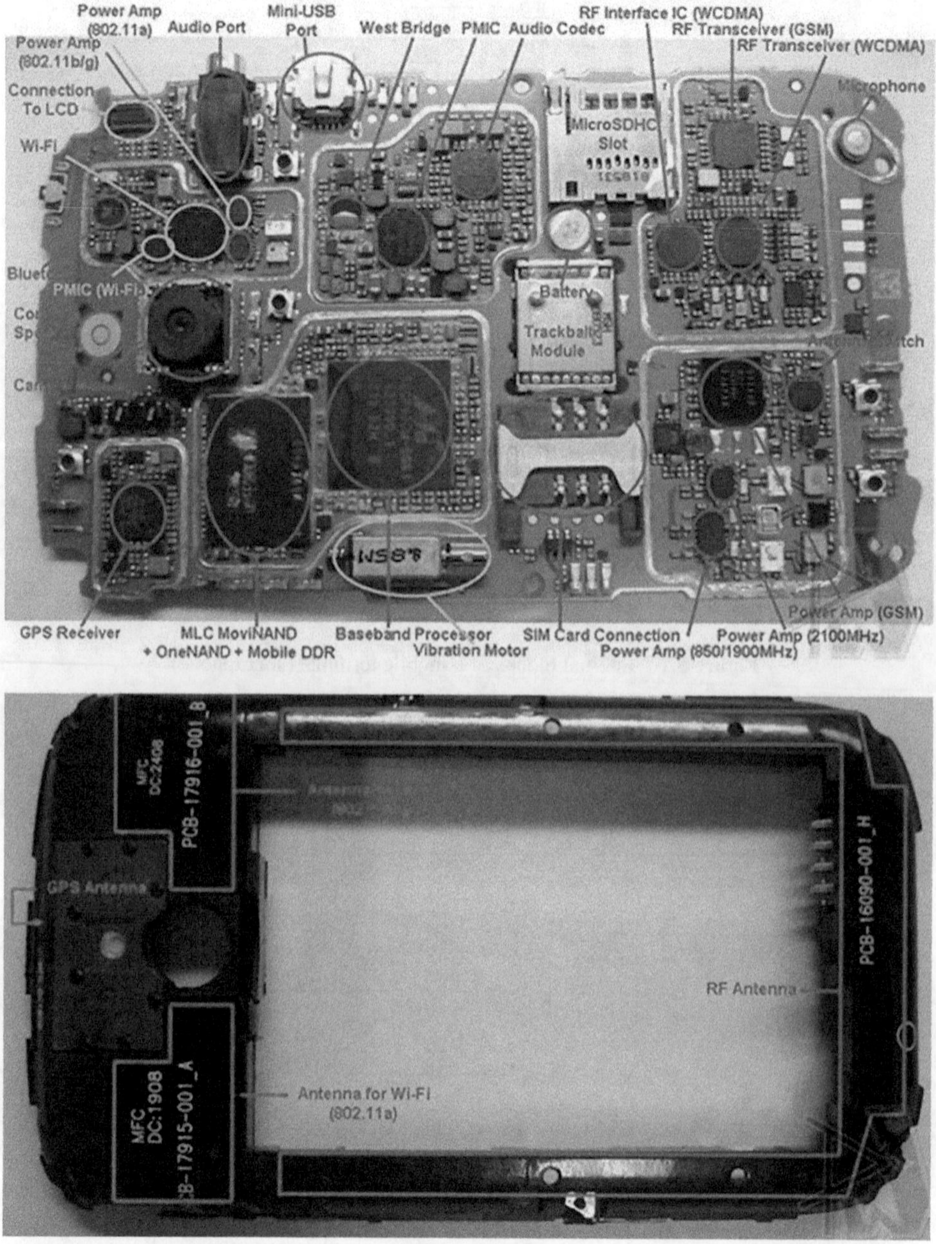

Figure 5.3 Internal view of upper and lower surface of Blackberry Bold mobile device

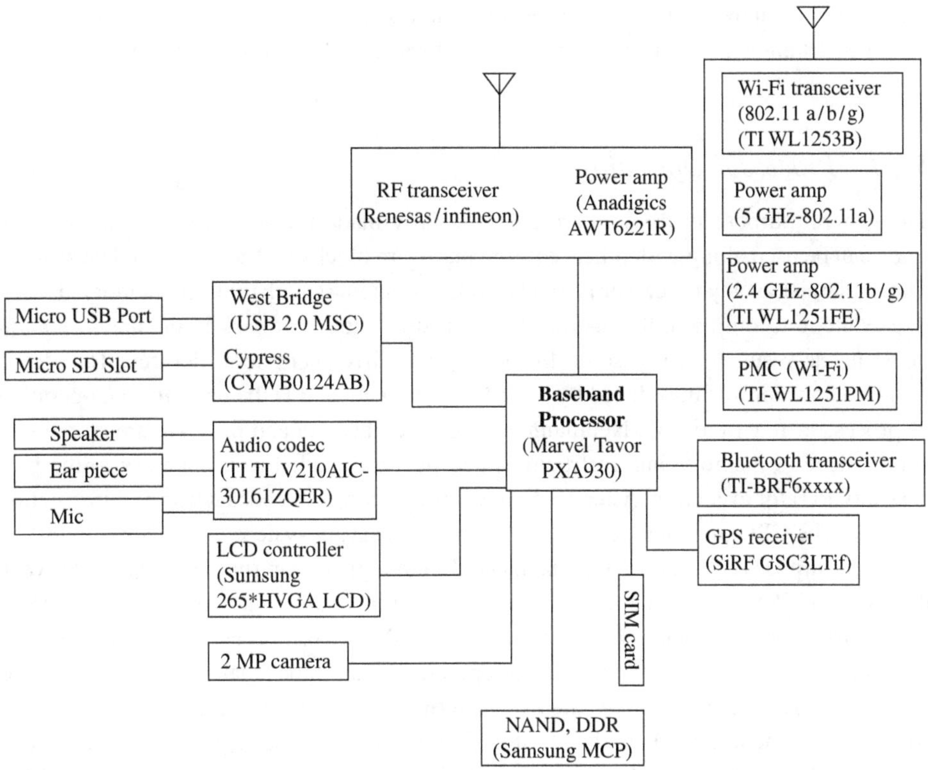

Figure 5.4 Internal blocks of Blackberry Bold mobile device

communication network, it uses baseband processors and for performing application specific computation it uses an application processor, which is a multicore general-purpose processor that helps to provide user interface and in running different applications. These processors could have their own peripherals, memory, and clocking. Sometime, due to demand for higher integration, these two processors are also integrated inside a single physical package and share common resources as much as possible.

The baseband modem manages radio communications, signal processing tasks, and protocol stack to enable the smartphone to access different types of wireless network technologies. To support receiver signal processing-related tasks, generally one DSP is used along with several hardware accelerators (HWA), and for protocol stack execution, one or more (in case of multi-RAT smartphone) processors, like ARM, are used. Apart from the modem signal processing-related tasks, a DSP is also used for noise suppression, echo cancellation or other such signal processing-related tasks.

The application unit relies mainly on a general purpose processor (GPP), most commonly a RISC processor, which provides the processing needed by the applications and provides user interfaces and overall command-and-control functions. However, nowadays, to drive application functionalities, special types of application processor or SoC are used, which can support a number of multimedia-related features such as Web browsing, e-mail, multimedia entertainment, and games, and also employs customized user applications.

5.2.1 Processor Operations

The general-purpose processors are not especially made for any particular application. They can be used along with other required hardware blocks to design a complete system. The major task of any processor is to read code / program (and data) from memory, decode it, execute it according to the instructions, and store the result back in the memory and / or write the result (data) in some I/O location to drive peripheral devices. The stored program concept was introduced by John Von Neumann and is the basis for the operation of processor. In case of Von Neumann architecture, the data and program are stored in a single piece of memory, and instructions and data are loaded from memory to the processor (CPU) for execution. After fetching an instruction, it decodes it using an instruction decoder. The CPU in general consists of a data-processing system, a number of registers (general-purpose registers) to hold data, and digital logic to control the sequence between the CPU and I/O devices and to manage the flow of information among the various units. The registers are fast memory, placed inside the CPU for fast access and used for operations like arithmetic, logical and control operations, and other manipulations as well as internal flag setting. As fetching data from external memory is time and power consuming, so, generally, the intermediate data is stored in general-purpose registers. The arithmetic and logical operation on the data is done in the arithmetic and logic unit (ALU). Accumulators are registers that can be used for arithmetic, logical, shift, rotate, and other similar operations. Often processors have only one accumulator with several other storage registers (general purpose registers). Generally, before transferring the result to the final destination, the data is stored temporarily in the ALU. Apart from that, the CPU must hold the instruction that it is executing and also should know from where the next instruction is to be fetched. For that it needs two essential registers – the instruction register (IR) and the program counter (*pc*). The IR holds the current instruction and the *pc* holds the address of the next instruction to be executed in memory. The instruction set consists of multiple pieces, including addressing modes, instructions, native data types, registers, memory architecture, interrupt and exception handling, and external I/O.

To serve any interrupt routine (or during a function call), the program jumps from the main routine to an interrupt service routine (ISR) (or subfunctions) and returns once it is served. The stack is used to remember the content of the *pc* during that jump, so that the processor can continue from where it left before the jump. The software or / and hardware stacks are used to store the *pc*. The hardware stack consists of a few additional registers identical to the length of the program counter. When a program calls a subroutine, the binary number representing the new memory location is loaded into the *pc* and jumps to the subroutine but the old value representing the memory address for the current sequence of the program instruction is pushed into the stack. When the subroutine is completed, the old number for the *pc* is again popped from the stack. Often one subroutine calls another subroutine; therefore the stack may be more than one level deep and operates on a last-in-first-out (LIFO) basis. The hardware stack is limited by the number of registers in the stack, so sometimes a software stack is used. The stack pointer (SP) register is used to indicate the location of the last item put onto the stack.

In some architecture, a barrel shifter unit is used for shifting n bits in a single cycle. It generally takes 2^n data inputs with n control signals producing n data outputs. The "multiply and accumulate" (MAC) unit provides special hardware support for multiplying and accumulating the result.

When a microprocessor is interrupted, it stops executing its current program and calls a subroutine program or process to serve the interruption. This routine is called an interrupt service routine (ISR). At the end of the service routine, the execution again returns to the normal flow of the main program. The source of interruption can be an external signal, which is applied to nonmaskable interrupt input PIN (NMI), or to any other interrupt input PIN of the processor chip. This is called hardware interrupt. Another type of interruption is called software interrupt, which arises due to special software instruction. There is also another type of interruption, which is generated when some exception occurs during the program execution. This is due to some condition produced by the execution of the instruction – for example, the program execution will be automatically interrupted if you attempt to divide an operand by zero. These are conditional interruptions, also be referred to as software interrupts. Generally the hardware interrupts are taken from the external world via device pin. For each of hardware interrupts there is a separate hardware pin. These interruptions can occur simultaneously. The hardware interrupts as well as the software interrupts are prioritized. The starting address of an interrupt service procedure is stored in a table of a nonvolatile memory, which is called the interrupt vector table and the starting address is called an interrupt pointer.

After powering on, the processor jumps to the reset vector location, which is loaded into the program counter. Generally, the content of the reset location tells the processor from where it has to load the boot code. Sometimes the boot code may be in the internal ROM (in the case of a microcontroller or DSP) or may be in the external ROM or flash memory (in a protected sector). The processor then starts the fetch-execute cycle to execute the program, which is in sequence leading to other required programs being loaded into the memory until the operating system kernel is resident in the memory and the processor is ready to execute tasks or application programs.

5.2.2 Processor Types

The processor architecture involves instruction architecture (ISA) design, microarchitecture design, logic design, and implementation. The ISA defines the machine code that a processor reads and understands. It also defines word size, memory address modes, processor registers, and data formats. The processor architectures, in general, have been evolved progressively towards greater complexity with larger instruction sets, more addressing modes, more computational power of the individual instructions, more specialized registers, and so on. The following equation is commonly used to express a processor's performance ability:

$$\text{Processing time (time / program)} = \text{(instructions / program)} * \text{(cycles / instruction)} * \text{(time / cycle)}$$

So, the processor performance is dependent upon: (i) instruction count per program: higher instruction count leads to more memory size requirement – for example, lower code density, higher processing time; (ii) cycles per instruction (CPI); (iii) time per cycle (clock cycle time).

Processors can broadly be divided into the categories of: complex instruction set computers (CISCs), reduced instruction set computers (RISCs), very long instruction words (VLIW), vector, hybrid, and special purpose:

- Complex instruction set computers. In earlier times, instruction access from the memory was very slow, so the instruction was designed so that a single instruction could execute several low-level operations. This means that it supplies many complex instructions, where each instruction contains many low-level operations. For this reason, the number of instructions per program reduces and code density and memory access reduces, but the complexity in the hardware for instruction decoding and operation increases. As a complex instruction is equivalent to several simple instructions, so normally it requires several clock cycles to complete. The CISC approach attempts to minimize the number of instructions per program, sacrificing the number of clock cycles per instruction.
- Reduced instruction set computers. In this case, the instruction set contains a limited number of simple, basic instructions, from which more complex instructions can be composed. So, here, the program code size increases – for example, code density decreases but the hardware complexity reduces. RISC chips require fewer transistors, which makes them cheaper to design and as the instructions are simple they require fewer cycles, which leads to faster operation. Simple instructions can be executed in a single cycle. So, RISC reduces the cycles per instruction at the cost of the number of instructions per program. RISC design requires the software and compiler to deal with greater complexity, whereas CISC puts more complexity on the hardware for instruction functionality. The processing of instructions is broken down into smaller units, which are executed in parallel by pipelines. Normally, in RSIC, an instruction has a fixed length to allow the pipeline to fetch future instructions before decoding the current instruction, whereas in CISC processors the instructions are often of variable size and take many cycles to execute. RISC processors have a large general-purpose register set, which can contain either data or an address. RSIC supports load / store architecture, where data-processing operations only operate on register contents, not directly on memory contents. In contrast, CISC processors have dedicated registers for specific purposes and data-processing operations can act on memory directly.

- Very long instruction words. These have multiple independent parallel execution units and an instruction scheduler determines which instructions will be executed on which execution unit, at what time.
- Vector processors, or array processors. These are processors that can operate on an instruction set containing instructions that operate on one-dimensional arrays of data called vectors. They reduce the fetch-and-decode bandwidth because the number of instructions fetched will be less.

5.2.3 Advanced Risk Machine (ARM)

ARM Holdings plc. is a British multinational semiconductor company that licensees processor IP cores to create microcontrollers and SoCs by adding some other required blocks, like a timer, USB device, or I2C in it. The ARM instruction set and processor architecture have evolved significantly from version v1 to v8 and this evolution is still continuing. The list of ARM cores is given in Table 5.1.

Classic ARM processors have been available since 1994. The Classic ARM processors consist of three processor families: (i) ARM7 Family (ARM7TDMI-S™); (ii) ARM9 Family (ARM926EJ-S™, ARM946E-S™); (iii) ARM11 Family (ARM1136J (F)-S™, ARM1156T2). A comparison between different ARM processors is given in Table 5.2.

ARM Cortex processors are mainly of three types: M (microcontroller), R (real time), and A (application). The basic differences between these are tabulated in Table 5.3.

The ARM architecture defines the ARM and Thumb® instruction sets, execution models, memory models (virtual memory, caches, tightly coupled memory (TCM), and memory protection), and debug models used by ARM processors. ARM architecture extensions define additional features such as floating-point support, single instruction multiple data (SIMD) instructions, security extensions, Java bytecode acceleration, and multiprocessing support. ARM architecture provides two types of memory management options: a memory protection unit (MPU) and a memory management unit (MMU). For example, ARM926EJ-S™, ARM11™ MPCore™, Cortex A9 supports MMU, whereas ARM946E-S™, Cortex-R4 support MPU.

As, smartphones have gone from single core to dual core and now quad core and it is only going to keep increasing, so multicore architectures are introduced. The ARM architecture v6K introduces the first MPCore processor supporting up to four CPUs and associated hardware. ARMv7 includes a hardware floating-point unit (FPU), with improved speed compared to software-based floating point.

5.2.3.1 ARM System Design

A typical embedded SoC based on an ARM core is shown in Figure 5.5. We can divide these into several hardware blocks. (i) ARM processor – it comprises a core plus the surrounding components like, memory management and caches that interfaces with the core

Table 5.1 ARM cores

Architecture	Architecture version	Bit width	Processor core	Main features
ARM Classic	ARMv1	32/26	ARM1	First ARM processor, 26-bit addressing
	ARMv2	32/26	ARM2, ARM3	Multiply and multiply accumulate, 32-bit multiplier, instruction, coprocessor support, 26-bit address space,
	ARMv3	32/26	ARM6, ARM7	Extended address range to 32 bits, separate cpsr and spsr, new modes for undefined and abort, MMU support
	ARMv4	32/26	ARM8	Half word load store instruction, in T variant instruction to transfer to Thumb state, new mode called system introduced
	ARMv4T	32	ARM7TDMI, ARM9TDMI	Improved ARM/thumb switching, add software breakpoint instruction
	ARMv5	32	ARM7EJ, ARM9E, ARM10E	Support for Java acceleration (Jazelle), VFPv2
	ARMv6	32	ARM11	V6 provides support for SIMD, TrustZone, Thumb 2
Cortex-M	ARMv6-M	32	ARM Cortex-M0, ARM Cortex-M0+, ARM Cortex-M1	
	ARMv7-M	32	ARM Cortex-M3	V7 provides support for NEON, Adv SIMD
	ARMv7E-M	32	ARM Cortex-M4, ARM Cortex-M7	
Cortex-R	ARMv7-R	32	ARM Cortex-R4, ARM Cortex-R5, ARM Cortex-R7	
Cortex-A	ARMv7-A	32	ARM Cortex-A5, ARM Cortex-A7, ARM Cortex-A8, ARM Cortex-A9, ARM Cortex-A12, ARM Cortex-A15, ARM Cortex-A17	
	ARMv8-A	64/32	ARM Cortex-A53, ARM Cortex-A57, ARM Cortex-A72	V8 support for crypto, scalar FP, Adv SIMD
	ARMv8.1-A	64/32	TBA	
Cortex-A	ARMv8-R	32	TBA	

Table 5.2 Comparison between ARM processors

Attributes	Pipeline depth	Typical MHz	MIPS/MHZ	Architecture	Multiplier
ARM7	3	80	0.97	Von Neumann	8×32
ARM9	5	150	1.1	Havard	8×32
ARM11	8	335	1.2	Havard	16×32

Table 5.3 Differences between ARM Cortex processors

ARM Cortex-A family (v7-A)	ARM Cortex-R family (v7-R)	ARM Cortex-M family (v7-M)
Application processor High-performance processors for open operating systems Applications: smartphones, notebooks, digital TV, home gateways	Real-time application processor Exceptional performance for real-time applications Automatic breaking systems, mass storage controller, networking, printing	Embedded processor Cost-sensitive microcontroller applications Mixed signal devices, smart sensors, IoT devices

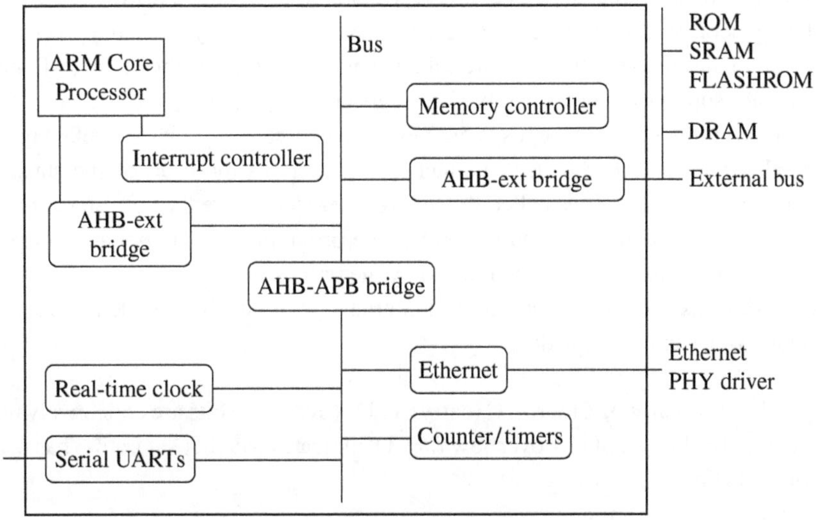

Figure 5.5 ARM-based system

via a bus. (ii) Controllers coordinate important functional blocks of the system. Two commonly used controllers are interrupt and memory controllers. (iii) Peripherals provide all the input-output capabilities external or internal to the chip, like serial ports, Ethernet, or timers. (iv) Bus – this is the interconnect system for communication.

5.2.3.2 ARM Processor

The ARM is a RSIC processor, having a simple architecture that helps for simpler implementations and very low power consumption. ARM architecture also helps for (i) maximizing the use of the ALU and shifter in every data-processing instruction; (ii) auto-incrementing / decrementing addressing modes to optimize loops; (iii) loading and storing multiple instructions to maximize data throughput; (iv) conditional execution of all instructions to maximize execution throughput. These features, on top of basic RSIC architecture, help ARM processors to achieve a good balance of high performance, good code density, low power consumption, and low silicon area.

5.2.3.3 ARM Core

As discussed earlier, ideally an ARM processor comprises a core (the execution engine that processes instructions and manipulates data) along with some associated components like memory management and caches that interface with it using a bus. The ARM core consists of 37 registers, each 32 bits wide; among these, 31 are general-purpose registers including a program counter (r0 to r15 in user mode, r8 to r15 in FIQ, r13 to r14 in IRQ, r13 to r14 in undefined, r13 to r14 in abort, r13 to r14 in supervisory mode) and six status registers – one current program status register (CPSR) and five stored program status registers (SPSRs). Out of 31 general purpose registers, 20 registers are available only when the processor is in a particular mode and these registers are called banked registers, as shown in the Figure 5.6. For example, supervisor mode has banked registers r13_svc, r14_svc and spsr_svc. All processor exception modes except system mode have a set of associated banked registers.

Generally, *r13* used as the stack pointer (*sp*) and stores the head of the stack in the current processor mode; *r14* is called the link register (*lr*), where the core stores the return address whenever it calls a subroutine; *r15* is the program counter (*pc*) and contains the address of the next instruction to be fetched by the processor.

The CPSR is used to monitor and control internal operations. It is divided into four fields (see Figure 5.7), each 8 bits wide:

- Flags: The N (negative), Z (zero), C (carry) and V (overflow) bits are collectively referred to as conditional flags. Sticky overflow flag, Q, indicates whether saturation has occurred or not (5TE/J only).
- Status: reserved for future use.
- Extension: are reserved for future use.
- Control: Mode bits (M [4:0]) indicates the mode in which processor operates. I bit is set to 1 to disable the IRQ and 0 to enable the IRQ. Similarly, F bit is set to 1 to disable FIQ. T- bit in is set to 0 for ARM state operation and set to 1 for Thumb state operation.

The CPSR is accessible in all modes. A privileged mode allows full read-write access to the CPSR and an unprivileged mode allows read-write access only to the condition flags

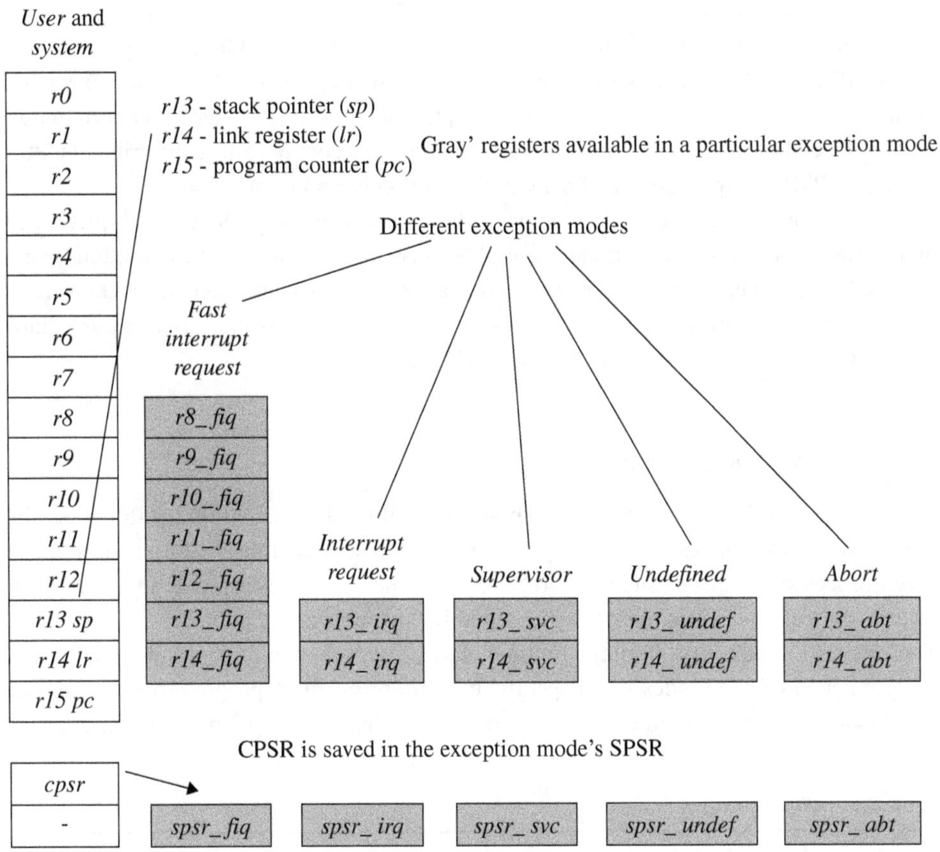

Figure 5.6 Different registers in ARM core

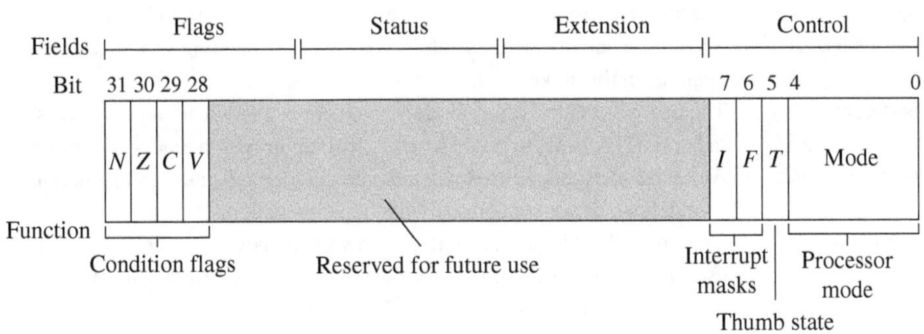

Figure 5.7 CPSR fields

but read only access to the control field in the CPSR. Every processor mode except user mode can change mode by writing directly to the mode bits of the CPSR.

An SPSR can only be modified and read in a privileged mode. When any exception occurs, the CPSR is saved in the occurred exception mode's SPSR. Again, to return to user mode a special return instruction is used that instructs the core to restore the original CPSR from the SPSR_<exp_mode>and bank in the user registers r13 and r14.

When the processor core is powered on, it starts in *supervisor* mode, which is privileged mode. Starting in a privileged mode is useful because initialization code can use full access to the CPSR to set up the stacks for each of the other modes. Any register can contain either data or an address. An ARM processor supports several data types like: Byte (8 bits), Half-word (16 bits – supported v4 and above), and Word (32 bits).

5.2.3.4 ARM Processor Modes

Ideally, in a good design the system should not allow the user program to access the protected memory or resources freely. So, in order to support this, ARM has introduced seven processor modes as shown in Table 5.4. Of these, only user mode, in which most of the applications run is unprivileged mode having limited or no access to resources. The other six modes are privileged modes, having access to memory and resources. A program enters these modes when specific exceptions occur. A program running in user mode cannot switch to another mode without generating an exception. So, mode changes

Table 5.4 ARM processor modes

Processor mode	Description
User<usr>	This is normal program execution mode, where most of the user program executes. It is unprivileged mode.
FIQ<fiq>	Fast interrupt request mode supports for fast interrupt processing.
IRQ<irq>	Interrupt request mode is used for general purpose interrupt handling.
Supervisor<svc>	The system enters this mode after reset and the operating system kernel operates in this mode.
Abort<abt>	The processor enters abort mode when there is a failed attempt to access memory. This helps to implements virtual memory or memory protection.
Undefined<und>	An undefined mode is used when the processor encounters an instruction that is undefined or not supported by the implementation.
System<sys>	System mode, which is present only in v4 and above, is a special version of user mode that allows full read-write access to the status register (CPSR). Here, exactly the same registers are available as in user mode; however, it is a privileged mode and therefore not subject to restrictions as in user mode. A program enters this mode, not due to exception, as the operating system tasks need access to system resources without using additional registers associated with exception modes. Avoiding such use ensures that the task is not corrupted by the occurrence of any exception.

can be made under software control, or can be caused by external interruptions or exception processing.

Switching between modes can be done manually through modifying the mode bits in the CPSR register.

5.2.3.5 Exceptions

Internal and external sources generate exceptions to cause the processor to handle an event. When an exception occurs some standard registers are replaced by registers specific to that exception mode. When an exception or interrupt occurs, the processor suspends the normal execution and jumps to the specific memory address location by setting the *pc* to that special address based on the exception / interrupt type. These special addresses for each exception types are stored in memory within a special address range called the vector table. Normally the vector table is located at 0x00000000 or can be optionally located at a higher memory address (starting at the offset 0xffff0000) which helps Linux or Windows OS (see Table 5.5).

So, once the pc is loaded with that vector address according to the exception type, the processor jumps to that respective memory location, where instruction is stored that branch to specific routines (ISR) designed to handle that particular exception / interrupt. When exception occurs, apart from pc some standard registers (banked registers r13, r14) are replaced with registers specific to the exception mode. The processor states just before handling the exception must be preserved so that the original program can be resumed when the exception routine is completed. So, when an exception occurs, r14 holds the return address after exception processing is over, r13 provides each exception handler with a private stack pointer. The FIQ mode also banks registers r8 to r12 so that interrupt processing can begin without the need to save or restore these registers. More than one exception can arise at the same time. The ARM supports seven types of exceptions in different processor modes.

When an exception causes a mode change, the core automatically saves the CPSR to the SPSR of the exception mode and saves the *pc* to the *lr* of the exception mode, sets the CPSR to the exception mode, and sets the *pc* to the address of the exception handler. Exceptions can occur simultaneously, so the processor has to adopt a priority mechanism.

When an exception occurs the following steps are performed:

R14_<exception_mode>= return link,
SPSR_<exception_mode>= CPSR,
CPSR [4:0]=exception mode number,
CPSR [5] =0 (execute in arm state),
If<exception_mode>== reset or FIQ then
CPSR[6]=1
Else CPSR[6] unchanged,
CPSR[7]=1 (disable normal interrupt),
PC=execution vector address,

After handling exceptions the SPSR is moved back into CPSR and R14_<exception_mode>is moved to the *pc* (return address).

Table 5.5 Exception vector table

Exception type	Description	Mode	Normal address	High vector address
Reset (prio-1)	When power is applied, the processor jumps to the reset vector location to execute the first instruction, which branches to the initialization code.	Supervisor	0x0000-0000	0xffff-0000
Undefined instructions (prio-6)	When the processor cannot decode an instruction it jumps to this vector location	Undefined	0x0000-0004	0xffff-0004
Software interrupts (SWI) (prio-6)	When SWI instructions are executed (the mechanism to invoke an operating system routine) the processor jumps to this vector location.	Supervisor	0x0000-0008	0xffff-0008
Prefect abort (prio-5)	This occurs when the processor attempts to fetch an instruction from restricted locations. The actual abort occurs in the decode stage.	Abort	0x0000-000c	0xffff-000c
Data abort (prio-2)	This is raised when an instruction attempts to access data memory without the correct access permissions.		0x0000-0010	0xffff-0010
IRQ (prio-4)	Interrupt	IRQ	0x0000-0018	0xffff-0018
FIQ (prio-3)	Fast interrupt	FIQ	0x0000-001c	0xffff-001c

5.2.3.6 Bus

A bus has two architecture levels: (i) physical level – this covers the electrical characteristics and bus width; (ii) bus protocol – the logical rules that govern the communication between the processor and a peripheral. The ARM mainly specifies the bus protocol part. The ARM AMBA (Advanced Microcontroller Bus Architecture) protocols are an open-standard, on-chip interconnect specification for the connection and management of functional blocks in a SoC, which facilitates the multiprocessor designs with large numbers of peripherals. The first AMBA buses, introduced in 1996, are: (i) Advanced System Bus (ASB) for high-performance system modules, and (ii) Advanced Peripheral Bus (APB) for low-power peripherals. Then in the second version (AMBA 2) AMBA High-performance Bus (AHB) for high-performance, high clock-requency system modules, which is a single clock-edge protocol. In the third version (AMBA 3), ARM introduced AXI (Advanced eXtensible Interface) for higher performance interconnects and ATB (Advanced Trace Bus) for on-chip debug and trace solution. AXI is most widespread AMBA interface and provide connectivity up to 100s of Masters and Slaves in a complex SoC. It provides high-frequency operation without using complex bridges. Next AMBA 4 was introduced with AXI4, which is extended later to system wide coherency with AMBA 4 ACE. In AMBA 5 CHI (Coherent Hub Interface) specification high-speed transport layer was redesigned and features were designed to reduce congestion.

5.2.4 DSP-Based Implementation

The first generation of mobile phones, the 1G system, were using analog transmission, so there were no digital baseband. DSP processors from one of the most important classes of mobile embedded processors in second-generation (2G) systems. DSP architectures are preferred over ASIC due to their shorter product lifecycle, and they are extensively used in GSM mobiles. Programmable DSPs provide a cost-effective and flexible architecture for mobile phones. AT&T introduced the first DSP in 1979, and subsequently Texas Instruments came up with other DSPs.TMS320C55 is a modern DSP architecture, which implements Harvard architecture, using one and three read buses for code and data, respectively. With TMS320C55 DSP architecture, features like programmable idle modes and automatic power saving were incorporated for better processor utilization at top speeds. Generally, the DSP performs the primary tasks in the receiver chain, such as channel estimation, Viterbi equalization, demodulation, decoding, forward error correction, error detection, and burst interleaving / deinterleaving. VLIW and SIMD architectures in modern cellular devices are becoming more popular; because they allow the frequency and voltage of the CPU chips to be reduced without losing performance. Modern DSPs can be more effective if they are able to support parallel processing.

5.2.5 SOC-Based Architecture

In SoC-based designs, system tasks can be managed by integrating microcontrollers, dedicated ASICs, or DSPs in a single chip. Highly integrated SoCs leveraging multicore technology have emerged for higher performance and low power designs. SoC packages integrating ARMs core include Nvidia Tegra's first three generations, CSR plc's Quatro family, Qualcomm's Snapdragon, ST-Ericsson's Nova and NovaThor, Silicon Labs' Precision32 MCU, Texas Instruments' OMAP products, Samsung's Hummingbird and Exynos products, Apple's A4, A5, and A5X, and Freescale's i.MX.

5.2.5.1 Qualcomm Snapdragon Processors

Snapdragon is a family of mobile SoC processor architecture provided by Qualcomm. The original snapdragon CPU, known as Scorpion, had many features similar to ARM Cortex-A8 core based on ARMv7 instruction set. This supports higher performance, utilizing SIMD operations. These SoCs are built around Krait processor architecture, which integrates LTE, GSM, and WCDMA modems to support seamless connectivity across 2G, 3G, and 4G LTE networks. This architecture supports a wider front end, with the ability to fetch and decode three instructions per clock. Qualcomm introduced Snapdragon 800 in 2013, which integrates 28 nm HPm quad core Krait 400 CPU, Adreno 330 GPU for graphics, Hexagon DSP for low power operation, and Gobi™ 4G LTE modem, and provides a 2.3 GHz clock speed.

Today, CMOS planar bulk technology era is ending – maybe 20 nm is the last CMOS bulk node. Continued transistor scaling does not give acceptable performance and power gains. Introduction of new technologies for future nodes are needed in order to continue transistor scaling with increased performance and lower power. Two new technologies have been introduced by industry:

- FDSOI (Fully Depleted Silicon on Insulator);
- FinFET (Tri Gate).

5.2.6 *Commonly Used Processors in Smart Phones*

As discussed earlier modern smart phones houses multiple (two, four, or eight) application processor cores. For example, Samsung Galaxy S4 i9500 is available with configurations: (1) 1.9 GHz quad-core ARM Krait + Qualcomm's Adreno GPU, (2) 1.6 GHz quad-core ARM Cortex-A15 + 1.2 GHz quad-core ARM Cortex-A7 + Imagination's PowerVR GPU. Apple iPhone 5 contains 1.3 GHz dual-core Swift (ARMv7-based) + PowerVR GPU. Nokia Lumia 920T contains 1.7 GHz dual-core Qualcomm Krait + Adreno GPU.

These multi-core application processors can deliver excellent performance with low power consumption, and cost.

5.3 LTE Smartphone Hardware Implementation

Figure 5.8 shows a reference hardware implementation of LTE mobile phone. Typical functions in layer 1 include forward error correction, interleaving and bit stream manipulation, constellation-modulation, MIMO encoding, OFDM signal modulation, and RFIC signal conditioning. All of the layer 1 functions could be implemented in a DSP processor and control and management functions could be implemented on an ARM processor. Commonly, upper layer protocol processing (layers 2 and 3 in Figure 7.8 in Chapter 7) is performed in another ARM Cortex-R7 processor. This processor will typically perform functions such as medium access control (MAC), packet data convergence protocol (PDCP), radio link control (RLC) and radio resource management (RRM). This processor (ARM Cortex-R7) is interfaced to the application processor, where the general-purpose OS (like Android) is running through some IPC/RPC (interprocess communication / remote procedure call) mechanism or common shared memory.

Today, some latest real time ARM processors (like the Cortex-R7 processor) has an extremely powerful superscalar architecture capable of delivering over 1500 Dhrystone

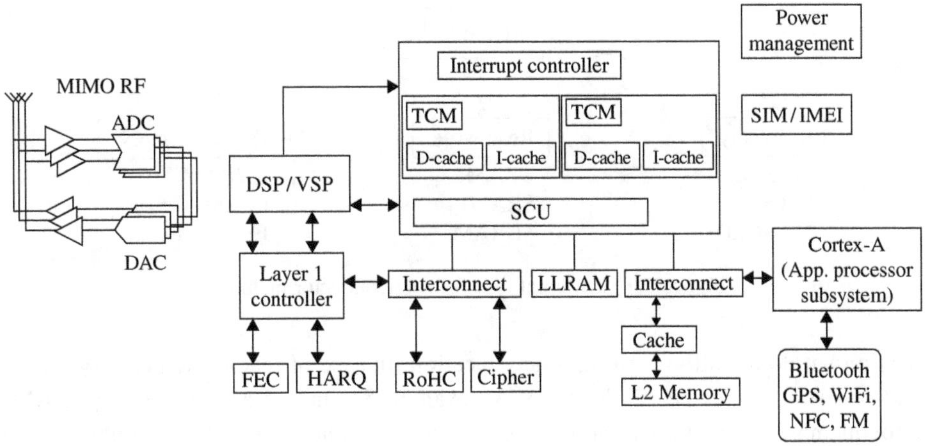

Figure 5.8 Typical implementation of a LTE Smartphone on ARM processor

MIPS from a square millimeter of 40 nm silicon with a clock frequency of 600 MHz. These can be used for LTE baseband system processing. In the LTE system, the interrupt arrives very fast (according to LTE symbol rate 14 interrupts per millisecond) and needs to be processed quickly to manage the modem and the radio as the channel environment changes very rapidly. So, the UE has to adapt constantly to changing signal conditions and download or upload data at a high rate. In the LTE system, there could be simultaneous data streams for audio, video and data communications together with control data from the network. So, in case of multicore systems, the individual core can have its own local L-1 cache for instructions and data. This helps to return more quickly to a power saving mode. Power consumption can also be reduced by minimizing costly access to off-chip memories and allowing the cores to spend a longer duration in power-saving modes. Tightly coupled memory (TCM) support is present, so time-critical interrupt routines can be placed in TCM memories. In some cases the high computation-intensive functions like, cipher and robust header compression (RoHC) can be offloaded to a hardware accelerator (HWA) and this can be interfaced easily with the Cortex-R7 processor low latency peripheral port (LLPP), which provides an optimized interface. In the LTE system, several such blocks can be implemented in hardware – for example, forward error correction (turbo code), hybrid automated resend request (HARQ), and robust header compression and ciphering (RoHC) for security. The modem is responsible for OFDMA coding and decoding of the radio signal, and could be implemented using DSPs or a vector signal processor (VSP).

5.4 Memory

Mobile-phone uses memory to store programs and data. Commonly, memory can be broadly classified into two categories – one is read-only memory and other is read-write memory (Figure 5.9). Read only memories (ROM) allow reading from any memory

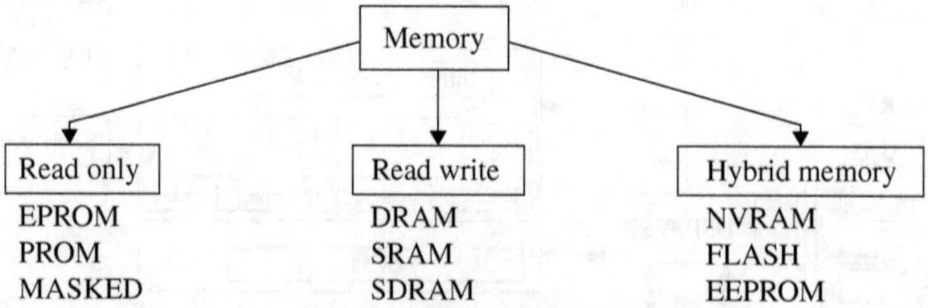

Figure 5.9 Different types of semiconductor memories

location but do not allow writing to any location. In case of read-write memory we can read or write to any specific location. An example is random access memory (RAM). Some memories contain dual properties and these are known as hybrid memories, such as flash or EEPROM. Again, depending on the storing property, we can broadly divide the memories into two categories – volatile memory and nonvolatile memory. In the case of volatile memory, the memory content vanishes when the supply power to this memory device is switched off. So, if the supply power is off, the content of the memory will be lost forever. RAM is the example of this type of memory. In the case of nonvolatile memory, as the name suggests, the memory content is nonvolatile. That means the memory contents are not lost, when the supply power to memory is off. ROM is an example of this type of memory.

5.4.1 Read-Only Memory (ROM)

ROM is read-only memory (no memory write is possible) and also nonvolatile memory, so once the data is stored or written into it, it remains there forever and does not lose the memory content once the supply power is off. It is low cost, high speed, nonvolatile memory, and is made up of arrays of memory cells like RAM. ROM can be used to realize arbitrary truth tables, generate characteristics, convert codes, or store system boot programs. The ROM is constructed by unipolar or bipolar devices. Some ROM devices are only programmable once – there is no way to alter the content or write the data into it – these are called one time (OT) programmable ROM. Other types of ROM, which can be written by using a special techniques like UV light or electrical signals, are called field-alterable ROM. These are not system programmable – in order to program these we have to use a special programming platform and device. Typical parameters of a ROM (uPD23C1000A) are given in Table 5.6. Different types of ROM are compared in Table 5.7.

5.4.1.1 Electrically Erasable Programmable ROM (EEPROM)

EEPROM is user-modifiable read-only memory, which can be erased and reprogrammed repeatedly by applying the higher electrical voltage. These are electrically programmable up to many thousand times. They can be packaged in simple plastic package, which reduces

Table 5.6 Typical parameters of a ROM (uPD23C1000A)

Operating voltage = 5 V	
Parameter	*Typical time in ns*
Address access time	200
Chip enable access time	200

Table 5.7 Comparison between different types of ROM

Memory device	Number of times programmable	Erasable from field or system?	Cost	Density
ROM	One time	NA	low	high
EPROM	Many times	Field erasable	high	high
EEPROM	Many times	Field erasable and system erasable	low	no

the cost of the device compared to the EPROM. This is commonly used for holding BIOS programs. An EEPROM chip has to be erased and reprogrammed entirely. It also has a limited life – that is, the number of times it can be reprogrammed is limited.

5.4.2 Flash Memory

Flash is a special form of EEPROM, which uses normal supply voltages (3.3–12 V) for erasing and reprogramming. So, it is easier to work with as it can be read or write with special techniques in the system. It is always desirable to have nonvolatile memory, as the power consumption reduces and there is no concern about the loss of information. The basic problem with EEPROM is that it needs special arrangements and high voltage to program – to write to it – whereas flash is system programmable and does not require any special platform to program it. It is a high density, truly nonvolatile, high-performance, read-write solution. The flash device also requires a higher voltage supply for programming, like a conventional EPROM. The electrical erase is by either hot electron or cold electron injection from a floating gate, with the oxide layer allowing the cell to be electrically erased through the source.

Flash memory is available on a separate chip with several blocks in it or it can also be integrated along with CMOS. The blocks or sectors may be of same size or different sizes. In some cases, where the boot code is stored in the flash, there are some small sectors on the top or bottom side of the flash memory. The boot block flash memory family has asymmetrically blocked memory array layouts to enable small parameters or boot code storage, along with efficient larger blocks for code and data-file storage. Symmetrically blocked memory arrays enable the best code and data file management, where all the sectors are of same size. This special top or bottom block is called a boot block or parameter block. The small boot block is designed to contain the boot code for the system and usually has some level of protection from accidental overwrite.

5.4.2.1 Flash Erasing and Programming

The flash chip contains several blocks. A fixed address range (like 0000-0FFFF) defines the blocks. When a flash memory is imported from the factory, all its locations contain same type of bit (either 1 or 0- depending on the flash type). This is called the erased state of the entire flash memory. During the write or program process, the flash cells changed from one binary voltage level to another (i.e. 1 to 0). That means, if the flash cells originally contain all 1s, then during writing, we can change the cell contents from 1 to 0 wherever required according to the input data pattern. But the reverse is not possible. That means, if the flash cells are not erased before writing, then the cells will contain 0s and 1s. So, writing to the cells, where it contains 0s will not be possible. It is therefore always required to erase the flash before writing. In the erase process, the flash cells are set back to their original binary voltage level or erase state. The erase process occurs in block basis or entire chip erase. When a block is erased, all address locations within a block are erased in parallel, independent of other blocks in the flash memory device. Flash components take a significant amount of time to erase a block or program compared to RAM. To modify any data content in a block, it is required to copy the original content of that block in some other memory location before erasing that block. Then modify the original data and again write back to the flash. There will be no problem of writing as it is already erased before freshly writing the modified data.

A flash memory device is very useful to reduce system costs as well as improving data reliability, providing easy update capabilities, increasing battery life, and providing stability after power loss.

Data for a Typical Flash Memory
AMDAm29DL400B, Write cycle time: 70–120 ns. Sector erase operation time: 0.7 s (64 kB). Chip erase time: 10 s. Read cycle time- 70–120 ns. Power consumption: single power supply: 2.7 to 3.6 V, Active read current: 12 mA, Active write current: 25 mA, Stand by current: 5 μA.

Recently Intel has developed a 1.8 V wireless flash memory (28F640W18), which is the highest performance solution for Internet phones.

5.4.2.2 Different Types of Flash Memories

Two main technologies dominate the nonvolatile flash memory market today: NOR and NAND. NOR flash was first introduced by Intel in 1988. NAND flash architecture was introduced by Toshiba in 1989. Most flash devices are used to store and run code (usually small), for which NOR flash is the default choice. Here, in Table 5.8, some differences between NOR and NAND flash memory are mentioned.

5.4.3 Random-Access Memory (RAM)

The name "random access" indicates that each cell in the memory chip can be read or written in any order. All RAM memories are read write memory. Some commonly used RAM memories are discussed below.

Table 5.8 Difference between NOR and NAND flash

Attributes	NOR Flash	NAND Flash
Capacity	Less (~1 MB–32 MB)	More (~16 MB–512 MB)
Performance	Very slow erase (~5 s) Slow write Fast read	Fast erase (~3 ms) Fast write Fast read
Reliability	Standard	Low
Erase cycles	10 000–100 000	100 000–1 000 000
Lifespan	Less than 10% the life span of NAND.	Over 10 times more than NOR
Interface	Full memory interface	I/O only, CLE, ALE and OLE signals must be toggled.
Access method	Random	Sequential
Ease-of-use (Hardware)	Easy	Complicated
Full system Integration	Easy	Hard. A simplistic SSFDC driver may be ported.
Ideal usage	Code storage – limited capacity due to price in high capacity. May save limited data as well. Some examples: simple home appliances, low-end set top boxes, low-end mobile handsets, PC BIOS chips.	Data storage only – due to complicated Flash management. Code will usually not be stored in raw NAND flash. Some examples:PC cards, compact flash, MP3 players, digital cameras.
Price	High	Low

5.4.3.1 Static RAM (SRAM)

"SRAM" is an acronym for "static random access memory"; this means that once the data is written into the memory cell it remains as it is as long the power is not switched off. SRAM is available in many varieties starting from the superfast bipolar and GaAs SRAM to the slow commodity CMOS variety. The early SRAM cells using NMOS technology consist of six transistors; out of that four enhance mode transistors and two depletion mode resistors. The CMOS SRAM cells are very low power, wide noise margin but low speed.

Typical SRAM Data

32K × 8 bit low power CMOS static RAM: K6T0808C1D family-manufactured by SAMSUNG electronics. Access time – read cycle time: 70 ns, write cycle time: 70 ns. Power: supply voltage: 4.5 V, operating supply current: 5 mA, standby current: 30 μA. Density: generally 4 to 6 transistors per memory cell.

5.4.3.2 DRAM

"DRAM" is an acronym for "dynamic random access memory," which means that, to remember the stored data, the memory chips requires every bit to be refreshed within a certain period of time. When the power is removed from the DRAM the data content is lost. The DRAM uses tiny built-in capacitors to store the data bits. The charge is stored in the capacitor C. When the transistor is turned on, the information is refreshed (or charged).

Typical DRAM Data
Mitsubishi Electric – M5M467400Dxx series. Fast page mode: access time, write cycle time: 90–110 ns. Access time from CAS: 13 ns. Access time from RAS: 50 ns. Read cycle time: 90–110 ns. Refresh cycle time: 64 ms. Power dissipation: 300 mW, Vcc: +3.3 V.

DDR stands for double data rate and is based on synchronous dynamic random access memory (SDRAM) design. Memories from this category transfer two data chunks per clock cycle (rising and falling edges of the clock signal).

The ideal memory should have fast access time, high density, high performance, low cost, low power dissipation, random access support, and be nonvolatile, highly reliable, easy to test, and standardized throughout the industry. So far a single type of memory cannot satisfy all these diverse requirements, and each memory has several advantages as well as disadvantages. So, a system uses various types of memories inside it according to its various requirements. Generally, inside a mobile phone, Flash memory is used to store the processor program and application program or data. An EEPROM series is used for storing the system and tuning parameters, user settings, and selections. The program is normally executed from SRAM or DDR after downloading it from flash memory. This is also used for scratch pad memory.

5.5 Application Processing Unit

As discussed earlier, today's mobile phones not only contains the voice or data modem part, but also contain a group of applications and associated hardware and software. Examples are audio players, video players, GPS, and connectivity modules like USB, Bluetooth, and IrDA. Figure 5.1 shows different hardware functional modules and their interface to a central application processor, which is controlling and executing the necessary software drivers for those. In Chapter 7, we will discuss the software details for audio and video players.

5.5.1 Application Processor Peripherals

In smartphones, application processors have the typical set of peripherals. These can be classified into: (i) processor core; (ii) multimedia modules (audio, speech, image, etc.); (iii) wireless modules (Bluetooth, IrDA, NFC, etc.); (iv) device interfaces (RTC, UART, IRDA, SPI, I2C, SD/MMC card controller, keypad scan controller, USB device, etc.). As discussed above, a smartphone application processor is an advanced RISC machine (ARM) specially optimized for low power consumption.

5.6 Multimedia Modules

Multimedia modules perform multimedia related functionalities. These are mainly:

- A speech and audio module. This includes a microphone, an A/D converter, a speech compression / decompression unit, a D/A converter, speaker, audio encoder / decoder, MP3 decoder, etc.
- Video module: this includes image encoder / decoder (JPEG), video encoder decoder (MPEG), A/D, D/A unit, LCD display and touch screen, digital camera, and so on. A JPEG unit is used for decoding pictures for viewing on the LCD screen, and encoding pictures taken with the camera for later viewing or for sending out on the network. An MPEG unit is used for decoding streaming live video, video on demand, and incoming video conferencing data, and encoding video taken with the video camera for later viewing, or for sending out via the network.

A smartphone also contains a graphics processing unit (GPU) for rapidly manipulating multimedia functions, where large blocks of data are processed in parallel. The role of this is to manage 2D and 3D graphics, video capture, playback, image compression, deliver mobile gaming, and provide better user interface. Different input / output components in a mobile phone are shown in Figure 5.12(b).

5.7 Microphone

A microphone is an acoustic-to-electric transducer or sensor that converts sound (air pressure variations) into electrical signals (current variations). Sometimes it is also referred to as a mike or mic.

5.7.1 Principle of Operation

A sound wave is generated by a source; it creates contractions and rarefaction in the air medium and propagates. When this strikes a microphone surface, it produces vibration and, from that, voltage / current is generated, which is proportional to the sound signal amplitude. A variety of mechanical techniques can be used to construct microphones. The two most commonly used designs are the magneto-dynamic and the variable condenser designs. Generally, for speech / audio applications, we use dynamic, ribbon, or condenser microphones (Figure 5.10).

- Dynamic microphones: a sound wave vibrates the attached coil of wire in the field of a magnet. This produces a voltage, which replicates the sound pressure variation – characterized as a pressure microphone.
  - Advantages: (i) relatively cheap and rugged; (ii) can be easily miniaturized.
  - Disadvantages: the uniformity of response to different frequencies is worse than the ribbon or condenser microphones.

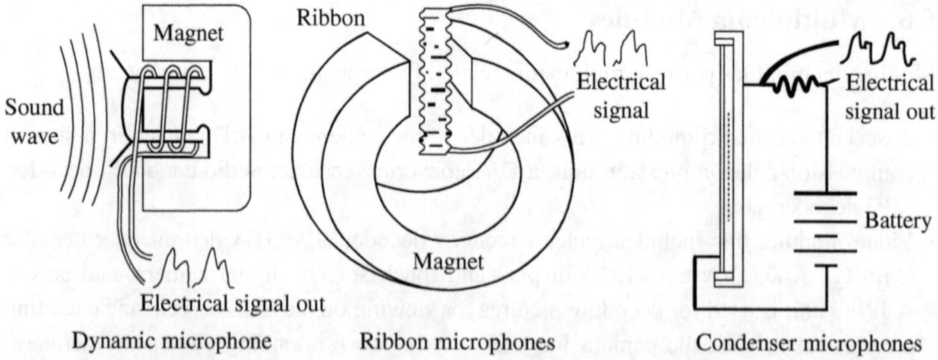

Figure 5.10 Dynamic microphone, ribbon microphone, and condenser microphone

- Ribbon microphone. In this type of microphone, the air movement associated with the sound wave moves the metallic ribbon in the magnetic field, which generates an imaging voltage between the ends of the ribbon. This voltage is proportional to the velocity of the ribbon, so, this type of microphone is characterized as a "velocity" microphone.
- Condenser microphone. In this type of microphone, the sound pressure changes the spacing between a thin metallic membrane and the stationary back plate. The plates are charged to a total charge:

$$Q = CV = [\alpha. \text{ (Area of plate). Voltage}] / [\text{plate spacing}]$$

  where, C is the capacitance and V the voltage of the biasing battery. A change in plate spacing will cause a change in charge Q and force a current through resistance R. This current replicates the sound pressure, making this a "pressure" microphone. A condenser microphones spans the range from cheap throwaways to high-fidelity quality instruments.
  - Advantages: this offers best overall frequency response, so it is selected for many recording applications.
  - Disadvantages: (i) expensive; (ii) may pop and crack when it is close to sound source; (iii) requires a battery or external power supply to bias the plates.
- Carbon microphone. This type of microphone is a capsule containing carbon granules pressed between two metal plates. A voltage is applied across the metal plates, causing a small current to flow through the carbon. One of the plates is a diaphragm, which vibrates when sound wave strikes it and produces varying pressure on the carbon. The changing pressure deforms the granules and this causes the change in contact area between each pair of adjacent granules, which leads to the change in electrical resistance of the mass of granules. The changes in resistance causes a corresponding change in the voltage across the two plates, and this is output from the mic as the electrical signal. Carbon microphones were formerly used in telephone handsets. This has extremely low-quality sound reproduction.

5.7.1.1 Characteristics of MIC

There is no inherent advantage in fidelity of one specific type of microphone over another. Condenser types require batteries or power from the mixing console to operate, and dynamic types require shielding from stray magnetic fields, which makes them a bit heavy sometimes. The most important factor in choosing a microphone is based on the application, size, and quality requirement. The following parameters must be considered for proper selection:

- Sensitivity. Sensitivity of MIC is a measure of how much electrical output is produced by a given sound.
- Overload characteristics. When it is over driven by loud sound, the microphone produces distortion. This is caused by various factors like coil pullout from magnetic field and amplifier clipping.
- Linearity or distortion. The distortion characteristics of a mic are determined mostly by the care with which the diaphragm is made and mounted.
- Frequency response. A flat frequency response is always desirable.
- Noise. Microphones produce a very feeble current, which requires amplification by a factor of more than 100. Now, any electrical noise produced by the microphone will also be amplified, so it should be noise free. Dynamic microphones are essentially noise free but the electronic circuit built into condenser types is a potential source of trouble and must be carefully designed and constructed of premium parts.
- Impedance matching. Microphones have electrical characteristic called impedance, which is measured in ohms (Ω) and this depends on the design. Typically it can vary from 600 Ω to 10 kΩ. To obtain the best sound, the impedance of the microphone must be matched with the load to which these are connected.

A typical GSM mobile phone implementation of a voice-sampling circuit is shown in Figure 5.11.

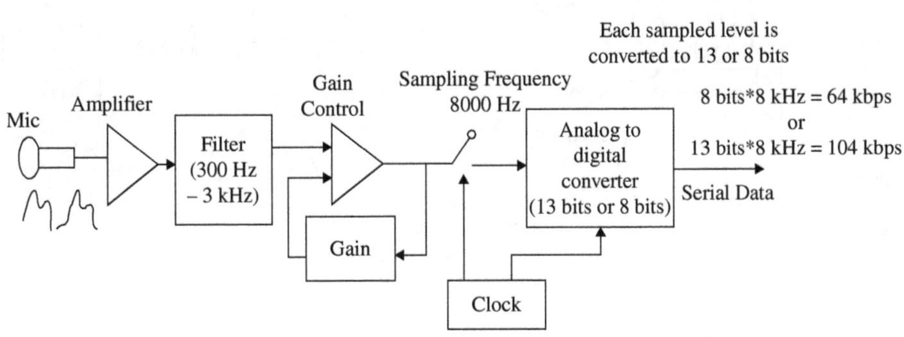

Figure 5.11 Microphone circuit and voice signal sampling

5.8 Loudspeaker

Generally, a speaker takes the electrical signal as input and translates it back into physical vibrations to create sound waves as shown in Figure 5.12. In 1876, Alexander Graham Bell patented the first loudspeaker as part of his telephone circuit. The modern design of loudspeakers based on moving-coil drivers was proposed by Oliver Lodge in 1898. Generally, a speaker uses a lightweight diaphragm connected to a rigid frame via flexible suspension. The flexible suspension part constrains a coil of fine wire to move axially through a cylindrical magnetic gap. The diaphragm is usually manufactured using paper, metal, or plastic in a cone or dome-shaped profile. The suspension system helps to keep the coil centered in the gap and provides a restoring force to make the speaker cone return to a neutral position after moving back and forth. A typical suspension system consists of two parts: (i) Spider – this connects the diaphragm or voice coil to the frame and provides the majority of the restoring force. The spider is usually made of a corrugated fabric disk. (ii) Surround: this helps to keep the coil in the center and also allows free movement. The surround can be a roll of rubber or foam or a corrugated fabric, attached to the outer circumference of the cone and to the frame. The narrow end of the cone is connected to the voice coil. The voice coil wire is round, rectangular or hexagonal in shape and usually made of copper, or aluminum, or silver. Running an electric current through the wire creates a magnetic field around the coil, magnetizing the metal it is wrapped around. The electromagnet and the permanent magnet interact with each other. The input alternating current

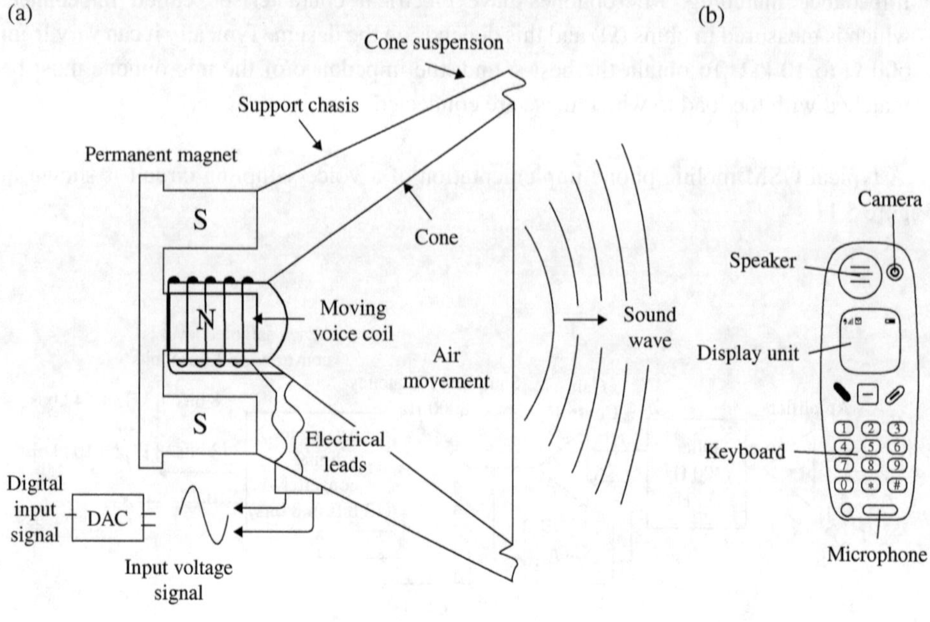

Figure 5.12 (a) Internal diagram of a loudspeaker. (b) Different input / output devices in a mobile phone

causes the polar orientation of the electromagnet to reverse itself many times a second; this constantly reverses the magnetic forces between the voice coil and the permanent magnet, and this pushes the coil back and forth rapidly, like a piston. This vibrates the air in front of the speaker, creating sound waves. A woofer is a driver, capable of reproducing low (bass) frequencies. A tweeter is a driver, capable of reproducing the high (treble) frequencies.

The efficiency of a loudspeaker is defined as the sound-power output (usually specified in dB, this is known as sensitivity of the speaker) divided by the electrical power input. The impedance of a speaker (typically 4 Ω, 8 Ω, etc.) is matched with the audio amplifier load to obtain the maximum power transfer. The rated power of a speaker is defined by two terms: the nominal power (continuous), and peak (or maximum short-term) power. These terms are important as they define the maximum input power that the loudspeaker can handle before it is thermally destroyed. Mobile phones use microspeakers due to smaller housing capacity. Designers use various techniques to increase the volume and improve the sound quality of these microspeakers but there is always risk involved as blown speakers are a common cause of failures in mobiles. Commonly, mobile phones have two speakers: one is used for the earpiece (to generate speech during call) and the second one is for sound reinforcement, for things like ringtones, music playback, and hands-free calling. These microspeakers have a permanent magnet and a voice coil, which is attached to a diaphragm. The diaphragm pushes the air to create sound. The speaker is enclosed in a protective box, which provides the "back volume" for pushing and projecting the sound from the speaker.

5.9 Camera

Today, most commonly, a digital camera is attached almost in every mobile phone. The digital camera is very similar to the conventional analog camera. It also contains most of the associated components that the conventional camera contains, like a lens and a shutter. Light falls upon an array of image sensors or photosensitive cells. Most commonly, an image sensor is a charged-couple device (CCD) converting light into electric charges, and is essentially a silicon chip used to measure light. These charges are stored as analog data that are then converted to digital via an analog-to-digital converter (ADC). The generated image data is too huge to store so, it is compressed using image coding techniques (as discussed in Chapter 7) and then it is stored on a memory card. The most common type of memory card is the compact flash card (CF card); apart from that, other popular formats are the memory stick (MS), the multi-media card (MMC), secure digital (SD), secure digital input output (SDIO), and so forth. The secure digital (SD) is a flash (nonvolatile) memory card. Some digital cameras use complementary metal oxide semiconductor (CMOS) technology based microchips as image sensors; these are cheaper and easy to integrate. For the camera application, there are three major external components – a camera, an external RAM and an LCD. The camera system and auto focus circuit inside a mobile phone is shown in Figures 5.13(a) and (b) respectively.

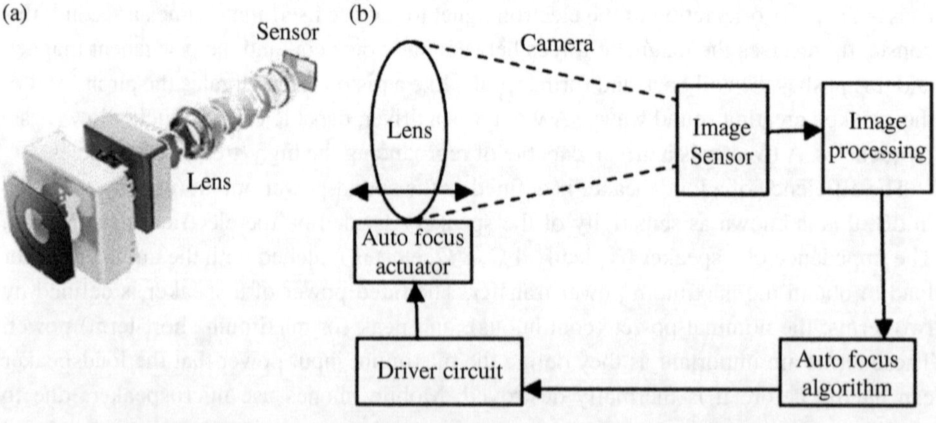

Figure 5.13 (a) Mobile camera system. (b) Autofocus circuit diagram

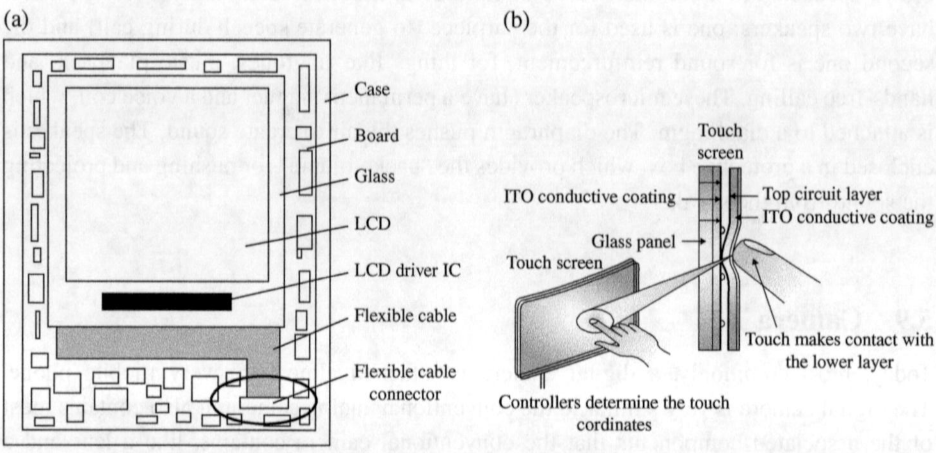

Figure 5.14 (a) LCD module and associated components. (b) Resistive touchscreen

5.10 Display

Nowadays, there are two technologies commonly used for smartphone display: liquid crystal display (LCD) and active matrix organic light-emitting diode (AMOLED). These have several variations, as discussed below.

(i) An LCD display is an electro-optical amplitude modulator realized as a thin, flat display device, made up of any number of color or monochrome pixels arrayed in front of a light source or reflector. It uses a very small amount of electric power for its operation. In Figure 5.14(a), different parts of an LCD subsystem are shown. In case of color LCDs, each individual pixel is divided into three cells, or subpixels, which are colored red, green, and blue, respectively, by additional filters (pigment filters, dye filters, and metal oxide filters). Here, each subpixel can be controlled independently by LCD driver software to yield thousands or millions of possible colors for each pixel. Active-matrix LCDs depend on thin film transistors (TFTs), which are basically tiny switching transistors and capacitors and these are arranged in a matrix on a glass substrate. To address a particular pixel, the appropriate row is switched on, and then a charge is sent down the correct column. Since all of the other rows that the column intersects are turned off, only the capacitor at the designated pixel receives a charge. The capacitor is able to hold the charge until the next refresh cycle. Passive-matrix LCDs use a simple grid to supply the charge to a particular pixel on the display. It starts with two glass layers called substrates. One substrate is given columns and the other is given rows made from a transparent conductive material. This is usually indium-tin oxide. (ii) Thin film transistor liquid crystal display. It uses thin-film transistor (TFT) technology to enhance the image quality. It is most commonly used in mobile phones. It offers better image quality and higher resolutions than earlier generation LCD displays. The limitations are narrow viewing angles, poor visibility in direct light or sunlight, and the fact that large TFT displays consume more power. But it is economical to manufacture, so it is most commonly found on budget phones, feature phones, and low-end smartphones. (iii) In-plane switching (IPS) LCD. This offers better display quality, wider viewing angles and consumes less power than TFT LCD display. But it is costlier than TFT LCD so it is only used on high-end smartphones. iPhone 4 uses IPS LCD (640×960 pixels), which is also called Retina Display, because its pixels cannot be individually identified by a human eye. (iv) Organic LED (OLED) technology uses anodes and cathodes for electron flow through a very thin film and color is controlled by the tiny red, green, and blue light-emitting diodes built into the display. Its brightness is determined by the strength of the electron current. (v) Active matrix OLED (AMOLED). Here, individual pixels (the active matrix) are lit separately on top of a thin film transistor (TFT) array that passes electricity through organic compounds (that's the OLED part). It is a newer technology than IPS LCD, helps for better battery life, and improves in some areas, but it suffers from a "burn in" issue, where pixel quality is degraded over time, costs more, and appears less sharp when viewed at very close range. Super AMOLED is an upgrade on plain old AMOLED. (vi) Super LCD (SLCD): Based on LCD technology, it gives warmer color tones and has better color definition than AMOLED.

Touchscreen (input device) and the actual LCD screen (output device) are two independent parts and touchscreen is discussed in the next section.

A backlight is used to provide the background light for the LCD display to see the screen or keypad buttons. Generally, LEDs are used for LCD backlighting. They are controlled by the signal voltage coming from the controller.

5.11 Keypad and Touchscreen

A keypad is a set of buttons arranged in a matrix form, which usually bear digits (0–9), letters (a–z), alphanumeric characters (*, #) and also some special symbols for accepting calls, rejectings call, cursor movement, and so forth. Figure 5.15 shows the internal circuit

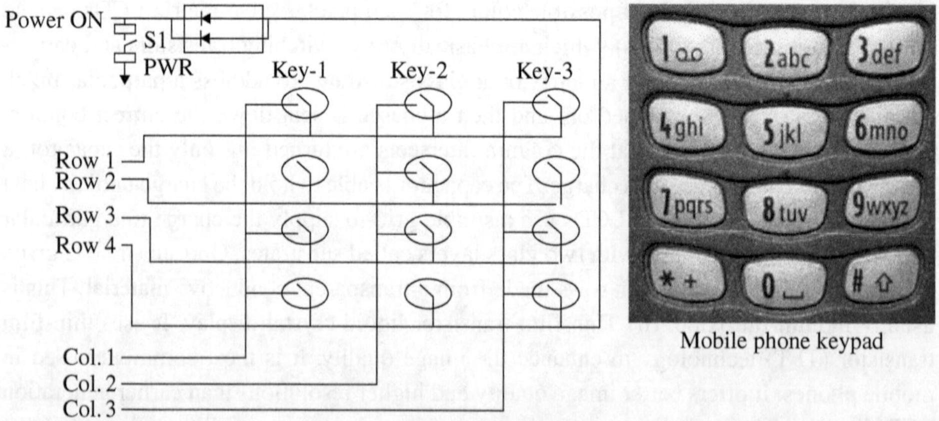

Figure 5.15 Keypad of a mobile phone. Source: Das. Reproduced with permission of John Wiley & Sons (Asia)

diagram of a keypad. The keypad hardware may be polled by the keypad software driver to check if a key has been pressed by the user. In this case, the hardware must have a memory of the last key pressed so that the driver will detect that a keystroke has occurred on the next polling interval. The driver must reset the keystroke memory so that additional keystrokes can be detected on future polls of the hardware. For this case, no interrupt service routine (ISR) is required. Alternatively, the hardware may be implemented so that it generates an interruption when any key is pressed. In this case, an interrupt service routine (ISR) is required. Generally, a keypad device will not use direct memory access (DMA) or shared buffers to transfer data, but will use a programmed I/O instead.

Nowadays most mobile phones are designed with touchscreen-based virtual keypad. A touchscreen is an electronic input device using single or multitouch gestures. Commonly, it is the thin transparent layer of plastic on the top of LCD display, which reads the signal from the touch and transports it to the processing unit. A touchscreen is a pointing device consisting of a specialized surface that can translate the motion and position of a user's fingers to a relative position on a screen and that position (coordinate) can be used to take decision on what key is pressed. It operates in many ways, like capacitance and conductance sensing. George Gerpheide created the matrix approach in April 1994, where a series of conductors is arranged in two layers in an array of parallel lines, separated by an insulator and crossing each other at right angles to form a grid. A high-frequency signal is applied sequentially between pairs in this two-dimensional grid array. The current that passes between the nodes is proportional to the capacitance. When a virtual ground, such as a finger, is placed over one of the intersections between the conductive layers, some of the electrical field is shunted to this ground point, resulting in a change in the apparent capacitance at that location. The capacitive shunt method senses the change in capacitance between a transmitter and receiver that are on opposite sides of the sensor. The capacitance value decreases when a finger is placed between the transmitter and receiver as some of the field lines are shunted away.

The most commonly used touchscreen types are:

- Resistive. The resistive touchscreen possesses a flexible top layer made of polyethylene (PET) and a rigid bottom layer made of glass with a very small gap between them, which acts as a resistance. Both the layers are coated with a conducting compound called indium tin oxide (ITO) and then spaced with spacers. While the monitor is operational, an electric current flows between the two layers. When the touchscreen is touched with a finger (or a stylus), the flexible screen presses down and touches the bottom layer. A change in electrical current is hence detected and the coordinates of the point of touch are calculated by the controller and parsed into the processor – see Figure 5.14(b). The Nokia N97, HTC Tattoo, and Sony Ericsson Satio use this. This has advantages like low cost, low power consumption, the ability to be activated with any object, and resistance to surface contaminants (moisture, dust, oil) but it also has disadvantages like lower image clarity and the fact that the outer polyester film is vulnerable to damage.
- Capacitive. Capacitive touchscreens are more responsive to human touch and most popular type of touch screens on the market. These are used in high-end smartphones with Gorilla Glass. A transparent electrode layer is placed on top of a glass panel and covered by a protective cover. Generally glass panel is coated with indium tin oxide (ITO). When an exposed finger touches the monitor screen, it reacts to the static electrical capacity of the human body and decreases in capacitance is detected by sensors located at the four corners of the screen, allowing the controller to determine the touch point. It durability is moderate, and it needs calibration during manufacture. A passive (nonconductive) stylus cannot be used for surface capacitive touchscreen.

Surface acoustic-wave touchscreen technology contains two transducers (transmitting and receiving) placed along the X-axis and Y-axis of the monitor's glass plate along with some reflectors.

5.12 Analog-to-Digital Conversion (ADC) Module

The analog to digital conversion unit is one of the important components in a digital mobile phone. Generally the RF front-end unit processes an analog signal, which needs to be converted to a digital signal for baseband processing, and for that purpose one ADC circuit is used. Whereas on the transmit side, the source signal from the microphone is the analog signal, which is also converted to a digital signal using ADC and is given to the source coder (speech codec). There are various types of ADC available, like dual slope (slow speed, medium cost), flash (very fast speed, high cost), successive approximation (medium to fast speed, low cost), sigma delta (slow speed, low cost).

Of these, the sigma delta converters have become very popular and are the most widely used in mobile phone receivers. The key feature of this converter is that it is the only low-cost conversion method, which provides both a high dynamic range and flexibility in converting low bandwidth input signals. A simple block diagram of a first-order sigma delta

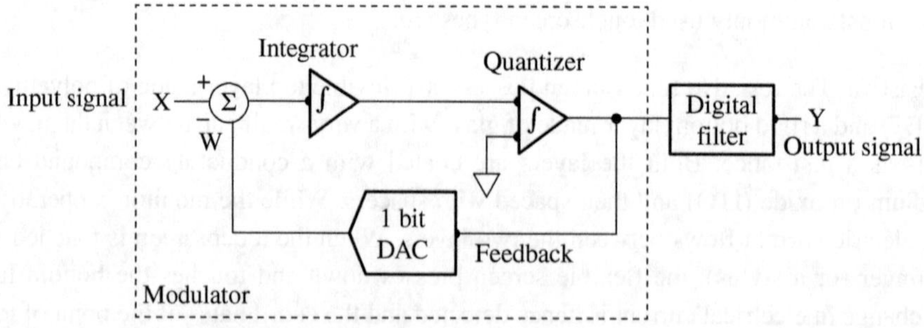

Figure 5.16 First order Sigma Delta ADC

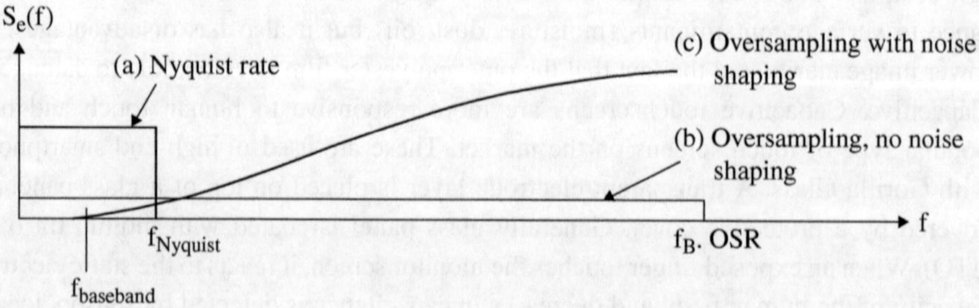

Figure 5.17 Spectral properties of quantized noise in case of (a) Nyquist rate sampling, (b) oversampling with no noise shaping, and (c) Nyquist rate with oversampling

analog-to-digital converter (ADC) is shown in Figure 5.16. The input signal X comes into the modulator via a summing junction. After that, it passes through the integrator, which feeds a comparator, which acts as a one-bit quantizer. Then, the comparator output is fed back to the input summing junction via a one-bit digital-to-analog converter (DAC). The same signal also passes through the digital filter and emerges at the output of the converter (Y). The feedback loop forces the average of the signal W to be equal to the input signal X.

- Noise shaping. Moving the quantization noise from the band of interest to outside this band is referred to as noise shaping. Assuming that noise type is additive white Gaussian noise (AWGN), we can use feedback to remove the noise from low frequencies (say 0-4 kHz for voice band) at the cost of increasing the noise at higher frequencies (out of the desired signal band). As shown in Figure 5.17, this is done by imbedding a filter and the D/A converter in a feedback loop. The noise-shaping filter or integrator of a sigma delta converter distributes the converter quantization error or noise so that it is very low in the band of interest.
- Oversampling. This is simply the act of sampling the input signal at a frequency much higher than the Nyquist frequency. Oversampling decreases the quantization noise in the band of interest.
- Digital filter. An on-chip digital filter is used to attenuate signals and noise that are outside the band of interest.

- Decimation. As the sigma delta converter samples at much higher rate, so the generated sampling data volume is much higher. The act of reducing the data rate down from the oversampling rate without losing information is known as decimation. This process is used in a sigma delta converter to eliminate redundant sampled data at the output. The sampling theorem tells us that the sample rate only needs to be two times the input signal bandwidth in order to reconstruct the input signal reliably without distortion. However, the input signal here was heavily oversampled by the sigma delta modulator in order to reduce the quantization noise. Therefore, there is redundant data that can be eliminated without introducing distortion to the conversion result. The decimation process simply reduces the output sampling rate, while retaining the necessary information.

Here, it should be noted that sigma-delta modulation only alters the spectral properties of the quantization noise by shifting the noise power to a high frequency domain but it still needs to be removed from the output signal by means of low-pass filtering. In fact, the total amount of quantization noise increases for higher modulation orders. As discussed, the filtering is achieved by means of a decimation filter, which also reduces the sampling rate and thereby reduces the number of samples to be processed inside the baseband processor.

5.13 Automatic Gain Control (AGC) Module

Due to noise and channel characteristics fluctuations, the received signal strength at the receiver changes from maximum to minimum or vice versa. Automatic gain control is widely used in communication systems to maintain a constant signal strength by varying the amplifier gain. Apart from the amplitude variation in the speaker volume, the variation of received signal strength causes several other issues like: (i) the performance of the amplifier (LNA and PA) circuit changes because, when signal strength is high this leads to saturation issue and when signal strength is low this leads to poorer amplification; (ii) the tolerances of power level, in different components of the transmitter-receiver chain are not the same, so this may cause damage to the circuit; (iii) ADC or DAC has a dynamic range – if the signal strength varies beyond that then there will be error (saturation) leading to signal clipping. Under extreme conditions of voltage and temperature the mobile phone should work properly with AGC loop circuitry and guaranteed component tolerances. In Figure 5.18(a), a typical AGC loop circuit is shown. This is a feedback system comprising a forward gain stage (A), feedback gain (β) and a signal comparison stage that generates a differential error signal. The AGC loop is analyzed in terms of its closed-loop gain (forward transfer function) and open-loop gain. R(s) and C(s) represent the input and output amplitude. There is a gain stage in the comparator and temperature compensation at the detector diode stage to compensate for diode detector forward voltage variation with temperature.

Automatic gain control plays an important role in the mobile receiver by adjusting the gain of the amplifier according to the received signal strength. The baseband AGC algorithm monitors the RSSI (received signal strength) and averages that over a period. Then, based on that, it programs the required gain value to the LNA, so that the received signal is amplified

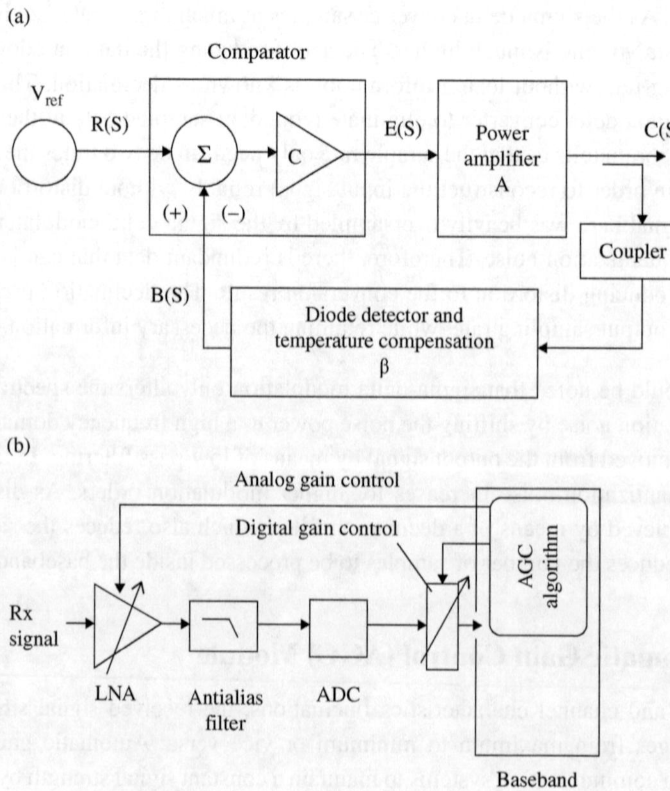

Figure 5.18 (a) AGC loop diagram and control loop components. (b) AGC circuit in mobile receiver

optimally without clipping. Today, mostly two levels of gain setting are used. Analog gain is applied by a feedback loop on an amplifier circuit before the signal is converted into digital by ADC, whereas the digital gain is just multiplication by a factor of the digitized value as shown in Figure 5.18(b).

Most commonly, in a receiver circuit, the AGC is implemented using an IF amplifier, a voltage-controlled amplifier controlled with analog voltage. The dynamic range is determined by minimum carrier-to-noise ratio and blocking signal (interferer) level at the ADC input. Generally, the dynamic range of a receiver is often limited by the dynamic range of the ADC. AGC can be used to dramatically reduce the required dynamic range. The system designers know that adding variable gain is much less expensive than increasing the dynamic range of ADC. The dynamic range of the receiver (reception window) is typically defined to be restricted above 15 dB and below 20 dB, a specific reference level. AGC function can be placed in the first stages of the receiver, after the RF conversion. Nowadays, the gain is controlled at several stages at the front end and also at the baseband module through a software algorithm to obtain a better result. In a typical mobile phone circuit premonitoring is used to maintain a constant output level. This premonitoring is done in three phases and determines the settling time for the RX AGC. The receiver is switched on approximately 150 μs before the burst begins and processor measures the received signal level and then adjusts the AGC-DAC in accordance with the measured signal level and / or switches on / off the LNA with the front-end amplifier control line. The requirement for the received signal level under static conditions is that the MS should measure and report to the BS over the range −110 dBm to −48 dBm.

5.14 Frequency Generation Unit

In a mobile terminal the clock frequency is supplied to different units, like the Rx local oscillator (for generating the LO frequency for the mixer for desired carrier frequency reception), the Tx local oscillator (for Tx carrier frequency generation), the ADC circuit (for sampling frequency), and digital baseband (for CPU frequency). In baseband, apart from supplying the clock signal to the processor (CPU), the clock generator unit must also provide other clock frequency signals to different peripherals and interfaces (such as DDR refreshers, video and graphics unit, and different IP blocks). As shown in Figure 5.19, in an RF unit there are several frequencies required: (i) the frequency for the LO in receiver mixer; (ii) the sampling frequency at the ADC; (iii) the transmitter LO frequency, and so forth. The simpler way to generate these frequencies is the use of dedicated crystals for each but that would increase the size, cost, and power consumption. So, instead of that, a single high-frequency crystal oscillator is used and different frequencies required in the system are derived from that master clock using a frequency synthesizer (as shown in Figure 5.19).

Generally, a frequency synthesis technique and frequency dividers are used to generate multiple frequencies from an accurate reference frequency (the input reference clock), commonly supplied by a crystal oscillator made of quartz crystals.

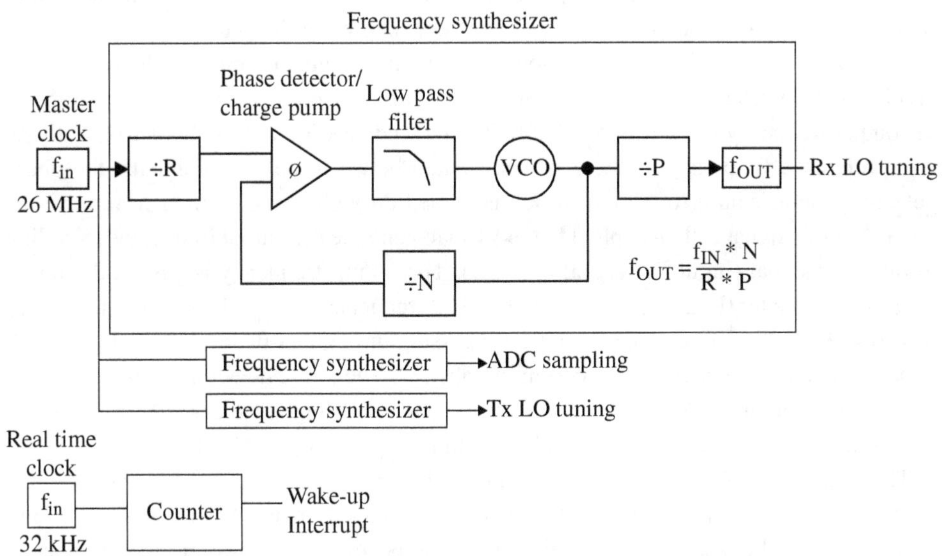

Figure 5.19 Frequency generation unit inside a mobile terminal

Appropriately cut quartz crystals are used for high-quality electromechanical resonators. Their piezoelectric properties allow them to be the frequency-determining element in electronic circuits. Quartz crystals are modeled electrically as a series LCR branch in parallel with a shunt capacitance that yields the well known result for LCR circuits with the natural frequency (ω_0) is the square root of the inverse of the product of the inductance (L) and capacitance (C) (see Figure 5.20(a). Note that crystals often have undesired mechanical resonances near the fundamental frequency (spurious frequencies, spurs).

There are various different types of frequency synthesizers available, each with merits and demerits.

- Direct. These are implemented by creating a waveform directly without any form of frequency-transforming element. These could be anlog or digital. The analog technique (a mix-filter-divide architecture) requires large amount circuitry, which makes it bulky and expensive. The digital technique creates the signal by having a stored version of the required waveform and then advancing the phase in fixed increments.
- Indirect. Here the output-required signal is generated indirectly by an oscillator, which is controlled by other signals based on phase locked loop (PLL) technology.

As shown in Figure 5.19, a typical digitally controlled analog PLL consists of a reference counter (R), feedback counter (N), postscaling counter (P), and the core analog blocks, which include a phase detector / charge pump, low-pass loop filter and voltage-controlled oscillator (VCO). The VCO is an electronic device whose output oscillation frequency is controlled by providing various input DC voltages to it, which means that the applied input voltage determines the instantaneous output oscillation frequency. Generally there are two type of VCOs: (i) harmonic oscillators, where the output is a sinusoidal waveform and (ii) relaxation oscillators where the output is a saw tooth or triangular waveform and output frequency varies with the time of charging and discharging of the capacitor.

In Figure 5.19, f_{in} is the reference input clock coming from an oscillator clock. It is then divided by R to generate a lower frequency (this is optional). The phase detector compares the output frequency (f_{out}), which is fed to it via a feedback loop (with divider N) with the input reference frequency. If there is a mismatch between these two signals, the phase detector generates an error signal, which is passed through low pass filter to remove the noise, and this signal is then applied to the VCO to generate the output frequency according to the error signal (input DC signal to VCO). The output frequency is passed through a postscaling counter (P) and used as output locked frequency (f_{clk_out}). This output frequency is also given to the phase detector unit after passing through a divider (N counter), which divides the output frequency by a certain number, N. The phase detector provides a signal to the charge pump to decrease or increase the control voltage of the VCO. In the above circuit the output clock frequency available will be: $f_{clk\_out} = (f_{in}) * (N / R*P)$.

The values of N, R and P are set according to the output clock requirement. So, the PLL allows the processor to operate at a high internal clock frequency derived from a low-frequency crystal clock input (f_{in}). This has several benefits: (i) the lower clock frequency input reduces the overall electromagnetic interference in the system; (ii) the ability to oscillate at different frequencies reduces the costs of the system by eliminating the need to add additional oscillators. Today PLL is best suited for frequency synthesis as a PLL-based clock generator provides a cost-effective solution for generating various frequencies, satisfying the tighter specification parameters like skew and jitter requirements.

Generally, in a mobile phone, there is a 26 MHz master clock oscillator, which provides the reference clock to other units. This is known as a master clock (see Figure 5.20). This clock runs when the device is in active mode but not in deep sleep mode. For power saving, whenever there is a long period of inactivity in the RF reception or transmission, then the device cuts the

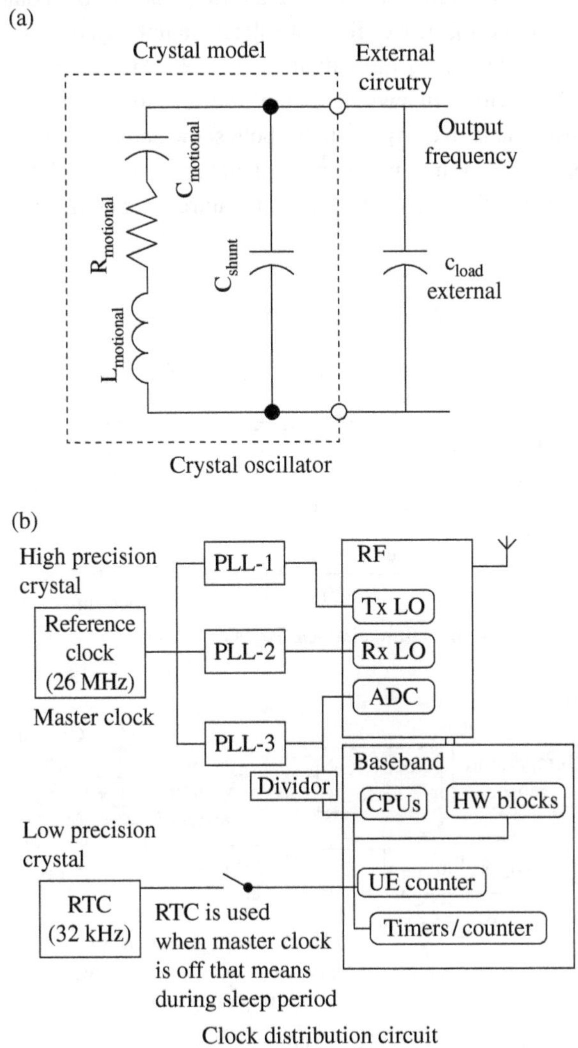

Figure 5.20 (a) Crystal oscillator circuit. (b) Clock signal distribution circuit in mobile terminals

clock to several modules (or all modules) and switch off this master clock. When the master clock is off, there is one more low frequency clock runs during that time. This is known as real-time clock (RTC). When device enters into sleep mode it switches off most of the modules including RF and programs a counter to provide wake-up interrupt. This counter and some other modules (which are required to be executed even during the sleep) run using the RTC.

5.15 Automatic Frequency Correction (AFC) Module

The primary requirement of the AFC is to keep the local transmitter (TX) frequency stable within certain limits, and on the receiver side the frequency error should be low enough to ensure a good receiver performance. To establish and maintain a robust wireless connection, the reference oscillator frequency must attain high levels of precision and accuracy. As accuracy over time is very critical in a wireless phone design, so the reference oscillator inside a mobile phone must be able to compensate for both static and dynamic errors. The frequency deviations can be caused primarily due to temperature drift, initial crystal offset, Doppler shifts, and aging. Figure 5.21(a) shows the temperature characteristics of crystal.

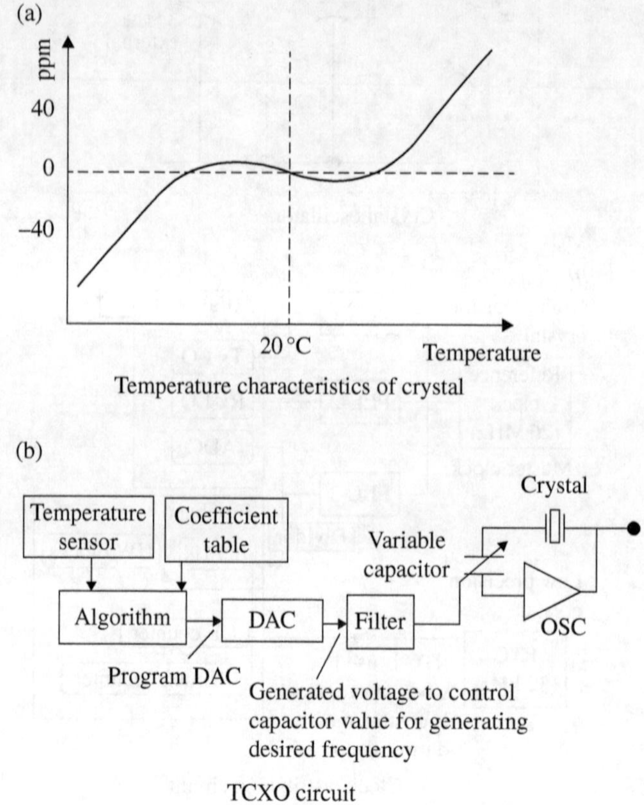

Figure 5.21 (a) Temperature characteristics of crystal. (b) Generic TCXO circuit

There are several types of digitally implemented frequency compensation circuits like: (i) temperature compensated crystal oscillator (TCXO); (ii) analog digital temperature compensated crystal oscillator (ADTCXO) and voltage controlled temperature compensated crystal oscillator (VCTXO); (iii) digital temperature compensated crystal oscillator (DTCXO); (iv) microprocessor (MCXO).

Compensated Crystal Oscillator.
Generally, handsets use voltage-controlled temperature-compensated crystal oscillator (VC-TCXO) modules as the system reference oscillator. The primary difference between a TCXO and a simple crystal oscillator is that the TCXO contains additional circuitry that compensates for the crystal's frequency versus temperature characteristics. Figure 5.21(b) shows the generic TCXO circuit. VC-TCXO modules use analog techniques to correct the frequency deviations using voltage control circuit. The problems are that VC-TCXOs bring a big price tag, a large footprint, and require some other external components. These issues are creating real design challenges in the competitive mobile handset market. Nowadays, one alternative solution to this problem is to use a digitally controlled crystal oscillator (DCXO). New mobile phone transceiver architectures are being developed that house digitally controlled crystal oscillators (DCXOs) and help to eliminate the headache of adding a VC-TCXO to mobile phone architecture.

5.15.1 The Analog VC-TCXO

The VC-TCXO module incorporates temperature compensation circuitry, which typically consists of a simple control loop using a thermistor circuit. The thermistor network biases an internal varactor diode to attain the correct crystal load capacitance to maintain the target oscillation frequency. The bias level of the varactor diode changes with temperature as the thermistor resistance changes. This helps to compensate for the temperature effect on the crystal's frequency.

As shown in Figure 5.22, the VC-TCXO also includes an external tuning voltage input to finely control the varactor diode (whose capacitor changes with applied voltage) to compensate for frequency errors other than temperature drift. To ensure the specified precision and accuracy, each resistor must be production trimmed to offset each crystal's unique static error. In the VC-TXCO, a low dropout (LDO) voltage regulator is used for control and to stabilize the supply voltage. Depending on the implementation, the supply voltage may "push" the oscillation frequency away from the target and induce errors. The additional internal components and manufacturing steps significantly increase the cost of the module.

5.15.2 Digitally Controlled Crystal Oscillators – DCXO

As we are moving towards the digital world, the DCXO is replacing the analog VC-TXO. In the past, many handset designers have developed discrete DCXOs to avoid using costly VC-TCXO modules and to reduce the overall bill of material (BOM) cost. DCXOs

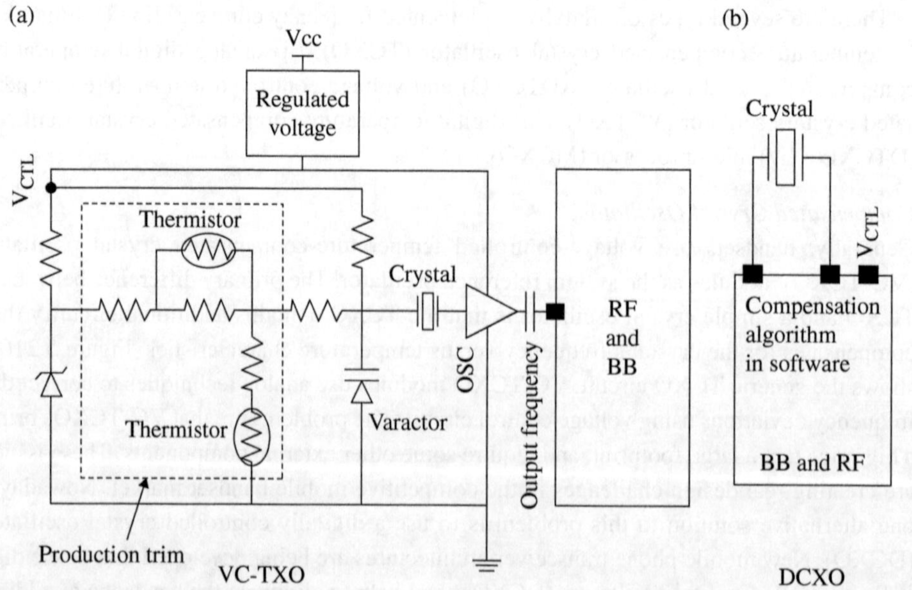

Figure 5.22 (a) Block diagram of a typical VC-TCXO circuit (direct compensation). (b) Block diagram of a typical DCXO implementation

compensate for frequency errors using a combination of digital and analog circuitry. In a GSM/GPRS mobile phone design, a DCXO can replace the VC-TCXO function with a standard AT-cut uncompensated crystal resonator. The digital circuit monitors the frequency deviation and constantly controls the deviation through circuit component adjustment. Frequency errors may be controlled by a software program using a control loop topology. Implementations are chip dependent and can be implemented using variety of configurations and using different methods.

5.15.2.1 Working Principle of DCXO

Conceptually, based on the frequency measurement calculations by a transceiver software program (the deviation of the local frequency to the network frequency), a DCXO "pulls" the crystal frequency to the required target value. As shown in Figure 5.22(b), a digitally configurable interface can be used to programmatically add or subtract load capacitance to the oscillator circuit to change the resonance frequency particularly useful for correcting static errors.

With the DCXO approach, the error compensation circuitry and voltage regulation are integrated into the IC. DCXOs must compensate for both dynamic and static errors. In addition, the DCXO must be able to adjust the frequency in incremental steps continuously so that a final frequency error of 0.1 PPM or lower is achieved throughout. As DCXO circuits do not include temperature sensors so they heavily rely on the frequency measurement in the digital controller. Generally, it involves three processes:

- Frequency estimation. This is the process of estimating the complex frequency components of a signal in the presence of noise or channel impairments. Different estimation methods can be used for frequency, like time-domain periodicity estimation, spectral pattern matching, and frequency-domain periodicity. Different algorithms used commonly for frequency estimations are maximum likelihood (ML), approximate ML, Fourier coefficient, filtering techniques, signal subspace, noise subspace, phase-weighted averagers.
- Frequency compensation. Once the frequency is estimated, next the deviation of the local clock frequency with the estimated frequency is computed and this computed frequency deviation is compensated by using different methods like changing the varactor diode voltage, and resistance value changes.
- Frequency tracking: once the frequency is estimated and deviation is computed, then it has to be constantly monitored and tracked to keep the deviation under a certain limit.

5.16 Alert Signal Generation

A buzzer is used to provide various alerting audio signals as an indication of incoming calls. The user key press and other responses beeps are also generated by the buzzer. The buzzer is controlled by a buzzer pulse width modulation (PWM) output signal from the baseband processor. The mobile phone uses a dynamic type of buzzer. The low-impedance buzzer is connected to an output transistor that obtains a drive current from the PWM output. The volume can be adjusted either by charging the pulse width, causing the level to change, or by changing the frequency to use the resonance frequency range of the buzzer.

A vibra alert device is used for giving a silent alert signal to the user about any incoming calls and this is controlled by a vibra PWM output signal from the baseband processor. Generally, an especially designed motor is used for a vibra alert device. A mass of improper weight distribution is attached to the motor's shaft / axis, so when the motor rotates, the irregular weight causes the phone to vibrate (there is an off-centered weight attached to the motor's rotational shaft that causes the motor to wobble, see Figure 5.23).

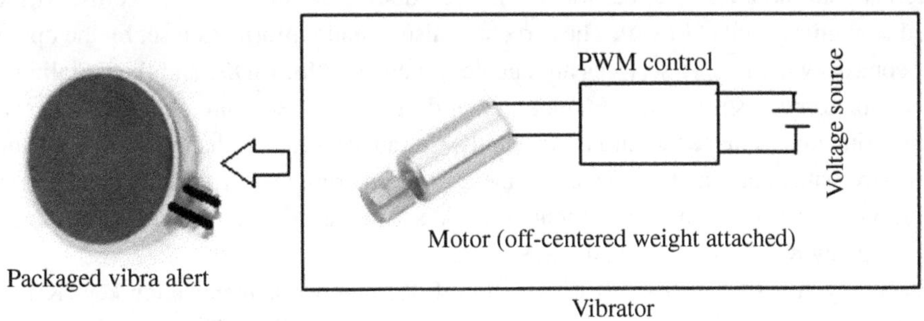

Figure 5.23 Cellphone vibrator

The vibra alert can be adjusted either by changing the pulse width of the input voltage or by changing the pulse frequency of the vibra PWM signal. The vibra device is placed inside a special vibra battery.

5.17 Subscriber Identity Module (SIM)

The subscriber identity module (SIM) is a small smart card, which contains microprocessor, memory, programs and information. The SIM card contains a microprocessor chip inside it, which stores unique information about the user's account, including the user's phone number, and identifies the user to the network. So it is not the cell phone, that determines the telephone number – rather it is the SIM card. The subscribers activate their phones by inserting SIM cards into them. Once the SIM is removed from the phone then the phone can't be used for making call except for some emergency calls. Generally, the phone is not tightly coupled with a SIM – a SIM can be moved from one mobile phone to another. This makes upgrades of mobile devices very simple for the mobile phone user.

The subscriber identity module serial number (SSN) is a 19 or 20 digit unique number that identifies an individual SIM card. Typically a SIM card has 16 to 64 kB of memory, which provides plenty of room for storing hundreds of personal phone numbers, text messages, and value-added services. SIM cards are available on a subscription basis. Generally network operators in specific geographic locations provide them. For this, a user can sign a contract with a provider and get a bill every month, or, alternatively, they are available on a prepaid basis, in which case users buy airtime as required. Generally, GSM specifies two types of SIM cards. An ID-1 card is the same size as a standard credit card and has embossing, a picture, lettering and a magnetic stripe similar to credit card. Some larger GSM phones use this. Another type is called a plug-in SIM card. An ID-1 card can be converted to a plug-in SIM card by removing the plastic holder. The ID-1 has a dimension of 54 mm × 85.6 mm whereas a plug-in SIM has a dimension of 15 mm × 25 mm.

The memory on a SIM card consists of a master file (MF), and this has two dedicated files (DFs). The DFs contain elementary files (EFs), which contain actual GSM data – each EF contains one record, which could be information such as a phone book or the IMSI (international mobile subscriber identity). Record sizes are measured in words, with one word containing 8 bits (1 byte). These records also contain information set by the operator that controls what feature services are enabled, such as SMS, ISDN, and fixed dialing.

The microprocessor-based SIM platform is designed to be secure. Attempts to reverse engineering may damage the card permanently. Certain data can be changed by the manufacturer only. Other data can be changed by the user by entering the correct PIN. The ciphering algorithms, A3 and A8, are implemented in the subscriber identity module (SIM) and the ciphering key K_c is also stored in the SIM.

Generally, the SIM carries this information: IMSI number, authentication key (Ki), subscriber information, access control class, cipher key (Kc), TMSI, additional GSM services, location area identity, forbidden PLMN, A3 and A8 algorithm., BCCH information. The SIM

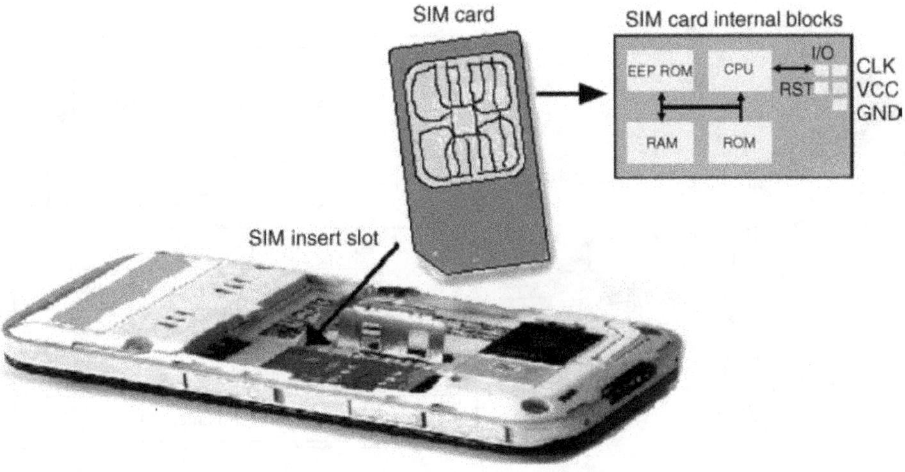

Figure 5.24 Blocks inside a SIM card

card provides the storage capability for administrative information, ID card identification, recently dialed numbers, SMSs, and so forth. Figure 5.24 shows the SIM internal blocks.

The interfaces between the mobile handset and the SIM card are fully standardized and there are already specifications in place. The SIM card readers or editors are hardware-software combinations that make it possible to gain access to the SIM card of a mobile phone on a PC. With a SIM card reader it is possible to view, create, edit, and backup phonebook entries using a PC. PIN codes, transfer data from one SIM to another, backup and export and import all phonebook entries can be managed. SIM card readers allow users to back up the SIM card phonebook data to local memory and avoid data losses in case the user loses or changes a SIM card or a GSM phone.

In GSM, the SIM consisted of the hardware and the software but in 3G (UMTS) this split into two parts: the SIM is now an application and hence only software whereas the hardware part was called the universal integrated circuit card (UICC). UMTS has introduced a new application: universal subscriber identity module (USIM). USIM brings several enhancements like mutual authentication and longer encryption keys, and an improved address book. (Refer to SIM Specification 51.011; SIM Specification 31.102; ETSI Recommendation GSM 11.11, Specifications for the SIM-ME Interface; TS 31.102, TS 31.101.)

5.18 Connectivity Modules

5.18.1 Bluetooth

Bluetooth is a telecommunications industry specification for wireless personal area networks (PANs), which is a short-range (32 feet ~ 10 m) radio frequency technology that operates at 2.4 GHz (ISM band) and is capable of transmitting voice as well as data – see Figure 5.25(a). The name Bluetooth comes from a Danish king, Harald "Bluetooth"

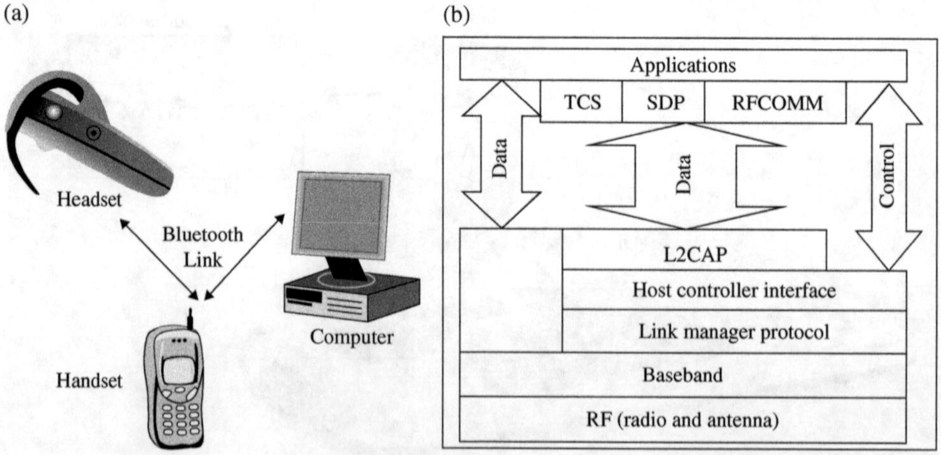

Figure 5.25 (a) Bluetooth-connected devices. (b) Bluetooth protocol stack

Blaatand, who unified Denmark and Norway. In the beginning of the Bluetooth wireless technology era, Bluetooth was aimed at unifying the telecom and computing industries. Today, Bluetooth provides a way to connect and exchange information between groups of devices such as mobile phones, headsets, laptops, PCs, printers, digital cameras, and video game consoles over a secure, globally unlicensed short-range radio frequency. Bluetooth radios use a fast frequency-hopping (FHSS) spread spectrum (FHSS) technique for the air medium access technique and time division duplex (1600 hops/s). Generally, it operates in the 2.4 GHz ISM band (79 MHz of spectrum = 79 channels) and uses the Gaussian frequency shift keying. The first version uses transmission power of around 1 MW–100 MS and provides data rate 1 mbps. Up to eight data devices can be connected in an ad hoc piconet. Each piconet supports up to three simultaneous full duplex voice devices (CVSD).

Bluetooth (BT) profiles are general behaviours through which Bluetooth-enabled devices communicate with other BT devices. Bluetooth technology defines a wide range of profiles that describe many different types of use cases, like advanced audio distribution profile (A2DP), audio / video remote control profile (AVRCP), basic printing profile (BPP), common ISDN access profile (CIP), cordless telephony profile (CTP), fax profile (FAX), file transfer profile (FTP), and so forth.

The protocol stack for Bluetooth is shown in Figure 5.25(b) (refer to IEEE 802.15.1). Transmitter and receiver blocks are shown in Figure 5.26. For more details please refer to the Bluetooth specifications [1]. RFCOMM is basically the cable replacement protocol, emulating serial ports over a wireless network. Service discovery protocol (SDP) allows applications to discover device information, services, and their characteristics. TCP/IP network protocols are for packet data communication and routing. L2CAP provides connection-oriented and connectionless data services to upper layer protocols; it is used to multiplex multiple logical connections between two devices. Link manager protocol is used for the setup and control of the radio link between two devices. The devices are addressed by a 48-bit IEEE

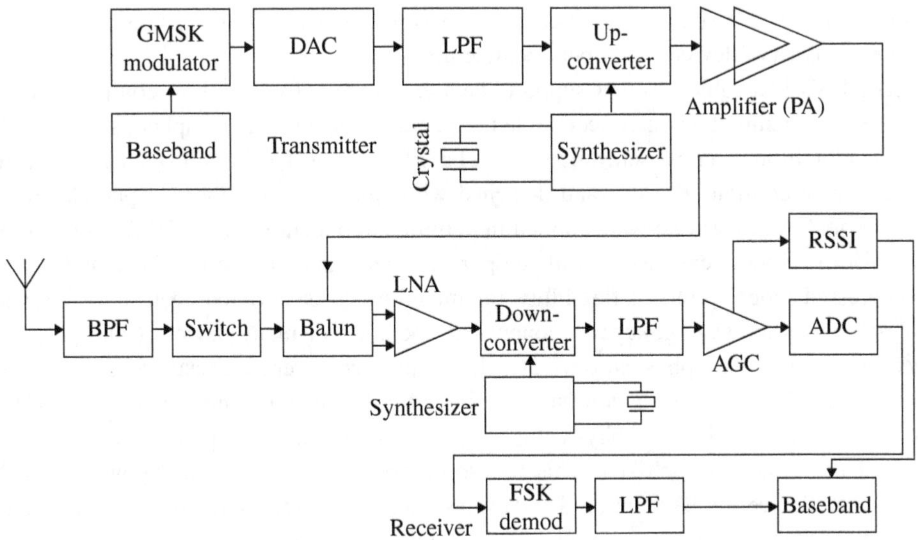

Figure 5.26 Bluetooth transmitter and receiver blocks

MAC address. Three bits are used for an active slave address, all zero for broadcast addresses and 8-bit parked slave addresses. Packets have 72 bits of access code, 54 bits of header, and 0-2744 bits of payload. For voice (speech-call) applications, use of forward error correction (FEC) and cyclic redundancy check (CRC) are optional, whereas for packet switch applications, automatic repeat request (ARQ) and FEC are optional. There are two main states and substates – standby (no interaction) and connection (working) – and seven more substates for attaching slaves and connection establishment. At any time, data can be transferred between the master and one other device, and the master chooses which slave device to address. The switching happens in round-robin fashion. Slave listens in each receive slot.

Bluetooth Low Energy was introduced in the 4.0 specification which uses same spectrum in a different mechanism. Bluetooth v4.2 was released on December 2, 2014, with features like LE Data Packet Length Extension, and an IPv6 connection option.

5.18.2 USB

Universal serial interface (USB) is a fast, bidirectional, isochronous / asynchronous, low-cost, dynamically attachable serial interface. The universal serial bus (USB) is specified to be an industry standard extension to the PC architecture with a focus on computer telephony integration (CTI), and consumer and productivity applications. An original intention of USB was to connect many devices to a PC host, as there was a shortage of serial and parallel ports in the PC. In 1994, an alliance of four industrial partners – Compaq, Intel, Microsoft, NEC started to specify USB. The main goals of defining the USB are

Plug-n-play, Port Expansion, low-cost high performance, enabling seamless integration of new classes of devices and open architecture.

The USB is a cable bus that supports data exchange between a host computer and a wide range of simultaneously accessible peripherals. The attached peripherals share USB bandwidth through a host-scheduled, token-based protocol. The bus allows peripherals to be attached, configured, used, and detached while the host and other peripherals are in operation. Several criteria were applied in defining architecture for the USB, like ease of use for peripheral expansion, full support for real-time data for voice, audio and compressed video, protocol flexibility for mixed-mode isochronous data transfers and asynchronous messaging, support concurrent operation of many devices (multiple connections)– up to 127 physical devices – lower protocol overhead resulting in high bus utilization. The USB system consists of a single USB host and a number of USB devices and interconnects. The USB physical interconnect is a tiered star topology, and there is a hub at the center of each star. This has some benefits – for instance, power to each connected device can be monitored and even switched off independently. High-, full- and low-speed devices can be supported. The main entities of an USB system are (i) USB host; (ii) USB function; and (iii) interconnections.

5.18.2.1 USB Host

There is only one USB host in the USB bus chain and the host is the master of the USB system. The USB interface to the host computer system is referred to as the host controller. The host controller may be implemented in a combination of hardware, software or firmware. The root hub is integrated into the host system to provide one or more attachment points. The USB host controllers have their own specifications. With USB 1.1, there were two host controller interface specifications: (i) the universal host controller interface (UHCI), developed by Intel, which puts more burden on software (Microsoft) and allows for cheaper hardware, and (ii) the open host controller interface (OHCI) developed by Compaq, Microsoft, and National Semiconductor, which places more burden on hardware and makes software simpler. In USB 2.0 standard, another host controller interface was defined: (iii) the enhanced host controller interface (EHCI) – developed by Intel, Compaq, NEC, Lucent and Microsoft – which is an enhanced version.

5.18.2.2 USB Device

USB devices are one of the following:

- A hub, which provides additional attachment points to the USB. Hubs are wiring connectors and enable multiple break attachment characteristics of the USB. Attachment points are referred to as ports. The USB 2.0 hub consists of three portions: (i) the hub controller, (ii) the hub repeater, and (iii) the transaction translator. USB specifications do not limit the number of downstream connectors that form a hub but seven seems to be practical limit; the most popular option is to have four.

- Functions. These are USB devices that are able to transmit or receive data or control information over the bus. They follow USB protocol. A function is typically implemented as a separate peripheral device with a cable that plugs into a port on a hub. However, a physical package may implement multiple functions and an embedded hub with a single USB cable. This is known as a compound device. A compound device appears to the host as a hub with one or more nonremovable USB devices.

5.18.2.3 Interconnects

To connect a host with a hub or USB function devices there is a need for a cable, which is called a USB cable. It carries power as well as data signal. The USB transfers signal and power through a four-wire cable, which is shown in Figure 5.27. The signaling occurs over two wires on each point-to-point segment. The power is transmitted over two wires – Vbus and Ground.

The colors of each wire are given in Table 5.9.

USB allows variable length cable up to 7 m. The cable length is decided considering the delay and voltage drop.

5.18.2.4 Speed of USB

USB version 1.1 supported two speeds, a full speed mode of 12 Mb/s and a low speed mode of 1.5 Mb/s. The USB 2.0 standard has introduced a high-speed mode, with speed of up to 480 Mb/S.

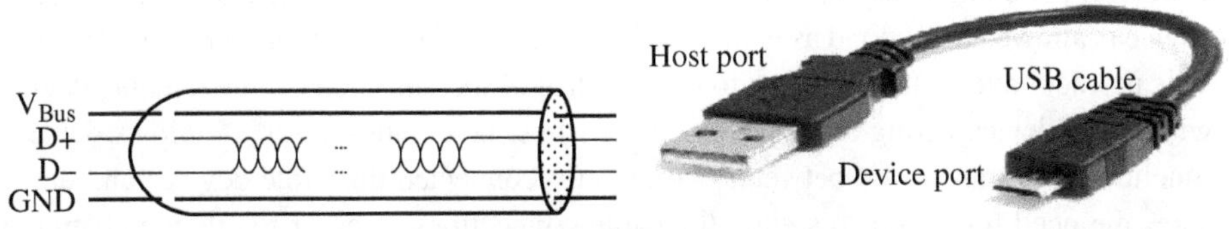

Figure 5.27 USB cable

Table 5.9 USB interconnects

Pin Number	Cable color	Function
1	Red	V Bus (+5 V)
2	White	D–
3	Green	D+
4	Black	Ground

5.18.2.5 USB on the Go (OTG)

As USB communication can only takes place between a host PC and a peripheral, this imposes a limitation of having a PC host in the connection. In order to enable this limitation, a supplement to the USB 2.0 specification was developed, allowing a portable device to take on role of a limited USB host, without the burden of having a PC. This new standard basically defines:

- A new type of device called a "dual role device." A dual-role device can operate (using the same connector) either as a normal USB peripheral or as a USB host. This single connector capability makes OTG especially useful for handheld or other small devices that do not have the space for multiple connectors.
- The standard also introduces two new receptacles and a new plug. So, an OTG-compliant device has only one USB connector (a mini-AB receptacle). The ID pin makes it easy for a dual-role device to determine if it should be the default host or the default peripheral. On a mini-A plug, the ID pin is shorted to ground, whereas on a regular mini-B plug, the ID is pin left open (or in the case of car kit it is shorted to ground through 102 kΩ).
- Full-speed operation as a peripheral (high speed optional) and full speed support as a host (low speed and high speed optional).
- It supports two new OTG protocols: (i) Session request protocol. This protocol allows the B device to request that the A device turn on VBus and start a session. Once the session is started the host negotiation protocol can be invoked to give control to the B device. The OTG supplement defines a session request protocol (SRP), which allows a B device to request the A device to turn on VBus and start a session. This protocol allows the A-device, which may be battery powered, to conserve power by turning the VBus off when there is no bus activity while still providing a means for the B device to initiate bus activity. Any A device, including a PC or laptop, is allowed to respond to SRP. Any B-device, including a standard USB peripheral, is allowed to initiate SRP. Any dual role device is allowed to respond as well as initiate SRP. (ii) Host negotiation protocol (HNP). This protocol allows the B device to take control of the bus and become the host device, with the A device acting as a peripheral. The host negotiation protocol allows the host function to be transferred between two directly connected dual-role devices and eliminates the need for a user to switch the cable connections in order to allow a change in control communications between the devices. The HNP will typically be initiated in response to input from the user or an application on the dual-role B device. HNP may only be implemented through the mini-AB receptacle on a device.

5.18.3 WiFi

Wireless local area network (WLAN) is a wireless network that links two or more devices using a short-range wireless medium access method. Wi-Fi means Wireless Fidelity, a trademark name used to brand products that belong to a category of WLAN devices based on the standards defined by IEEE 802.11. In 1999, the Wi-Fi Alliance was formed. Wi-Fi is a physical and link layer interface, as is Ethernet.

System Architecture

All wireless devices that join a Wi-Fi network, whether mobile, portable or fixed, are called wireless stations (STAs) or nodes, and are connected to an access point (AP). The STAs and AP in same radio coverage form a basic service set (BSS) and this is controlled by a single coordination function (CF). A distribution system connects several BSSs via AP to form a single network. This extends the coverage, and is known as extended service set (ESS). Two operating modes are specified in IEEE 802.11 standard: infrastructure mode and ad hoc mode. A BSS operating in ad hoc mode is isolated and there is no connection to other Wi-Fi networks or to any wired LANs, whereas, the infrastructure operating mode requires that the BSS contain one wireless access point (AP). An AP is an STA with additional functionality, a major role of which is to extend access to wired networks for the clients of a wireless network. Routers allow wireless clients access to multiple networks but APs allow access to a single network.

Protocol and System Parameters

IEEE 802.11 (Table 5.10) defines two layers in the ISO model: the physical layer and the data link layer.

The physical layer (PHY) is divided into two sublayers: (i) a physical medium-dependent sublayer, which includes wireless medium access techniques based on infrared, or radio transmission using frequency-hopping spread spectrum (FHSS) or direct-sequence spread spectrum (DSSS) in the ISM frequency band and methods for transmission and (ii) a reception physical layer convergence procedure (plcp), which converts data into frames and listens to the medium. There are two signaling frequencies currently used by Wi-Fi networks: (i) 2.4 GHz, which comprises 14 channels, each with a bandwidth of approximately 20 to 22 MHz, and (ii) 5 GHz, which comprises 13 channels, each with a bandwidth of approximately 20 MHz.

Table 5.10 IEEE 802.11 standard

Standard	Frequency band	Bandwidth	Modulation	Max data rate
802.11	2.4 GHz	20 MHz	DSSS, FHSS	2 Mb/s
802.11b	2.4 GHz	20 MHz	DSSS or complementary code keying (CCK)	11 Mb/s
802.11a	5 GHz	20 MHz	OFDM	54 Mb/s
802.11g	2.4 GHz	20 MHz	DSSS, OFDM	54 Mb/s
802.11n	2.4 GHz, 5 GHz	20, 40 MHz	OFDM	600 Mb/s
802.11ac	5 GHz	20,40, 80, 90, 160 MHz	OFDM	6.93 Gb/s
802.11ad	60 GHz	2.16 GHz	SC, OFDM	6.76 Gb/s

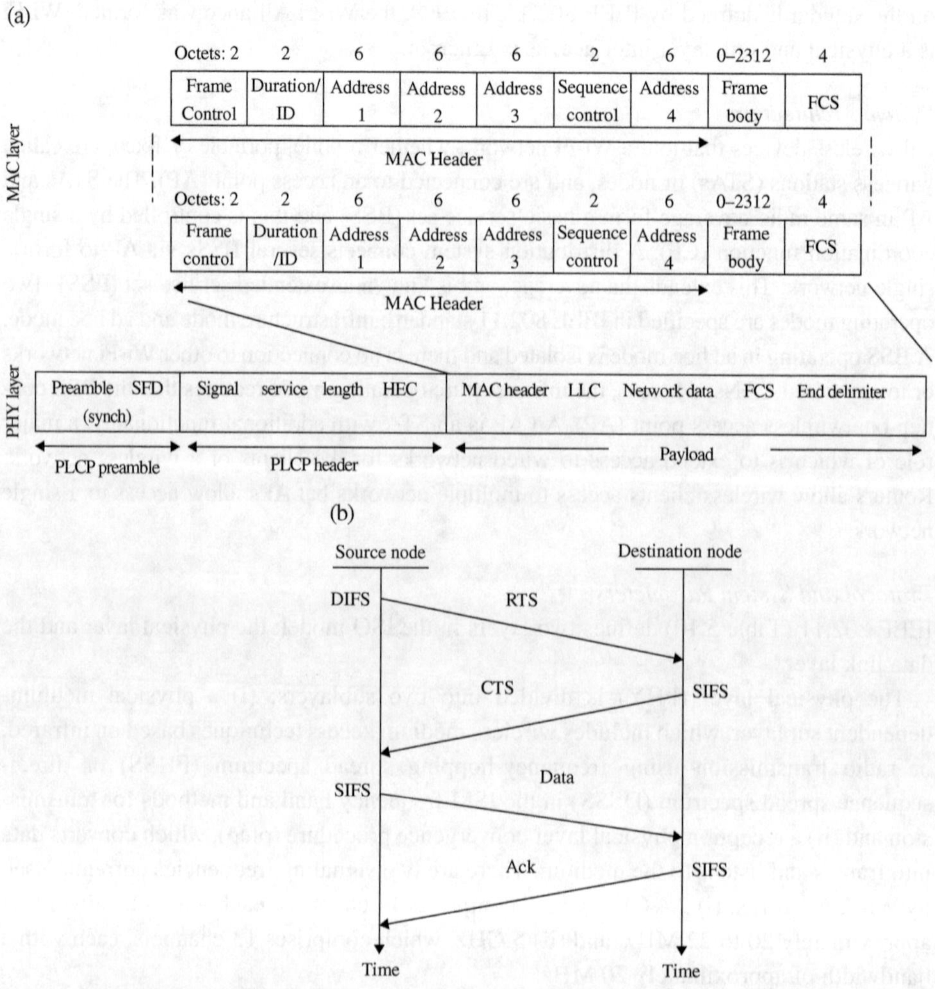

Figure 5.28 (a) MAC and PHY layer frame structure (b) RTS and CTS signaling

Medium Access Control Layer (MAC)

This is responsible for accessing the medium, establishing wireless links, authentication, and power management. To move data packets across a shared channel, the MAC layer uses carrier sense multiple access / collision avoidance (CSMA/CA), which is very similar to the strategy used in 802.3 MAC layers: collision detection (CSMA/CD). This random access scheme is carrier sense collision avoidance through random backoff – Figure 5.28(b). The node senses whether the medium is free or busy. If the medium is found to be idle (free) for more than the DCF interframe space (DIFS) duration, then only a node is allowed to access (send) through the medium. But if the medium is busy, nodes have to wait for the DFIFS duration, then repeat the same process. In case of collision, each node chooses a random backoff time and then tries again. In the case of DFWMAC-DFC with the

RTS/CTS method, the standard defines an additional technique, which uses the request to send (RTS) and clear to send (CTS) mechanism. In this case, after waiting for DIFS the sender can send a RTS control packet, where the RTS includes the receiver of the data transmission to come and duration of the whole data transmission. After receiving this, other nodes (except intended receiver) set the net allocation vector (NAV) so that they will not transmit for that duration. The intended receiver, after receiving the RTS, sends the CTS. The MAC frame is composed of: (i) a MAC header, which consists of a frame control field, the duration, address fields 1–3, a sequence control field; (ii) a frame body – this field varies in size and consists of information based on the frame type to be carried; (iii) a frame-check sequence (FCS) – this is a 32-bit CRC part as shown in Figure 5.28(a).

In the physical layer, a frame consists of two basic parts: PLCP (preamble and header), and payload part – see Figure 5.28(a). Generally the PLCP preamble part has a synchronization and start-frame delimiter (SFD) and is transmitted at a fixed rate (1 mbps). The synchronization part is used for synchronization, gain setting, energy detection, and frequency offset compensation. The SFD indicates the start of the frame. The header part contains a signal, which indicates the data rate of payload, the header check sequence (HEC), and so forth.

The transmitter and receiver blocks are shown in Figure 5.29(a) and (b).

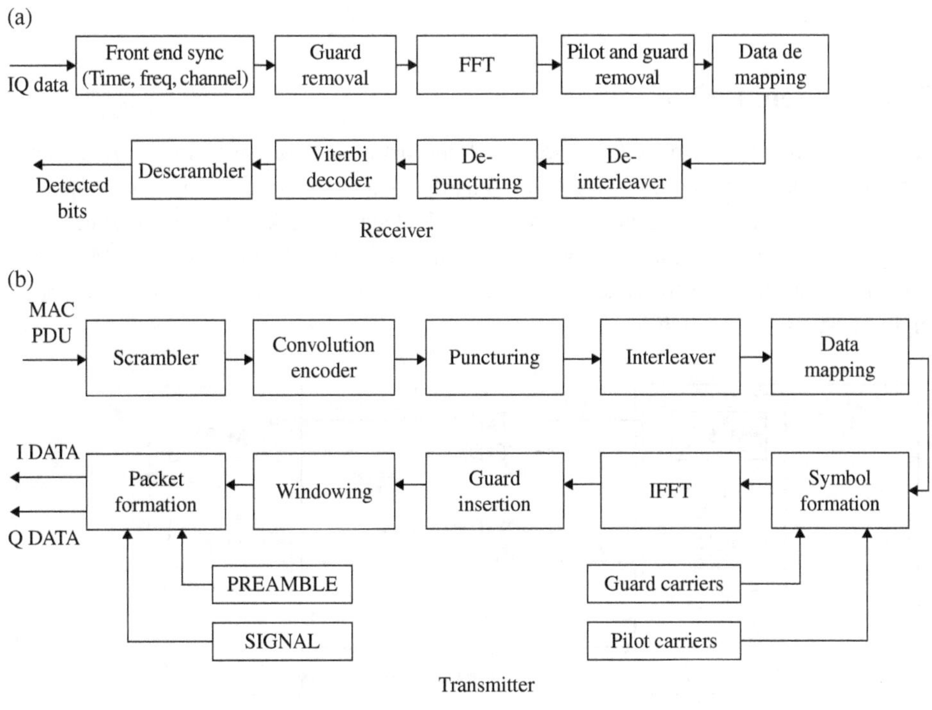

Figure 5.29 (a) WLAN receiver and (b) WLAN transmitter blocks

Operations

- Creating or joining a network. Each wireless station, both clients and access points, must be configured for operation.
- Operating mode – there are two options for the operating mode: infrastructure and ad hoc.
- Operating channel – the 802.11 extension in use (a, b, g, n …). Country regulatory agencies determine the channels available to the network.
- Network name – service set identifier (SSID) is the name of a Wi-Fi network and some networks broadcast their SSIDs to wireless devices in range.
- Scanning: devices search for available networks within the range of the scanning device. The device can be directed to search for a particular SSID or a particular channel or all channels. AP transmits beacons typically at every 100 ms.

5.19 RF Baseband (BB) Interface

A mobile phone system vendor may prefer to buy RF and BB blocks from different vendors, so a standard interface is required between the BB and RF. Most commonly, a DigRF interface is used for the RF-BB interface. The IQ bit sequence is passed to the BB from the RF during reception or passed to the RF from the BB during transmission via the DigRF interface protocol. The DIGRF uses the MIPI standard of DigRF interface version 4 or 3. Figure 5.30 shows the signals on the DigRF interface.

5.20 System Design

A mobile phone is an embedded system and not just a piece of hardware or collection of software; rather, it is a combination of both hardware and software. In today's competitive market, developing a good product is a real challenge. The key factors to success are designing a system that can work with minimum resources (like memory size, MIPS etc.) and that offers high performance in terms of execution speed, low power consumption, high robustness, and reliability.

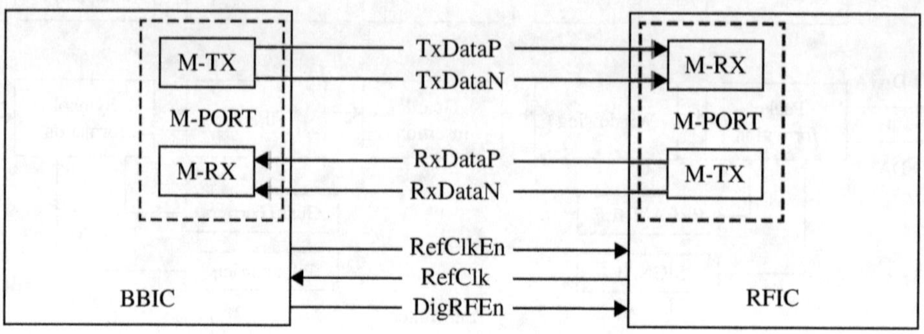

Figure 5.30 DigRF interface signals between RF and baseband

5.20.1 System Design Goal and Metrics

The primary goal of a system designer should be to design a system that must confirm size and weight limits, consume a lot less power, satisfy safety and reliability requirements, meet tight cost targets and, above all, guarantee real-time operation reacting to external events. Some requirements are functional and some are nonfunctional.

- Functional requirements. Output as a function of input. There are set of defined requirements that need to be fulfilled. In case of the modem, these requirements are clearly defined in the specifications, and conformance test cases are available for verification.
- Nonfunctional requirements. These are requirements like the time required to compute output; size, weight, power consumption, reliability, and so forth.

The mobile phone system is a complex embedded system and the design goals of an embedded system varies based on the system's application area, like:

- Real-time operation – real-time operation means that the computation of the data should be completed within a time limit. So, during the design of the system, the worst-case situation should always be taken into account. In case of reactive computation the software executes in response to external events. These events may be periodic; in this case scheduling of events to guarantee performance may be possible.
- Portability – the size and weight of the designed system plays an important role in the system design. In many cases especially in mobile environment, it is always desirable that the system should be as small as possible in size and weight.
- Power consumption. Power consumption is a vital issue for mobile devices. Care should be taken in the system design to reduce the power consumption in all possible ways. This is discussed in more detail in Chapter 8.
- Safety and reliability. The systems have risks associated with failure but the probability of failure will be reduced if the system is designed properly and tested rigorously before delivery. A safe and reliable product enhances the customer's faith and satisfaction. After taking all precautions, failure may still happen so there should be some way to recover quickly or to debug and make it work again.
- Cost. The system may be designed to meet all the challenges and satisfy all the requirements but it will not draw the customer's attention if the cost is not low enough. Design at any particular stage can't drastically reduce the cost, so it has to be done at each and every stage of the system design. A good designer has to manage all the issues to deliver a cost-effective system at the right time with right price. The costs can be divided into two types: nonrecurring engineering cost (NRE cost) – this is one time monetary cost of designing the system and unit cost – this is monetary cost of manufacturing each copy of the system.
- Time to market. This is the time required to develop a system to the point that it can be released and sold to customers.
- Flexibility. The ability to change the functionality of the system without incurring heavy NRE costs.

- Maintainability. The ability to modify the system after its initial release.
- Correctness. The system should be functioning and performing as expected.

5.20.2 *System Architecture*

A single baseband processor capable of supporting multimode technologies like 2G, 3G and 4G LTE offers the advantages of a smaller device size, reduced power consumption, and a reduced bill of materials. A combination of digital baseband along with an RF transceiver from the same company facilitates a lower integration burden to OEMs as everything has been previously proven and verified to function properly in a reference design format provided by smartphone platform provider companies.

Due to an increase in data traffic volume and support for various complex applications, the load on the application processor has also increased tremendously. So, the performance

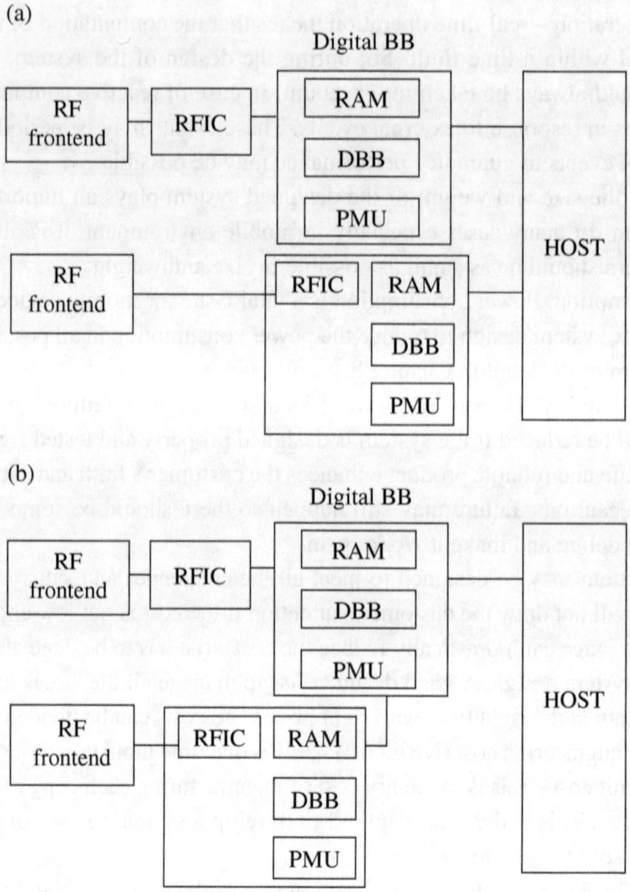

Figure 5.31 (a) Host-centric DSDA architecture. (b) Modem centric DSDA architecture

of application processors is a key consideration for any smartphone design and the application processor is becoming an integral part of any smartphone design. Several companies, such as Qualcomm, Mediatek, Intel, ST-Ericsson, Samsung, and Nvidia offer very powerful application processors with multicore technology that will be able to handle LTE requirements.

Data cards (CPE, Hotspot Routers) use similar architecture to the smartphone, except that instead of requiring a complex application processor, a simple processor will be enough. This will send the data from the modem to the host device like a laptop or PC. This common design helps the designated OEMs to reduce device costs and to expedite the device development process.

Recently, simultaneous handling of multimode 2/3/4G technologies through single / multiple baseband chipset has been imposing a lot of demands on dual SIM devices. Figure 5.31(a) shows an architecture that provides simultaneous support for 2/3G and LTE. This comes with the extra burden of requiring a larger device space and higher power consumption due to support for two simultaneously active modems (LTE/HSPA/GSM/TD-SCDMA + GSM modems). Actually there are various options available – like, dual SIM dual active (two simultaneous CS/PS connections) (DSDA), dual SIM dual call (DSDC), and dual SIM dual standby (one active and one standby or two standby modems) (DSDS) – architecture corresponds to these is shown in Figure 5.31.

Reference

[1] Coulton, P., Edwards, R., and Clemson, H. (2007) *S60 Programming: A Tutorial Guide*, John Wiley & Sons, Ltd.

Further Reading

Adams, J. E., Jr. and Pillman, B. (2013) *Digital Camera Image Formation: Introduction and Hardware*, Springer.

ARM (n.d.) *ARM Related Books*, http://www.arm.com/support/resources/arm-books/index.php (accessed May 6, 2016).

Axelson, J. (2009) *USB Complete: The Developer's Guide*, 4th edn. Lakeview Research.

Das, S. K. (2010) *Mobile Handset Design*, John Wiley & Sons, Ltd.

Eargle, J. (2011) *The Microphone Book: From Mono to Stereo to Surround – A Guide to Microphone*, 3rd edn. Focal Press.

Ericsson (n.d.) *Mobile Learning: A Practical Guide*, Leonardo da Vinci Programme of the European Commission,http://www.ericsson.com/res/thecompany/docs/programs/incorporating_mobile_learning_into_mainstream_education/book.pdf (accessed May 6, 2016).

Guo, Y., McCain, D., Cavallaro, J. R., and Takach, A. (2006) *Rapid Industrial Prototyping and SoC Design of 3G/4G Wireless Systems Using an HLS Methodology*, Nokia Networks Strategy and Technology.

Heydon, R. (2012) *Bluetooth Low Energy: The Developer's Handbook*, Prentice Hall.

Roshan, P. and Leary, J. (2004) *Wireless LAN Fundamentals*, Cisco Press.

6

UE RF Components and System Design

6.1 Introduction to RF Systems

As described in Chapter 1, the RF (radio frequency) unit is one of the most important blocks in the mobile phone. It helps to transmit or receive signals wirelessly over the air. Today, apart from cellular system modems, mobile phones house several other wireless systems such as WLAN, GNSS, broadcasting receivers, Bluetooth, FM receivers, and NFC. All these systems have RF front-end units and digital baseband units. This chapter primarily covers the RF systems for a cellular system's modem. As shown in Figure 6.1, the basic building blocks of RF system are:

- RF front-end module (FEM) – this is the front-end part of the RF unit and generally consists of an antenna, antenna switch, impedance-matching unit, and band-pass filters.
- RF transceiver unit – this consists of an analog receiver and a transmitter circuit.
- Frequency generation unit – this generates and synthesizes different frequencies required for the RF unit.

6.2 RF Front-End Module (FEM)

6.2.1 Antenna

For wireless communication, information has to be carried via air or free space. So, the air or free space medium will be used for transmitting and receiving the information. Electrical signals cannot be transmitted directly through air or free space but electromagnetic (em) waves can travel through free space and air, so an electrical signal is converted to em waves

Mobile Terminal Receiver Design: LTE and LTE-Advanced, First Edition. Sajal Kumar Das.

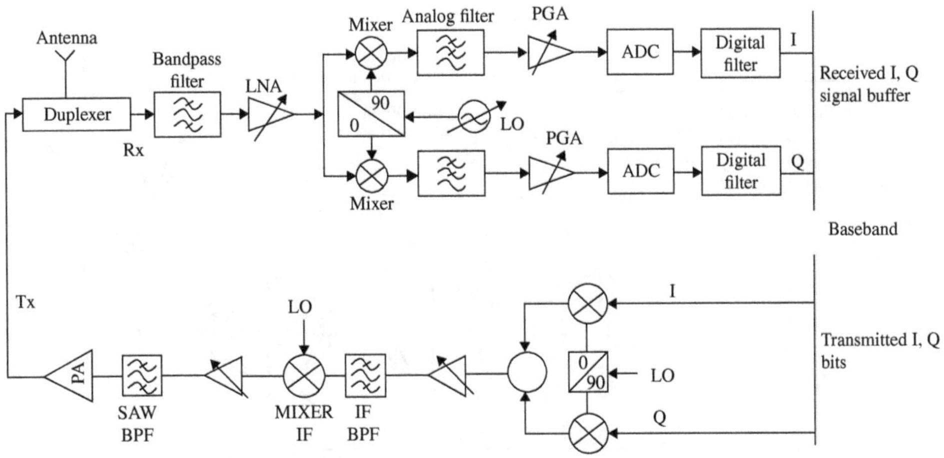

Figure 6.1 Block diagram of RF frontend unit

and used as carrier to send and receive information over the channel / medium. Antennas are transducers, which are used to convert electrical signals into electromagnetic waves for transmission and electromagnetic waves to electrical signals for reception – they convert bound circuit fields into propagating electromagnetic waves, and collect power from electromagnetic waves that they receive. An antenna acts as a bridge between the air / free-space medium and the communication radio device. Physically, an antenna is a metallic conductor; it may be a small wire, a slot in a conductor or piece of metal, or some other type of device.

6.2.1.1 Action of an Antenna

Transmission of a Signal Using an Antenna

Let us take two wires or conducting rods (two collinear conductors) and connect one end of an AC signal source to conductor A and another one to B as shown in Figure 6.2. The current in the lower conductor is 180° out of phase with the current in the top conductor:

1. When the AC signal is positive – the signal is in the positive half (Figure 6.2(a)) – conductor A of the antenna will be positive due to a lack of electrons (charge), whereas at that same moment conductor B will be negatively charged due to the accumulation of electrons. As the end points of A and B form an open circuit, a charge will accumulate there. We know that electrical flux lines always start from the positive point and end at the negative point, and form a closed loop. In a 3D space (free space) these flux lines will actually be spherical in shape. If the radius of the sphere is r, and the power of the AC signal (V * I) during the time when these flux lines were generated is P_t, then the power P_t is spread over the surface of a sphere whose radius is r, so the power density at each point in that sphere will be $P_t/4\pi r^2$. The driving current can be written as, $I = I_0e^{-j\omega_t}$ and it accelerates the charges in the antenna conductors. The electric current is defined

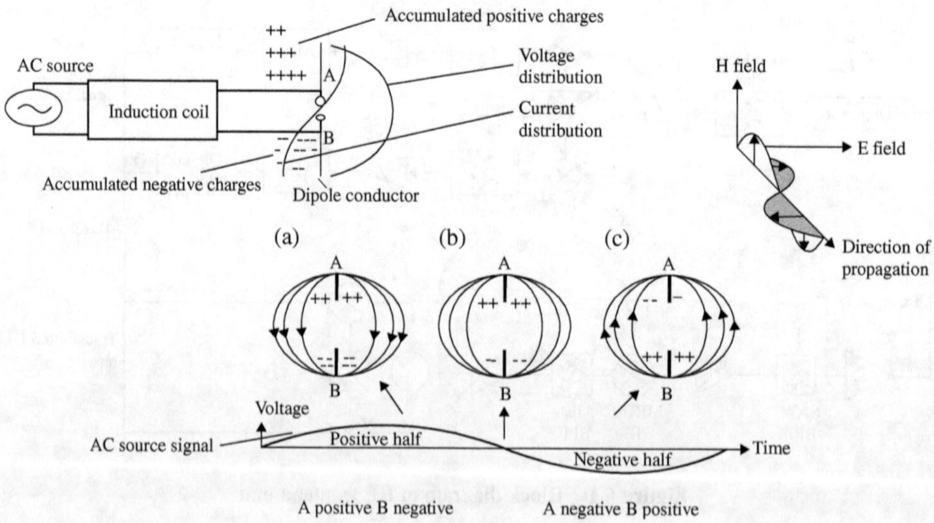

Figure 6.2 Creation of flux lines in the space with the RF input signal variation (positive to negative)

as the time derivative of electric charge (I = dq/dt). A current flowing through a conductor produces a magnetic field and voltage produces an electric field. So, lines of electric and magnetic fields will be established across the antenna conductors.

2. At the next point in time – see Figure 6.2(b) – when the applied input AC signal decreases to zero, there is no energy to sustain the closed flux loops of the magnetic and electric field lines, which were created at the previous point in time. So, these will detach from the antenna and remain self-sustained.
3. At the next point in time – see Figure 6.2(c) – when A becomes negative and B positive, the electric and magnetic flux lines will be created again, but this time the direction will be reversed, as the positive and negative points are interchanged. Next, when the RF signal goes to zero again, these flux lines again become free and self-sustained. The flux lines that were created in the first instance, when A was positive and B was negative, will have the opposite direction to the flux lines that were created when A is negative and B is positive. Thus, these flux lines will repel each other, and will move away from the antenna as shown in Figure 6.2.

Maxwell's four equations are given below:

$$
\begin{aligned}
&\nabla \cdot \mathbf{E} = \frac{\rho_v}{\varepsilon} && \text{(Gauss' law)} \\
&\nabla \cdot \mathbf{H} = 0 && \text{(Gauss' law for magnetism)} \\
&\nabla \times \mathbf{E} = -\mu \frac{\partial \mathbf{H}}{\partial t} && \text{(Faraday's law)} \\
&\nabla \times \mathbf{H} = \mathbf{J} + \varepsilon \frac{\partial \mathbf{E}}{\partial t} && \text{(Ampere's law)}
\end{aligned}
\tag{6.1}
$$

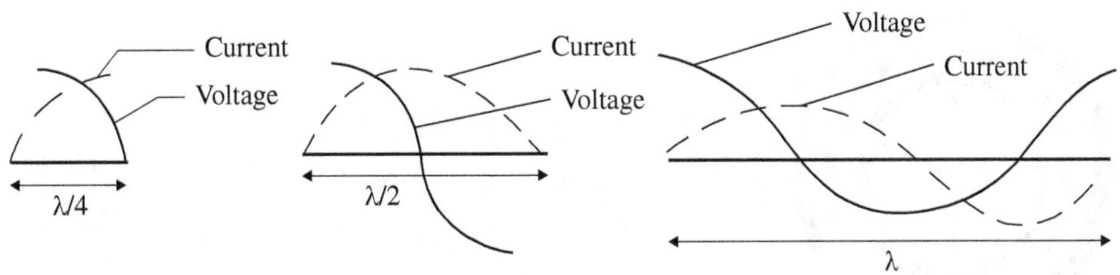

Figure 6.3 Voltage and current distribution across the antenna conductor for different lengths

The curl operator represents the spatial variation of the fields, which are coupled to the time variation. As shown in last two equations in Eq. (6.1), when the E-field travels, it is altered in space, which gives rise to a time-varying magnetic field; similarly, a time-varying magnetic field gives rise to a time-varying electric field. As shown in Figure 6.2, the em wave has its two fields oriented orthogonally, and it propagates in the direction normal to the plane defined by the perpendicular electric and magnetic fields. As explained earlier, antennas radiate spherical waves that propagate in the radial direction for a coordinate system centered on the antenna. The Poynting vector describes both the direction of propagation and the power density of the electromagnetic wave and is expressed as: $S = E \times H^*$ W/m^2, where H^* is the complex conjugate of the magnetic field phasor. The propagating wave field intensity decreases by $1/r$ away from the source, whereas power decreases by $1/r^2$.

The electric field results from the voltage changes occurring in the RF antenna that is radiating the signal, and the magnetic changes result from the current flow. When an AC signal is applied to the input of an antenna, the current and voltage waveform established in the antenna of length λ/2 is as shown in Figure 6.3. If the length of the antenna changes then the distribution of charge on the wire will also change and the current–voltage waveform will also change. In Figure 6.4, various waveforms for different antenna lengths are shown. Wherever the voltage is a maximum, then the current is a minimum as these are 90° out of phase.

Taking energy P_t, the flux lines will move away from the transmitting antenna. Now, as they move away from the antenna, the size of the sphere increases, r increases but the same power (Pt) is contained within that sphere. Thus the power density Pt $/4\pi r^2$ decreases as it travels far from the transmitting antenna.

Reception of a Signal Using an Antenna

An antenna also helps for reception of an em wave. It transforms the received em wave into an electrical signal. When the transmitted wave arrives at the receiving end, it tries to penetrate via the metallic wire (conductor) of the antenna. We know that the em wave consists of an electric field and a magnetic field and these are perpendicular to each other, and also they are perpendicular to the direction of propagation. Thus when the em wave touches the metallic antenna (from Maxwell's third equation) the magnetic field (H) will generate a

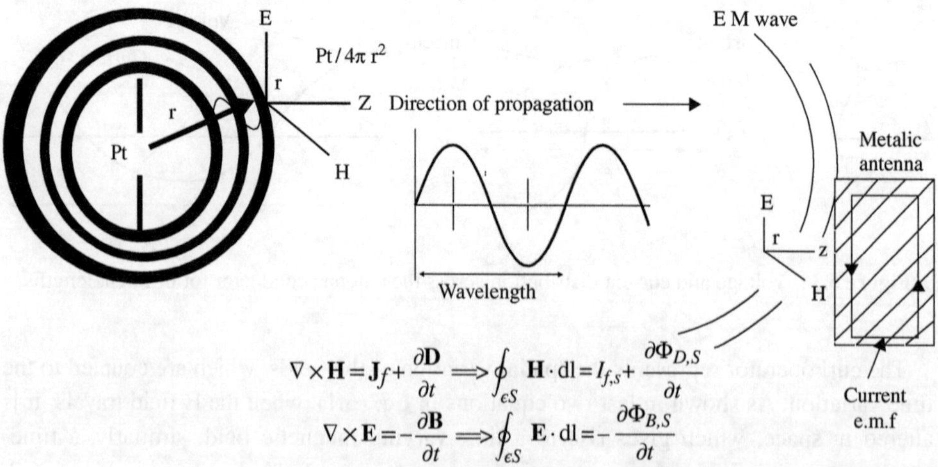

$$\nabla \times \mathbf{H} = \mathbf{J}_f + \frac{\partial \mathbf{D}}{\partial t} \Longrightarrow \oint_{\partial S} \mathbf{H} \cdot \mathrm{d}\mathbf{l} = I_{f,S} + \frac{\partial \Phi_{D,S}}{\partial t}$$

$$\nabla \times \mathbf{E} = -\frac{\partial \mathbf{B}}{\partial t} \Longrightarrow \oint_{\partial S} \mathbf{E} \cdot \mathrm{d}\mathbf{l} = -\frac{\partial \Phi_{B,S}}{\partial t}$$

Figure 6.4 Signal reception by the antenna

surface current on the metallic antenna, when it tries to penetrate via the metal (as metal is a good conductor). However, it will die down after traveling a thickness of skin depth (δ), and, the em wave will generate an electrical current in the metal body of the antenna. Similarly (from Maxwell's fourth equation), the electric field will generate an electric voltage in the antenna, as shown in Figure 6.4. This phenomenon can be experienced by placing a radio inside a closed metallic chamber and finding that it does not play, as the em wave cannot penetrate via the thick metallic wall (as the thickness is greater than skin depth). However, it can penetrate through a concrete wall (as concrete is not a good magnetic or electric conductor). For the same reason, a mobile telephone call disconnects inside an enclosed metallic lift or chamber due to the degradation of the signal strength inside the chamber.

Thus with the help of an antenna again we are able to convert the transmitted energy (transmitted using em waves) back into the electrical signal. So the antenna helps for transmitting as well as receiving the information through the air.

As the user wants to send as well as receive the information, so, ideally, the user device should have both transmitting and receiving antennas. But, in general, in a mobile device, the same antenna is used for transmission as well as reception purposes as explained in section 6.3.

6.2.1.2 Antenna Parameters

There are several critical parameters that affect an antenna's performance and can be adjusted during the design process. These include resonant frequency, impedance, directivity, gain, aperture or radiation pattern, polarization, efficiency, and bandwidth.

Resonant Frequency

Due to presence of parasitic elements, the effective length of an antenna becomes slightly larger than its physical length. The electrical length of an antenna is usually the physical length of the wire / dipole divided by its velocity factor (the ratio of the speed of wave propagation in the wire to the speed of light in a vacuum). The "resonant frequency" and "electrical resonance" is related to the electrical length of the antenna. Typically an antenna is tuned for a specific frequency and is effective for a range of frequencies (BW) usually centered on that resonant frequency. However, the other properties of the antenna (especially radiation pattern and impedance) change with frequency, so the antenna's resonant frequency may merely be close to the center frequency to satisfy those other important properties. Antennas can be made resonant on harmonic frequencies with lengths that are fractions of the target wavelength. Some antenna designs have multiple resonant frequencies, and some are relatively effective over a very broad range of frequencies. The most commonly known type of wide-band aerial is the logarithmic or log periodic, but its gain is usually much lower than that of a specific or narrower band aerial.

Polarization

The polarization of an electromagnetic wave is defined as the orientation of the electric field vector with respect to the ground. The electric field vector is perpendicular to both the direction of propagation and the magnetic field vector. The *polarization* of an antenna is the orientation of the electric field of the radio wave with respect to the Earth's surface and is determined by the physical structure of the antenna and its orientation. Generally four types of polarization are used – horizontal, vertical, circular, and elliptical. A vertical antenna will have vertical polarization and a horizontal antenna will have horizontal polarization. In circular polarization, the antenna continuously varies the electric field of the radio wave through all possible values of its orientation with regard to the Earth's surface. Circular polarizations, like elliptical ones, are classified as right-hand polarized or left-hand polarized using a "thumb in the direction of the propagation" rule. In practice, regardless of any other parameters, it is important that linearly polarized transmitter and receiver antennas should be matched – for example, horizontal should be used with horizontal and vertical with vertical. Transmitters mounted on vehicles with large motional freedom commonly use circularly polarized antennas so that there will never be a complete mismatch with signals from other sources.

Impedance

When an incoming electromagnetic wave or an outgoing RF signal travels through the different parts of the antenna system, it may encounter differences in impedance (E/H, V/I, etc.) between the input and output port. Then, depending on the impedance match, some fraction of the input energy will reflect back towards the source, forming a standing wave in the feed line. The ratio of maximum power to minimum power in the wave is called the standing wave ratio (SWR). Although a SWR of 1 : 1 is ideal, a SWR of 1.5 : 1 is also considered to be marginally acceptable in low-power applications. The complex impedance of an antenna is related to its electrical length at the wavelength in use. The impedance of an antenna can be matched to the feed line and radio by adjusting the impedance of the feed line, using the feed line as an impedance transformer. More commonly, the impedance is adjusted at the load with an antenna tuner, a balun, a matching transformer, and matching networks composed of inductors and capacitors or matching sections.

Radiation Pattern

The radiation pattern of an antenna is a graphical representation of the radiated fields or power along different directions in space. When radiation is expressed as field strength E volts / meter, it is called a field strength pattern. If the power per solid angle is plotted in 3D space, it is called a power pattern. A power pattern will be the product of electric and magnetic field patterns or proportional to the square of the field strength pattern. The radiated field pattern looks like a conical lobe. This lobe has an angular width. These are measured as null-beam width and half power-beam width. As shown in Figure 6.5, the beam width goes to 0 at point p. Now if we draw a tangent to this beam surface at both side from point p, the angular separation between these two tangents (θ) will be the measure of the null-beam width. The half power points are the points in the radiated field pattern, where the field reduces to $1/\sqrt{2}$ times its peak value (power becomes half). The angular separation between the two half power field points is called half power-beam width.

There are many types of antenna patterns: an isotropic radiation pattern is symmetric in all directions (omnidirectional); directional patterns focus towards one particular direction; and a plane pattern, which consists of the electrical field plane and magnetic field plane.

Efficiency of an Antenna

The radiation efficiency of an antenna is defined as the ratio of radiated power to the total input power:

$$\eta = P_{radiated} / P_{input}$$

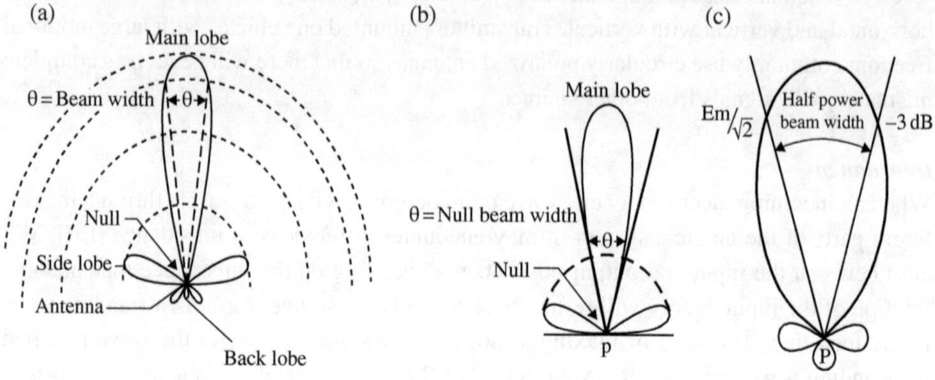

Figure 6.5 (a) Radiation pattern. (b) Null beam width. (c) Half power beam width

Directivity and Gain

The directive gain, D (θ, Φ) of an antenna is a measure of the concentration of the radiated power along a particular direction θ, Φ. If U (θ, Φ) is the radiation intensity, which is the measure of concentrated radiation, then the radiated power will be $P(rad) = \oint_S U(\theta,\phi)d\Omega$, where $d\Omega = \sin\theta d\theta d\Phi$ is an element of the solid angle Ω. The total radiated power is therefore the integral of the radiation intensity over a solid angle 4π. So, the average radiated intensity $U(avg) = P(rad)/4\pi$. An isotropic antenna radiates em power uniformly in all directions but a nonisotropic antenna concentrates power along a particular direction. If U (θ, Φ) is the radiated power intensity along the direction (θ, Φ) then the directivity D (θ, Φ) of antenna is defined as the ratio of maximum radiation intensity along this particular direction to the average radiation intensity. $D(\theta,\Phi) = U(\theta,\Phi)_{max} / U(avg) = (P(rad)/\Omega_A)/(P(rad)/4\pi) = 4\pi/\Omega_A \approx 4\pi/(HP_E * HP_H)$. Here, HP_E and HP_H are the half-power beamwidth of electric and magnetic fields respectively. The power gain is defined as the product of the directivity and efficiency of the antenna, $G(\theta,\Phi) = \eta D(\theta,\Phi)$. It accounts for the losses of the antenna. Generally, gain is expressed in dB, where $G(dB) = 10\log_{10} G$. Equivalently, we can say that the gain of an antenna is the gain over an isotropic antenna: $G(dB) = 10\log_{10}(G/Go)$.

Specification Absorption Rate (SAR)

The SAR is the measurement of the energy that has been absorbed by the human body during the transmission of the radio frequency electromagnetic field. This causes two issues: if the human body absorbs the RF energy, then we will lose some energy; secondly, it will affect the human body badly. It can be calculated by integrating or averaging over 1 g or 10 g:

$$SAR = \int \frac{\sigma(r)|E(r)|^2}{\rho(r)} dr \tag{6.2}$$

The above equation explains that the SAR is a function of the induced electrical field. The radiated energy E can be measured by volts / meter, the electrical conductivity σ can measured in Siemens / meter, and the mass density ρ can be measured in g/m^3. The unit of the SAR is W/kg. The value of SAR varies based on the design of the mobile phone and placement of the antenna in the mobile phone, so most commonly it is placed at the bottom of the phone. A lower SAR value is desirable. The SAR also affects the quality of the power by reducing the power level due to radiation absorption.

The Techniques of Reducing SAR

Reducing the em field of the antenna will reduce the value of the SAR in the human head. This can be done by RF shielding the front side of the phone – RF shielding made of ferromagnetic material can help to reduce the value of the SAR as it reduces of the surface current at the front of the mobile phone. Table 6.1 shows the level of SAR in different types of antennas.

Table 6.1 The level of SAR in different types of antennas

Type of antenna	Level of SAR
Helix	High
Slot	Low
PMA	High
PIFA	Low

6.2.1.3 Conventional Mobile Phone Antennas

Antennas are of different types like – wire antennas, aperture antennas, printed antennas, array antennas, reflector antennas, and so forth. A wire antenna is the basic type of antenna, widely used on top of the buildings, automobiles, ships, and spacecraft. These antennae are made into different shapes such as a straight wire (dipole), loop, and helix. Aperture antennas are in the form of a slot or aperture in a metal plate and are used in aircraft and spacecraft applications. Printed antennas are fabricated using a standard photolithography technique. The most common version of the printed antenna is the microstrip antenna. The shape and size of the patch determines the frequency of operation of the antenna and its performance. Due to low cost and easy integration, these types of antenna are more popular and today are most commonly used in mobile devices. In an array antenna, several antennas are placed in an array and separated from each other. They are geometrically arranged to give the required radiation characteristics. In earlier times, conventional mobile phones used either whip or helical antennas that extended from the top of the mobile handset, or else it they were contained within the upper part of the handset. Today, several different types of antenna are used in mobile phones for cellular communication and some of these are discussed below. Apart from the cellular system's antenna, mobile phones also house several other antennas for WLAN, GPS, FM, and so forth.

Dipole Antenna

One of the most commonly used antennas today in radio communications is the dipole antenna. The dipole antenna consists of two terminals or "poles," or two identical conductive elements such as metal wires or rods, which are usually bilaterally symmetrical. Radio frequency current flows into this. The driving current from the transmitter is applied, or, for receiving antennas, the output signal to the receiver is taken, between the two halves of the antenna. The length of the dipole elements is determined by the wavelength of the radio waves used. The most commonly used dipole antenna is the half-wave dipole, in which each of the two elements is 1/4 wavelength long, so the whole antenna is a half-wavelength long. The radiation pattern of a vertical dipole is omnidirectional; it radiates equal power in all azimuthal directions perpendicular to the axis of the antenna. Several different variations of the dipole are also used, such as the folded dipole, the short dipole, the cage dipole, the bow tie, and the batwing antenna.

Planar Inverted F antennas (PIFA)

For many years, planar inverted F antennas (PIFAs) were used in mobile phone handsets. Figure 6.6(a) shows a simple single-band PIFA. This has a low profile resonant element of about ¼ wavelength long. During the operation, currents oscillate in the inverted L section. The impedance of this type of antenna is determined by the position, where the feed is connected along the L section.

Helical Antenna

A helical antenna consists of a conducting wire wound in the form of a helix as shown in the Figure 6.6(b). Generally, helical antennas are mounted over a ground plane. It radiates when the circumference of the helix is of the order of at least one wavelength. The radiation along the axis of the helix is found to be the strongest. Generally, this type of antenna is directional. This is a simple antenna type and offers high-gain and broad band frequency characteristics. The radiation from a helical antenna is circularly polarized (clockwise or counterclockwise). Generally, a helical antenna has two operating modes: normal and axial. (i) In normal mode (broadside), the dimensions of the helix are small compared to the wavelength. The far field radiation pattern is similar to an electrically short dipole or monopole. These antennas tend to be inefficient radiators and are typically used for mobile communications where reduced size is a critical factor. (ii) In axial mode (end fire), the helix dimensions are at or above the wavelength of operation. This is like waveguide antennas, and produces true circular polarized waves. These antennas are best suited for space communication, where the orientation of the sender and receiver cannot be controlled easily, or where the polarization of the signal may change, but, due to large antenna size, these are not very popular in mobile handsets. Terminal impedance in axial mode ranges between 100 and 200 Ω. The resistive part is approximated by $R \approx 140$ (C/λ). Where R is resistance in ohms, C is the circumference of the helix, and λ is the wavelength. Impedance is matched with the cable C by a short strip-line section between the helix and the cable termination. The maximum directive gain can be expressed as $D_0 \approx 15\,N\,(C^2S/\lambda^3)$. The approximate bandwidth for a helical antenna is $0.75 * \lambda$ to $1.3 * \lambda$.

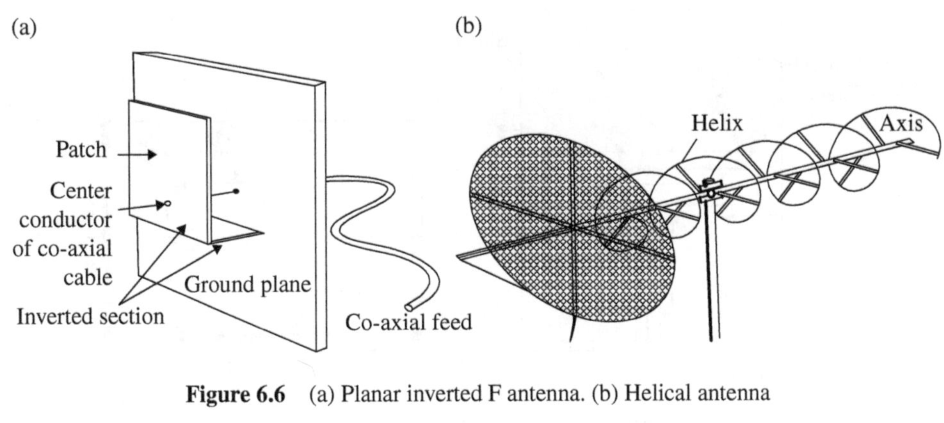

Figure 6.6 (a) Planar inverted F antenna. (b) Helical antenna

Whip Antenna

A whip antenna is a single-element antenna that can be used with an unbalanced feed line such as coaxial cable, or attached directly to a wireless transceiver – see Figure 6.7(a). This antenna is mostly used on handheld two-way radios and cell phones. These are usually attached to a vehicle and are designed to be flexible, so that they do not break when struck. Its name is derived from their whiplike motion when perturbed. The whip resembles a ground-plane antenna without the radial system. Generally, whip antennas are a short, flexible "rubber-duck" type, whereas in some cases, long, flexible, stainless-steel material is also used. The whip antenna's electrical and mechanical design is very simple and it is very easy to install but these are not very efficient as most whip antennas are operated with a poor electrical ground system. The whip antenna is stiff but flexible, wire mounted, and generally vertical orientation is used, with one end adjacent to a ground plane. This can also be called a half-dipole antenna, and this creates a toroidal radiation pattern, where the axis of the toroid centers about the whip. The length of the whip determines its wavelength, although it may be shortened with a loading coil anywhere along the antenna. Whips are generally a fraction of their actual operating wavelength. This type of antenna may cause uncertainty about the biological safety.

Slot Antenna

Many planar inverted F antenna (PIFA) and monopole antenna designs have been developed for use in mobile phones; however, these designs cannot guarantee adequate radiation performance for general mobile phone use, especially when antenna reception is blocked by conductive housing in the ground plane. Conductive blocking structures and other components adjoining an antenna significantly degrade radiation performance as these components serve as em field scatterers and create unwanted parasitic inductance and capacitance. To avoid this problem, slim and compact device configurations have begun to incorporate antennas that are more compatible with metal enclosures. Most such designs feature a loop,

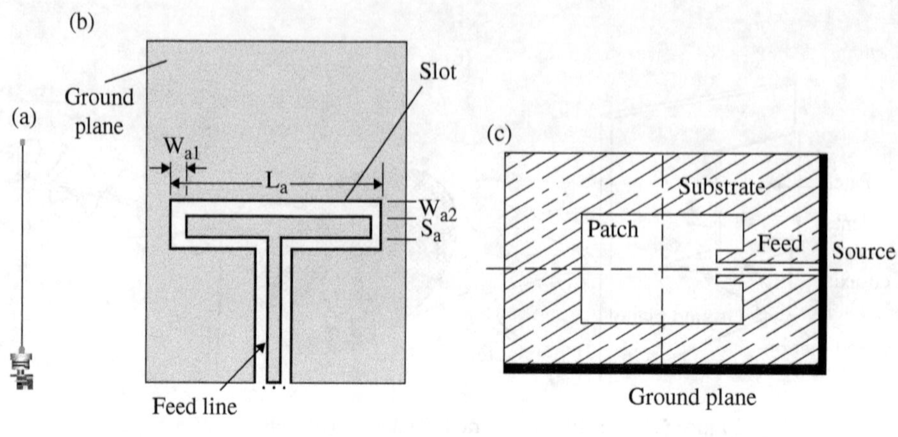

Figure 6.7 (a) Whip antenna. (b) Slot antenna. (c) Microstrip antenna

slot, or an open-slit antenna that utilizes the metal enclosure as a radiator component – see Figure 6.7(b). Generally, slot antennas are half-wavelength antennas that use the first resonant frequency. The physical length of a slot antenna will be determined by its operating frequency and by the materials surrounding the slot. Slot antenna radiation performance depends on the shape of the slot, the shape and size of the ground plane, and on the matching elements and supporting materials.

Microstrip Patch Antennas

Microstrip patch antennas are commonly used in mobile communications terminals due to their many attractive features, such as simple structure, low production cost, light weight, smaller size and robustness. These antennas are planar resonant cavities that leak from their edges and radiate as shown in Figure 6.7(c). Microstrip consists of a metal strip on a dielectric substrate (ε_r) covered by a ground plane on the other side. Unlike stripline, the single ground plane shields the circuit on only one side and on other side air is there. In this inhomogeneous type of structure a pure transverse electromagnetic (TEM) mode cannot exist, as it is not possible to satisfy the boundary conditions for a TEM mode at the surface of the dielectric and air. The em field lines in microstrip are not contained entirely inside the substrate. An impedance match occurs when a patch resonates as a resonant cavity and, when it is matched, the antenna achieves peak efficiency. A normal transmission line radiates little power because the fringing fields are matched by nearby counteracting fields. Power radiates from open circuits and from discontinuities such as corners, but the amount depends on the radiation conductance load to the line relative to the patches. Without proper matching, a small amount of power radiates to space.

A dielectric slab on a ground plane will support transverse magnetic (TM) modes when thin and transverse electric (TE) modes when thick. The TM mode is polarized normal to the slab surface, whereas the TE mode is polarized parallel to the slab surface. Today's standard mobile terminals operate on many frequency bands, e.g. GSM850/900/1800/1900, so multiband antenna elements are required. This is easier to achieve using microstrip antenna; this type of antenna can also be fabricated easily by a photolithographic process and is easily integrated with other passive and active microwave devices. Table 6.2 provides different mobile phone models and their antenna types.

Table 6.2 Mobile phone models and their antenna types

Model	Antenna types	Frequency band
Apple iPhone (2G)	Planar monopole	GSM850/900/1800/1900
Apple iPhone (3G)	Planar monopole	GSM850/900/1800/1900 + 3G
Apple iPhone 4	Planar monopole	GSM850/900/1800/1900 + 3G
Blackberry 8100	Planar monopole	GSM850/900/1800/1900
Motorola E398	PIFA	GSM850/900/1800/1900
Motorola L2000	Helix	GSM850/900/1800/1900
Nokia 6108	PIFA	GSM850/900/1800/1900
Nokia 5500	PIFA	GSM850/900/1800/1900

6.2.1.4 Antenna Tuner

Ideally an antenna can be considered as a reactive load having finite resistance and reactance (impedance). From network theory it is known that if the terminated impedance is matched with the input port, only then will the maximum power will be transferred to the terminating load, otherwise a significant fraction of it will be reflected back to the source port. Basically, the antenna is connected to a communication device port, so the impedance should be matched to transfer maximum power; similarly, on the other side, the antenna is also connected to the air / free space, so the impedance should be matched on that side to transfer maximum power. An antenna tuner unit (ATU) matches the antenna impedance with a fixed impedance of the communication device (typically 50 Ω for modern transceivers) as shown in Figure 6.9(b). Coupling through an ATU allows the use of one antenna on a broad range of frequencies. An antenna tuning system can be an open or closed loop. In open-loop antenna tuning, the matching network element is fine-tuned to optimize the antenna performance at different frequencies, modulation schemes and modes of operation (hands free, slide open, closed, and so forth). This configuration is stored in a look-up table in the nonvolatile memory of the handset at the time of production. Based on the information provided by the higher layer software, the tuning algorithm selects the appropriate setting for the matching network. In case of a closed loop the antenna impedance is adaptively tuned using a feedback circuit element.

6.2.2 Baluns

A balun is used to transform a signal between balanced and unbalanced modes. An unbalanced signal is referenced to a ground plane, as in a coaxial cable or microstrip. A balanced signal is carried on two lines and is not referenced to a ground plane. Each line can be considered as carrying an identical signal but with a 180° phase difference.

6.2.2.1 Tx–Rx Path Separation Block

Generally the same antenna is used for transmission and reception purposes. So we have to have a mechanism to multiplex the same antenna between transmit and receive path. There are several techniques available for tx-rx separation based on the duplexing technique used:

- Tx–Rx switch. Here the same antenna is time switched between the Rx and Tx paths – see Figure 6.8(a). Systems where the Tx and Rx path signals are not present simultaneously (for example, a half-duplex system like a GSM system and in TDD systems), then a Tx–Rx switch can be used. Diodes can be used as switching elements, and switching is controlled by the microprocessor to connect the Tx or Rx path with the antenna.
- Diplexer. If the Tx and Rx frequency bands are different then they could be separated by filters and then connected to the Rx or Tx path – Figure 6.8(b). This type of filter based technique for Tx and Rx path separation is known as a diplexer. It works for systems with separated Rx and Tx frequency (FDD systems).

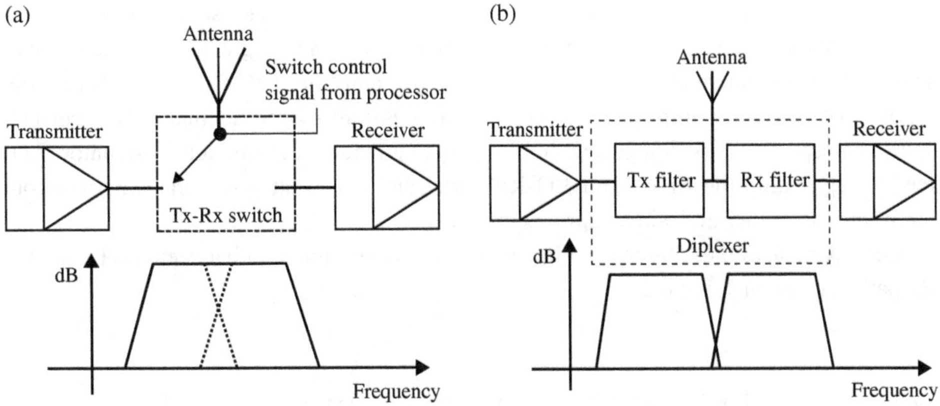

Figure 6.8 Antenna multiplexing using (a) Tx-Rx Switch. (b) Diplexer

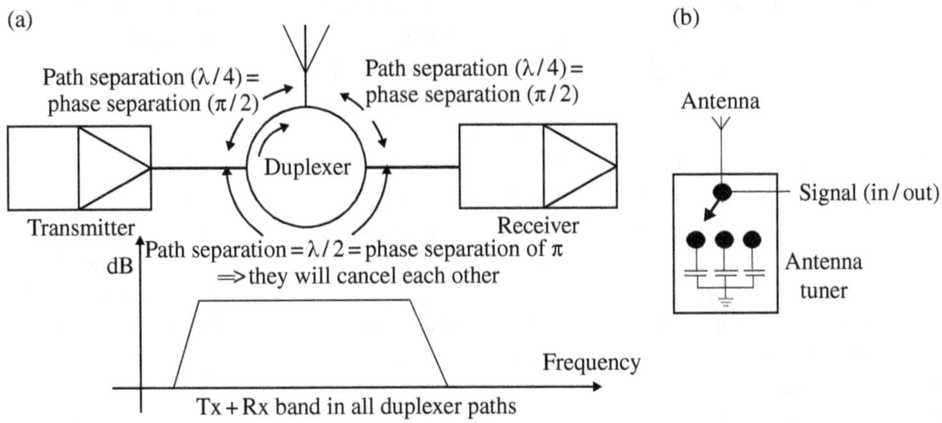

Figure 6.9 (a) Duplexer. (b) Antenna tuner

- **Duplexer** – A duplexer is an electronic device that isolates the receiver from the transmitter while permitting them to share a common antenna. The Tx circuit, Rx circuit, and antenna are connected to the three-port duplexer – see Figure 6.9(a). Tx and Rx are separated by a path difference of $\lambda/2$ = phase difference of π, so, they are opposite in phase (positive and negative) and will cancel each other out. Thus the Tx port is isolated from the Rx port, but the signal from Tx and Rx will arrive at the antenna at same phase ($\pi/2$), as the path difference between Tx or Rx with an antenna port is $\lambda/4$. Electrically a duplexer is a device that uses sharply tuned resonant circuits to isolate the transmitter circuit from the receiver circuit. This allows both of them to use the same antenna at the same time without the transmitter RF frying the receiver circuit. The separation or isolation between the transmit and receive signal paths is mandatory in order to avoid any destruction of the receiver when the Tx signal is injected, or at least to avoid any degradation of the receiver sensitivity due to the frequency proximity of the high-power signal from the transmitter block.

In the cellular systems band, the Tx-Rx separator's jobs are: (i) to isolate the transmitted signal from the received signal in the receive band to avoid any degradation of the receiver sensitivity; (ii) to attenuate the transmit path's power amplifier (PA) output signal to avoid driving the low-noise amplifier (LNA) into compression; (iii) to attenuate the receiver's spurious responses (first image and others); (iv) to attenuate the first local oscillator (LO) feedthrough using the first mixer LO-RF ports; and (v) to attenuate transmitter output harmonics and other undesired spurious products.

A comparison of these three techniques for antenna multiplexing between the Tx and the Rx path is given in Table 6.3.

Table 6.3 Comparison between different antenna multiplexing techniques

Duplexer	Diplexer	Tx-Rx switch
This allows a transmitter and receiver to use the same antenna with same or very near tx, rx frequencies	This allows tx and rx path to share one common communications channel (when tx, rx are separated by frequency)	This helps to switch the antenna between tx and rx path
This is a passive device, so no power supply is needed	This is a passive device, so no power supply is needed	Power supply may be required
Power-handling capability is less	Power-handling capability is medium	Power-handling capability is good but it is switch dependent
High-isolation, low-insertion loss is critical for its operation (typical isolation: 20 dB)	Isolation is dependent on the filter performance	Isolation and insertion loss is not very critical
Size can be smaller than diplexer	Filters are normally bulky	This is space-saving design
Permanent VSWR matching	Lowest third intermodulation products	Less problem
As isolation is dependent on wavelength, so it offers narrower frequency bands	Narrower frequency band (as filters are designed for particular frequency band)	Wider frequency bands
Tx and Rx can be of same frequency, and they can be transmitted simultaneously, can be used for both TDD and FDD systems; good for multimode receiver	Tx and Rx should be different frequencies, cannot be used for a FDD system	Tx and Rx should not be operational at the same instant; can be used for a TDD system
Most commonly used in radar systems	Most commonly used in WCDMA FDD, OFDMA FDD based cellular systems	Most commonly used in GSM and other TDMA based systems

Band-Pass Filter

This is used to extract the desired band of signal from the entire band of the received signal. Whatever em waves impinge on the antenna out of those some (based on their wavelength in relation to the antenna length) will be converted into RF electrical signals. In the reception path there will be many such RF signals with different frequencies (wavelength), which will be mixed up and appear in the receiver circuit. Of these, we need to take only the desired frequency band by using an appropriate band-pass filter before amplification.

Available Analog Filtering Technologies: Advantages and Tradeoffs

The most commonly used RF filtering technologies include LC filters, surface acoustic wave (SAW) filters, ceramic filters, bulk acoustic wave (BAW) filters, and low temperature cofired ceramic (LTCC) filters. Although LC filters can be used at high frequencies and can be integrated inside SoC but major bottlenecks are their larger size requirement and limited quality factor (Q). Ceramic filters offer low insertion loss (IL about 1.5–2.5 dB), high out-of-band rejection (>30 dB) and low cost but a drawback is that its larger size penalizes the integration. SAW filters are smaller than LC and ceramic filters but have limitations in the frequency domain (up to 3 GHz) and their maximum output power rating. For these types of filters the typical IL varies between 2.5 and 3 dB and out-of-band rejection ~30 dB. The SAW filters are not compatible with silicon integration. However, used as off-chip components. LTCC filters are a multilayer technology that offers integration of high Q passive components along with low IL, high maximal operation frequency and acceptable out-of-band rejection. LTCC filters are smaller than LC and ceramic filters. BAW filters use Film Bulk Acoustic Resonators (FBAR), which are characterized by a high-quality factor Q. These have low IL (1.5–2.5 dB), significant out-of-band rejection (≈40 dB) and high maximal operation frequency (up to 15 GHz).

6.2.2.2 Low-Noise Amplifiers

The amplitude of the received signal from the antenna is very feeble. So, we need to boost this signal without adding extra noise signal to it. A low-noise amplifier (LNA) is placed in the receiver circuit, which will amplify the received signal. In a mobile receiver, two LNA topologies are used in the RF circuit: (i) wideband LNAs that cover all the frequency bands of interest; (ii) tunable narrowband LNAs that are tunable over a wide range of frequency bands. The parameters that affect LNA performance at the device and board design levels are captured below:

- Device-level variables: feedback, transistor geometry, process technology, package parasitics, stability, biasing, and power dissipation.
- Board-level variables: temperature, input matching, layout and grounding, em shielding, supply decoupling, output matching.

Table 6.4 LNAs made from different materials

Year	Type	Node (mm)	Gain (dB)	NF (dB)	IIP3 (dBm)	Power (mW)	Supply (V)	Area (mm²)	BW (GHz)
1991	GaAs	300	11.5	2.5	9	13.8	3	3	0.2
1992	MESFET	1000	19.6	2.2	6	10	5	8	0.1
1995	GaAs HBT	300	11.1	1.9	11	4	2	0.861	2
2013	CMOS	130	15.4	1.74	4.09	5.16	0.6	0.691	0.9
2014	CMOS	180	10.8	5.5	−6.4	6.4	1.1	0.97	7.5

The primary parameters that influences the performance of LNAs are the noise figure (NF), gain, and linearity. Noise is mainly due to thermal and other sources, with typical noise figures in the 0.5 to 1.5 dB range. Typical gain is between 10 and 20 dB for a single-stage design. Some designs use cascaded amplifiers with a low-gain, low-NF stage followed by a higher gain stage, which may have higher NF.

Earlier, gallium arsenide was mainly used for LNA, but this is expensive. Today, due to the attractive scaling of complementary metal oxide semiconductor (CMOS) technology, low standby power requirements, low cost, and the fast development of technology, CMOS transistors are becoming very popular for both analog and RF circuits. This has created an excellent opportunity to integrate analog, RF, and digital circuits on the same die, which makes CMOS an excellent technology for future system-on-chip implementations. Essentially there are three silicon technologies for the integration of RF circuits – CMOS (bulk CMOS), BiCMOS where both MOS and bipolar transistors are available, and the third one is silicon on insulator (SOI) where the MOS transistors are built on an insulator (usually sapphire or an oxide layer). Table 6.4 shows LNAs made from different materials, and their parameters.

Noise Figure

The NF is the ratio of the signal to noise at the input to the signal to noise at the output of a device. This indicates how much extra noise has been added to the signal by the device. The NF is represented as

$$\mathrm{F} = \frac{\mathrm{SNR}(\text{at input port})}{\mathrm{SNR}(\text{at out put port})} = (\mathrm{S/N})_{\mathrm{in}} / (\mathrm{S/N})_{\mathrm{out}} \tag{6.3}$$

The NF is a measure of degradation of the signal-to-noise ratio (SNR), caused by the components in the device circuit. It is always greater than 1, and the lower the NF the better the device. When T is the room temperature represented by T_o (290 K) and the noise temperature is T_e, then the factor $(1+T_e/T_o)$ is the of an amplifier.

The NF is the decibel equivalent of the noise factor. It is a ratio of power so it is expressed in dB. The minimum NF value is 1 dB. $F=10^{NF/10}$; $NF=10\log(F)$. If several devices are cascaded, the total noise factor can be found with Friis formula:

$$F = F_1 + \frac{F_2 - 1}{G_1} + \frac{F_3 - 1}{G_1 G_2} + \frac{F_4 - 1}{G_1 G_2 G_3} + \cdots + \frac{F_n - 1}{G_1 G_2 G_3 \cdots G_{n-1}} \tag{6.4}$$

where F_n is the NF for the nth device and G_n is the power gain (numerical, not in dB) of the nth device. This indicates that the NF of the first block should be minimal to keep the noise under control. That is why, in the receiver circuit, at the first stage, one LNA is placed to boost the signal strength without increasing the overall noise level.

Gain

The RF and microwave circuits are optimized for power gain instead of voltage or current gain. The unit of power used to specify absolute power level is the dBm, or decibels referenced to 1 mW. Gain is expressed as:

$$G = \frac{P_{out}}{P_{in}} = \frac{V_{out}^2 / R_L}{V_{in}^2 / 4R_s} = 4\frac{R_s}{R_L}\frac{V_{out}^2}{V_{in}^2} \tag{6.5}$$

Power gain in dB: dB = 10 log $_{10}$ (P_r/P_t), log ratio of two signal power levels.

This is named after Alexander Graham Bell. It can also be expressed in terms of voltages: 20 $\log_{10}(V_r/V_t)$, as $P=(V^2/R)$. dBm (dBmW) is power relative to 1 mW = 10 $\log_{10}$ (P_r in W/1 mW) = 10 $\log_{10}$ (P_r in W/10^{-3} W) = 30.$\log_{10}$(P_r in W). So, W = $10^{\text{dB mw/10}} \times 10^{-3}$ and 0 dBm is 1 mW. dB μW = 10 log10(P in Watt / 1 μW) = 10 $\log_{10}$(P_r in W / 10^{-6} W) = 60 $\log_{10}$(P_r).

6.2.3 Mixers

Mixers perform the mixing operation by multiplying the two input signals. This translates the signals from one frequency band into another, by mixing frequencies. It allows the conversion of signals between a high frequency (the RF frequency) and a lower intermediate frequency (IF) or baseband. The output, intermediate frequency (IF), is the product of the two input signals coming from RF antenna and LO, and it contains the sum and difference of the two input frequencies. Of these two output frequencies, in receivers, the lower frequency component is usually the desired one and can be obtained by lowpass filtering the mixer output signal.

As explained above, the nonlinear behavior of mixing devices (diodes, field effect transistors (FETs) and bipolar transistors) is used to realize the mixing function. If the diode is excited by two sinusoids, Cos (ω_1t) and Cos (ω_2t) the current through the diode can be expressed as given in Eq. 6.6:

$$I = a_1\left(Cos\left(\omega_1 t\right) + Cos\left(\omega_2 t\right)\right) + a_2\left(Cos\left(\omega_1 t\right) + Cos\left(\omega_2 t\right)\right)^2 + \ldots \tag{6.6}$$

When expanded, this contains the term 2a_2 Cos (ω_1t) Cos (ω_2t). It can be expanded as shown below:

$$2Cos\left(\omega_1 t\right)Cos\left(\omega_2 t\right) = Cos\left(\left(\omega_1 - \omega_2\right)t\right) + Cos\left(\left(\omega_1 + \omega_2\right)t\right)$$

It is either the sum or difference term that is the desired output of a mixer. This means that if excited correctly they should be able to produce a strong mixing product. So, the basic mixer design entails injecting the signals to be mixed and extracting the desired mixing product whilst maximizing the efficiency of the conversion.

In a mobile receiver, local oscillators generate an RF signal locally and mix the incoming and locally generated LO signal in order to down convert the incoming RF signal. Suppose the received RF signal is $A_r \sin\omega_r t$ and the local oscillator signal is $A_o \sin\omega_o t$. If these are mixed (that is, multiplied) the resultant signal will be $A_r \sin\omega_r t.\ A_o \sin\omega_o t = (A_r A_o/2)\{\cos 2\pi(f_o - f_r).t + \cos 2\pi(f_o + f_r)t\}$. That means it generates two frequencies $(f_o - f_r)$ and $(f_o + f_r)$. This signal is passed through a channel select filter, which will filter out the frequency $(f_o + f_r)$, so that only $(f_o - f_r)$, known as the intermediate frequency (IF) will be passed forward. In some receivers, two or more such IF stages are used. Thus the RF signal is downconverted in several steps and finally it arrives at the last stage, where the $(f_o - f_r)$ becomes equal to the baseband signal frequency, $f_{baseband}$. This is then sampled at a minimum rate of $2 * f_{baseband}$ = Nyquist rate, to recover the baseband data signal, and then digitally demodulated. Similarly, the mixer is used for frequency upconversion.

6.2.3.1 Different Types of Mixer

Generally, two types of mixers are implemented – single balanced mixer (SBM) and double balanced mixer (DBM) as shown in Figure 6.10(a) and (b). The term balanced mixer is used to imply that neither of the input terms will appear at the mixer output. However, in practice, suppression of these input components is never perfect in an analog mixer circuit.

Mixers can be classified into two broad categories: passive or active. The most commonly available and used are passive diode mixers. Active mixers, on the other hand, involve transistors and the most popular ones are built from the basic Gilbert cell structure.

A balanced mixer can be implemented using a transformer coupled diode arrangement, or by using an active transistor based design. Both types of mixer produce signals at odd harmonics of the carrier frequency, particularly the diode ring mixer. In most instances, these can be easily filtered out. One disadvantage of balanced designs is that they require a higher LO drive level.

The SBM performs multiplicative mixing of RF and LO signals which are applied to different ports. Either the LO drive or the RF signal is balanced (applied in antiphase), adding destructively at the IF port of the mixer and providing inherent rejection. This is more

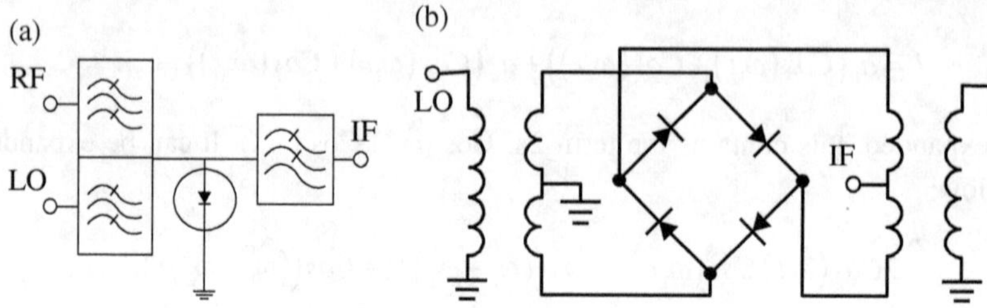

Figure 6.10 (a) Single-ended mixer. (b) Double-balanced mixer

commonly seen in two diode mixer configurations, a balanced transformer drives the diodes out of phase for the LO and in phase for signals present at the RF port.

Adding two more diodes and another transformer to the SBM results in a double balanced mixer (DBM), as shown in Figure 6.10(b). It normally uses four diodes in a ring or star configuration with both the LO and RF being balanced and all ports of the mixer are inherently isolated from each other. Its frequency response is largely determined by the frequency response of its transformers. As second-order harmonics are the most difficult to suppress, so double balanced mixers are the favored solution.

Apart from diodes, mixers are also implemented using FET. FETs can be used in mixers in both active and passive modes. They have the advantage of providing the possibility of conversion gain rather than loss and can also have lower noise figures than passive designs.

Mixer Performance

The mixer performance is dependent on parameters like conversion loss, isolation, dynamic range, DC offset, DC polarity, two-tone third-order intermodulation distortion, and intercept point. Although a mixer works by means of amplitude-nonlinear behavior, we generally want it to act as a linear frequency shifter. The degree to which the frequency-shifted signal is attenuated or amplified is an important parameter in mixer design. A mixer also contributes noise to the output frequency shifted signals. The degree to which a mixer's noise degrades the SNR of the signals is evaluated in terms of noise factor and noise figure. The load presented by a mixer's ports to the outside world can be of critical importance to a designer for VSWR matching. Isolation between ports plays a major role in reducing DC offset in a mixer. Generally, the mixer performance is measured using the following parameters:

- Conversion loss. This is the ratio of the wanted output signal level to the input, normally expressed in dB.
- Noise figure. This was discussed in the previous section.
- Compression. As shown in Figure 6.11, for small input signal levels, each dB increase in signal level results in a dB increase in the output signal level. As the input signal level continues to increase, the conversion loss of the mixer will eventually start to increase. The 1 dB compression point is the input signal level at which the conversion loss has increased by 1 dB. So, the 1 dB compression point (CP1) is the point where the output power of the fundamental crosses the line that represents the output power extrapolated from small-signal conditions minus 1 dB. Mixers should be used "backed off" from the 1 dB compression point as, in addition to the distortion of the wanted signal, operation at or close to it would give rise to significant increases in the level of spurious outputs.
- Third-order intercept (OIP3). A third-order intercept is used to characterize the linearity of a two-port, RF, circuit-like mixer. It is defined as the point at which the third order intermodulation product becomes equal to the ideal linear, uncompressed output. So, the third-order intercept point (IP3) is the point where the third-order term, as extrapolated from small-signal conditions, crosses the extrapolated power of the fundamental. Both

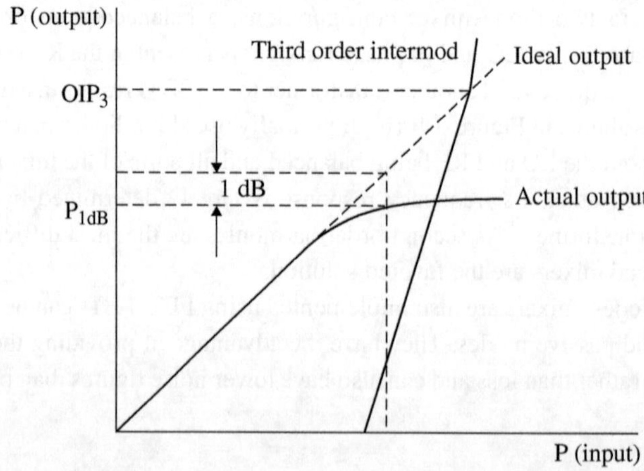

Figure 6.11 Typical pin-pout characteristics of mixer circuit

CP1 and IP3 are illustrated in Figure 6.11. Mixer distortion limits the sensitivity of a receiver, if there is a large interfering signal within the bandwidth of the RF input filter. Intermodulation distortion occurs when signals at frequencies f1 and f2 mix together to form the response at 2f1–f2 and 2f2–f1. If f1 and f2 are close enough in frequency, then the intermodulation products 2f1–f2 and 2f2–f1 will be in the band and so these will interfere with the reception of the input signal. Distortion of the output signal occurs because several of the odd-order intermodulation tones fall within the bandwidth of the circuit. Intermodulation distortion is typically measured in the form of an intercept point. As shown in Figure 6.11, one can determine the third-order intercept point (IP3) by plotting the power of the fundamental and the third-order intermodulation product versus the input power. Both input and output power should be plotted in some form of dB. Extrapolate both curves from low power level and identify where they cross – that is the intercept point.

- Linearity. The linearity of a mixer refers to its signal-level handling ability. Higher linearity produces lower distortion in output signal. The dynamic range of any RF / wireless system can be defined as the difference between the 1 dB compression point and the minimum discernible signal.
- Spur (spurious product). Spur refers to any unwanted mixing product that generates other unwanted frequencies from the higher order of terms of equation E1. These spur frequencies increase the interference level in the system and degrade the receiver performance.

Difference between Mixer and Amplitude Modulator

The mixer and AM modulator work in the same fashion and generate three output signal frequencies. The only difference is that in the mixer the two different signals are multiplied (A_r Sin $w_r t$. A_o Sin $w_o t$), whereas in the AM modulator the amplitude of the RF carrier signal is varied according to the input signal $v(t) = A(1 + m.\ \sin w_m t)\ \sin w_c t$.

6.3 RF Downconversion

The mixer circuit converts the RF signal to low frequency baseband signal, so that it can be sampled by the ADC. In the previous section, we have discussed about the mixer block of a radio receiver. Now, the conversion from the RF frequency to baseband frequency can be done in one or several stages based on the receiver downconversion technique.

6.3.1 Different Types of RF Downconversion Techniques

Typically, radio communication systems operate with carrier frequencies at many hundreds of MHz to several GHz. If we want to sample the received signal at the antenna itself (this will help to bypass RF hardware block and will help to process everything in digital domain, which is particularly desirable for software-defined radio) then the minimum sampling frequency requirement will be 2*fs (~2 * 1 GHz), and the sampled data will be too huge to handle by any present-day digital signal processor (DSP). Directly converting the antenna signals to digital form in an integrated ADC would also require prohibitively large sensitivity, selectivity, linearity, and very high conversion speed. As of today, an analog-to-digital converter does not exist that can offer this service. For that reason the received RF signals need to be converted to lower frequencies (baseband frequencies) for signal-processing steps like channel selection, amplification, and detection. This conversion is accomplished by a mixing process, producing a downconverted (in receiver block) and an upconverted (used in transmitter block) component.

Now, based on the mixing of the local oscillator (LO) frequency with the desired incoming RF frequency, there are several downconversion techniques. We can classify these into two broad categories – heterodyne and homodyne receivers. (i) In heterodyne receivers, the LO frequency and desired RF frequency are set to be different – an example of such architecture is the super heterodyne, low intermediate frequency (IF), wide IF receiver. It uses several intermediate stages to convert the RF frequency to baseband frequency. (ii) In the case of the homodyne (same mixing) receiver, the LO frequency and the desired RF frequency are set to be the same, so the IF is 0.

Prior to the selection of optimum receiver architecture, here different types of RF downconversion receiver architectures are reviewed and compared.

6.3.1.1 Heterodyne Receivers

Conventional radio receivers use the so called heterodyne architecture (hetero = different; dyne = mix). This architecture, translates the desired RF frequency to one or more intermediate frequencies before demodulation.

What is heterodyning? "Heterodyne" means a mix of two different frequencies together (one incoming signal frequency from the antenna and other locally generated from the local oscillator), to produce a beat frequency, namely the difference between the two and sum of two. For example, $A_r \mathrm{Sin}\,\omega_r t\, A_o \mathrm{Sin}\,\omega_o t = (A_r A_o/2)(\mathrm{Cos}2\pi(f_o - f_r)t + \mathrm{Cos}2\pi(f_o + f_r)t)$.

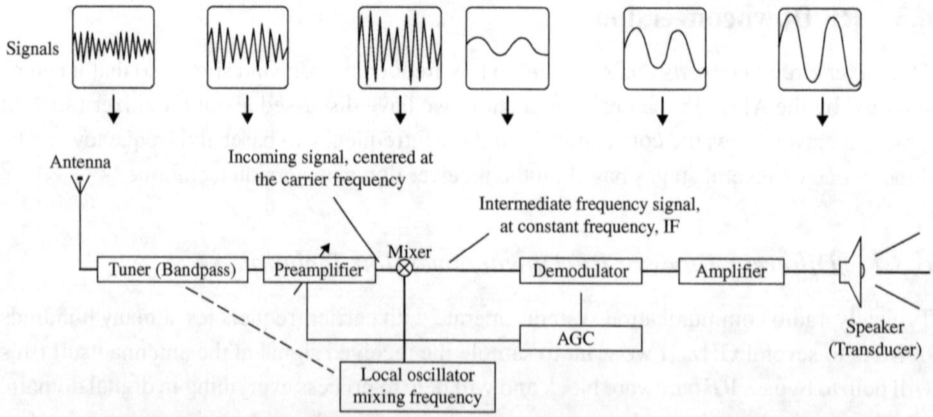

Figure 6.12 Basic block diagram of a superheterodyne receiver (analog receiver)

What is superheterodyning? When we use only the lower side band (the difference between the two frequencies), we are superheterodyning. Strictly speaking, the term "superheterodyne" refers to the creation of a beat frequency that is lower than the original signal. Edwin Armstrong came up with the idea of converting all incoming frequencies to a common frequency. The superheterodyne receiver, invented in 1917, has enjoyed a long run of popularity.

Superheterodyning is simply reducing the incoming signal frequency by mixing. In a radio application (Figure 6.12), we are reducing the incoming AM or FM signal frequency (which is transmitted on the carrier frequency) to some intermediate frequency, the IF = $f_o - f_r$.

This is essentially the conventional receiver with the addition of a mixer and local oscillator. The local oscillator is linked to the tuner because they both must vary with the carrier frequency. Let us look at a specific example. An FM radio is tuned to a station operating at 89.9 MHz. This signal is mixed with an LO signal at a frequency of 100.6 MHz. The difference frequency at the output of the mixer is 10.7 MHz. This is the IF signal. If the FM radio is tuned to a different station at 107.9 MHz, the LO frequency is also retuned to 118.6 MHz. The mixer once again produces an IF signal of 10.7 MHz. In fact, as the FM radio is tuned across the band from 87.9 to 107.9 MHz, the local oscillator (LO) is tuned from 98.6 to 118.6 MHz. No matter what frequency the radio is tuned to (in its operating range), the mixer's output will be 10.7 MHz.

The superheterodyne overcomes the variable sensitivity and selectivity problems of the RF receiver module by doing most of the amplification at the intermediate frequency, where the gain and selectivity can be controlled carefully (Figure 6.13). However, the superheterodyne introduces some new challenges. First, the LO signal must always differ from the input signal by exactly the IF frequency, regardless of whatever input frequency is selected. This is known as "tracking." Second, there are two different frequencies that can mix with the LO signal to produce the IF signal. One of those frequencies is our input

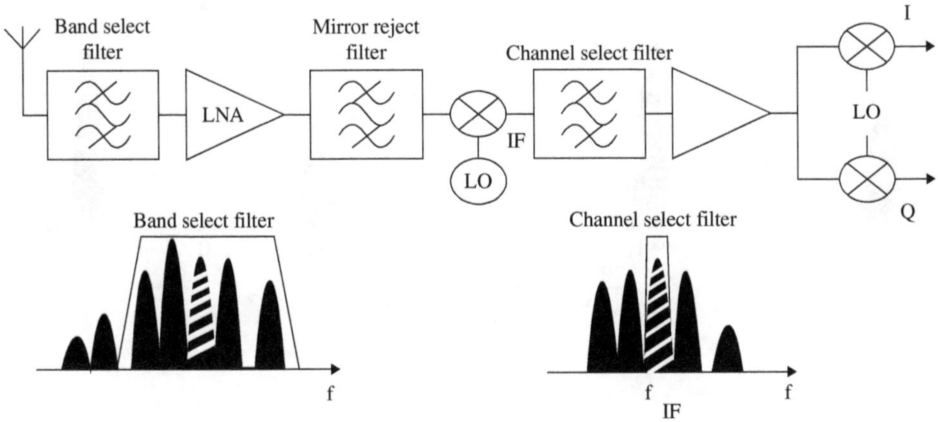

Figure 6.13 Superheterodyne receiver architecture (digital receiver – I/Q modulated)

signal frequency; the other is known as the "image frequency." The image, input, IF and LO frequencies are related as follows:

Image is another incoming frequency, which is mistakenly treated as desired input signal.

$$\text{IF} = \text{LO} - \text{Input and IF} = \text{Image} - \text{LO. So, Image} = \text{IF} + \text{LO} = \text{IF} + (\text{IF} + \text{Input}). \\ \text{Image} = \text{Input} + 2 * \text{IF} \tag{6.7}$$

Here let us take an example that references our earlier discussion about the FM radio. When the receiver is tuned to 89.9 MHz, the 89.9 MHz signal can mix with the LO signal of 100.6 MHz to create a 10.7 MHz IF signal. However, a signal (or noise) at 111.3 MHz can also mix with the LO signal to create a 10.7 MHz IF. Therefore any incoming noise or interference signals at this frequency have to be rejected. One filter is used to stop $(f_0 + fi)$ and pass $(f_0 - fi)$ to the demodulator unit. $f_{if} = f_0 - fi$, which is allowed to pass, but there may be another frequency, which is $(f_0 + f_{if})$. When this is input to the mixer, then the resultant will be $(f_0 + f_{if}) - f_0 = f_{if}$. So, this will also pass via the filter. But here, the IF is not because of the desired input signal (fi) rather it is another input signal $(f_0 + f_{if})$. This is known as image frequency as discussed earlier $(f_0 + f_{if}) = f_{image}$. This also needs to be rejected as it causes the following issues.

Image Frequency Problem

The desired RF input signal frequency $= f_{if} = f_0 - fi$ (which is allowed to pass via filter). But there is another RF input signal frequency $f_0 + f_{if} = f_{image}$, and when this is fed to the mixer, the resultant frequency $= f_0 - f_{image} = f_0 - f_0 + f_{if} = f_{if}$. As this resultant frequency is same as the IF frequency, so this will also pass via the filter (Figure 6.14). But this is derived from an unwanted RF signal, and this is called the image frequency, which needs to be stopped, unless this will corrupt the original signal and appear in the detector.

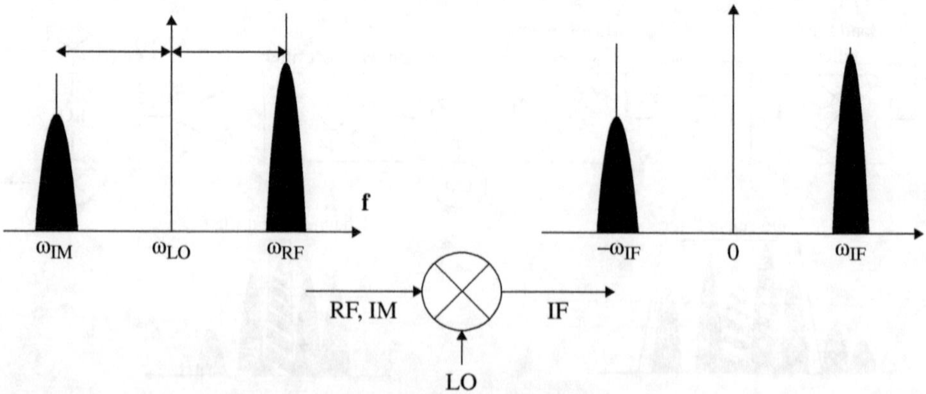

Figure 6.14 Image frequency

To reject the mirror / image frequency signal, an additional filter is often applied in front of the mixer. This is known as the image rejection filter (IR). It is bulkier and makes on-chip integration difficult.

On-Chip Superheterodyne Receiver Architecture

The on-chip architecture of the superheterodyne receiver discussed above is shown in Figure 6.15. A passive band pass filter limits the input spectrum provided by the antenna. The mixer introduces some noise, so the signal is first amplified by a low noise amplifier (LNA) before mixing. Mixers translate the RF signal to IF frequencies. The LO signal is tuned at a particular spacing above or below the RF signal, is injected into the mixer circuits. Hence, these bands must first be removed by an image reject filter. For that, the signal goes off-chip into an image rejection (IR) filter using passives with a high quality factor. Then, mixing with a tunable LO signal, the selected input channel frequency is downconverted to the IF. This LO1 output needs to be variable in small frequency steps for narrow band selection. To alleviate the sensitivity-selectivity tradeoff in image rejection mentioned above, an off-chip, high-Q bandpass filter performs partial channel filtering at a relatively high intermediate frequency. A second downconversion mixing step translates the signal down to the baseband frequency, which can be treated in the digital domain. This reduces the requirement for the final integrated analog channel selection filter, as now it can be done digitally.

In digital modulation, a signal is often expressed in terms of in-phase (I) and quadrature (Q) elements. On a polar diagram, the I axis lies on the zero degree phase reference, whereas the Q axis is rotated by 90°. In rectangular representation, the signal vector's projection onto the I axis is its "I" component and the projection onto the Q axis is its "Q" component. As shown in Figure 6.15, both the components (I and Q) can be generated in the second mixing stage. As the channel of interest is already selected by the first mixer tuning the LO, the frequency of the second LO can be fixed.

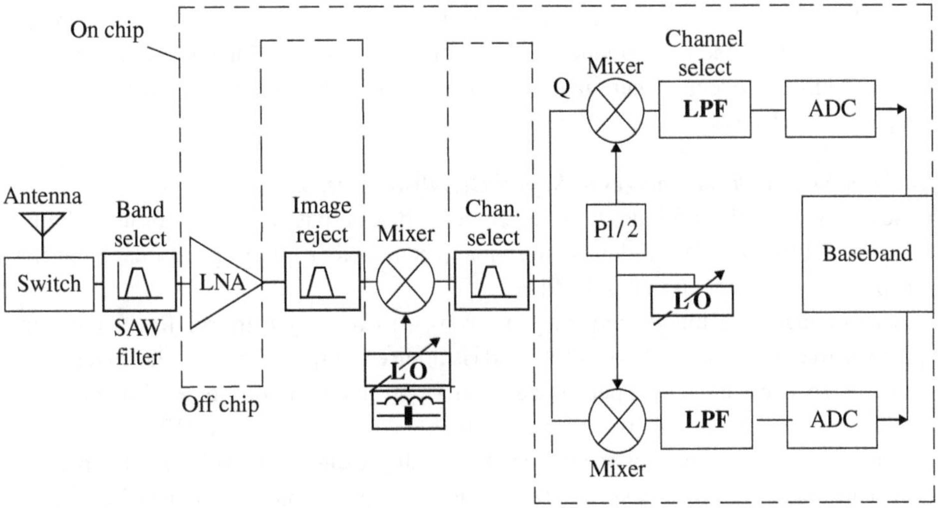

Figure 6.15 Heterodyne RF converter (with on-chip and off-chip components)

Off-chip passive components provide filters with a high Q-factor and this results in good performance for both sensitivity and selectivity, and this makes the heterodyne architecture a common choice. Furthermore, noise introduced by the local oscillator is less problematic as it is filtered by subsequent channel selection. Image rejection and adjacent channel selectivity is better in this type of architecture. The filters can be manufactured in different technologies like surface acoustic wave (SAW) – an external filter, bipolar, and CMOS. However, off-chip filtering comes at the price of extra signal buffering (driving typically 50 Ω loads), increased complexity, higher power consumption, and larger size. Narrow bandwidth passive IF filtering is typically accomplished using crystal, ceramic, or SAW filters (these are passive filters). These filters offer better protection than the zero-IF receiver's gyrator filters (active filters) against signals close to the desired signal because passive filters are not degraded by the compression of signal resulting from large signals. The active gyrator circuit does not provide such protection.

However, undesired signals that cause a response at the IF frequency in addition to the desired signal are known as spurious responses. In the case of the heterodyne receiver, spurious responses must be filtered out before reaching the mixer stages. One spurious response is known as an image frequency. An RF filter (known as a preselector filter) is required for protection against the image unless an image-reject mixer is used. Additional crystal-stabilized oscillators are required for the heterodyne receiver.

Generally, superheterodyne receivers cost more than zero-IF receivers due to the additional oscillators and passive filters. These items also require extra receiver housing space, which increases the size. However, a superheterodyne receiver's superior selectivity may justify the greater cost and size in many applications.

The benefits of the superheterodyne architecture are enormous. Most of the filtering and gain takes place at one fixed frequency, rather than requiring tunable high-Q band-pass filters or stabilized wideband gain stages. In some systems, multiple IFs are used to distribute the gain and selectivity for better linearity.

Advantages and Disadvantages of Superheterodyne Receiver

- Advantages: (i) this provides high selectivity and sensitivity as a result of using high-Q filters and the double RF downconversion scheme; (ii) it offers good image-rejection capability due to the use of an IR filter.
- Disadvantages: (i) image frequency problem – to reject the mirror frequency signal (image frequency), an additional filter (IR) is often employed in front of the mixer, which makes the size bulkier; (ii) poor integration – port integration becomes difficult in this architecture as it uses high-Q devices, a double conversion scheme, IR, and a passive channel filter – so, it cannot be integrated in small packages; (iii) as integration becomes difficult, so, it becomes larger in size and weight; (iv) it consumes more direct current (DC), hence power consumption is higher; (v) it offers only a fixed signal bandwidth; (vi) there is a higher cost for an increased number of components; (vii) there is improved protection and larger physical size, which requires an extra printed circuit board (PCB) or silicon real estate.
- Applications: the superheterodyne receiver is typically used in radio receivers and satellite receivers.

6.3.2 Homodyne Receivers

6.3.2.1 Zero-IF Receiver (DCR)

The homodyne (homo = same, dyne = mix) architecture uses a single-frequency translation step to convert the incoming RF channel directly to baseband, without operations at intermediate frequencies. It is therefore also called zero-IF or direct conversion (DCR) architecture. Here, the IF frequency is chosen as 0 Hz (DC frequency) by selecting the local oscillator frequency that is the same as the desired RF input signal frequency. So, after mixing both at the I and Q channels, the generated frequency components will be $(f_0 - fi) = 0$, and $(f_0 + fi) = 2fi$ as $f_0 = fi$. This is shown in Figure 6.16. The portion of the channel translated to the negative frequency half axis, becomes the image to the other half of the same channel translated to the positive frequency half axis. After the downconversion, the input signal has a bandwidth of B Hz around center 0. Figure 6.17 shows this architecture in the case of a quadrature downconversion (I-Q demodulation) receiver. In this architecture (like in heterodyne architecture) an off-chip RF filter first performs band limitation before the received signal is amplified by an integrated LNA. Channel selection is done by tuning the RF frequency of the LO to the center of the desired channel, making the image equal to the desired channel. So, here the problem of images is not present and the off-chip IR filter can be omitted. A subsequent channel-selection low-pass filter (LPF) then removes nearby channels or interferers prior to A/D conversion.

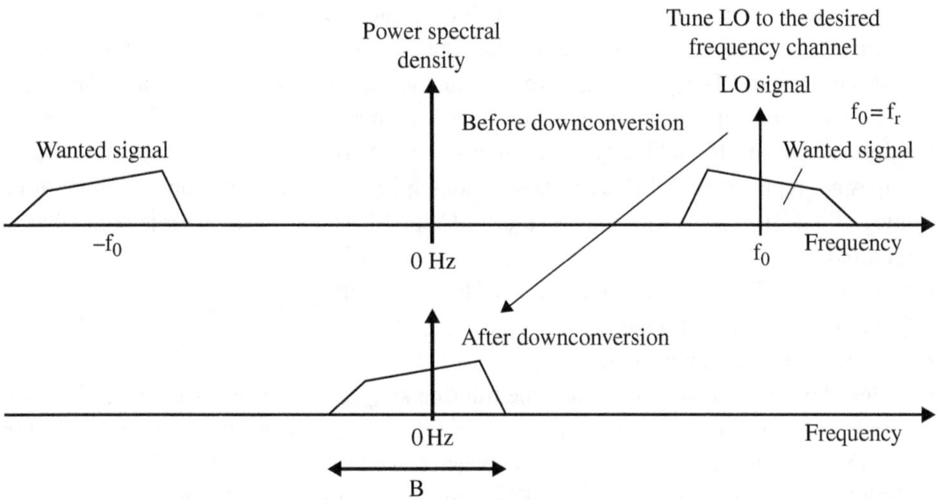

Figure 6.16 Direct conversion technique

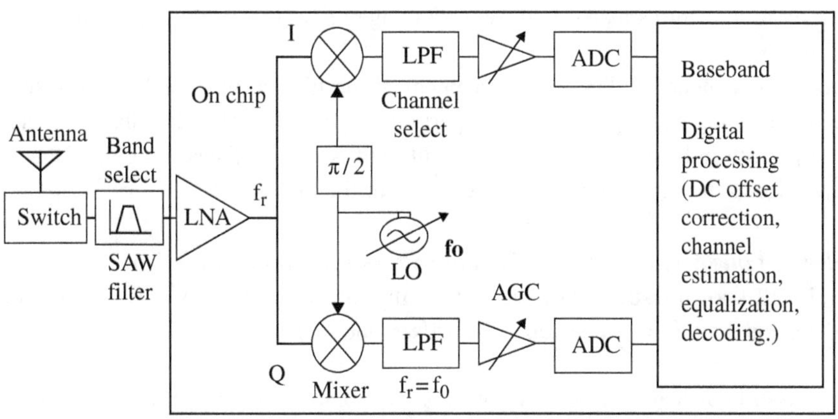

Figure 6.17 On-chip zero-IF direct conversion RF converter

Due to direct conversion to DC, homodyne receivers are more susceptible to disturbances arising from I/Q phase mismatches, nonlinearities, and flicker noise, than heterodyne designs. To control the performance loss, additional circuitry and design effort is required. However, there is no need for image rejection or other off-chip filters, which helps to save power and reduces the total receiver size.

Advantages of this Architecture

- No image frequency problem. An advantage of zero-IF receiver is that no image exists and an image-reject filter (or image-reject mixer) is not required.
- The LPF can be integrated – channel filtering is now possible entirely on chip as, after the downconversion, the frequency is now at baseband frequency. The zero-IF receiver

can provide narrow baseband filtering with integrated LPFs. Often, the filters are active op-amp-based filters known as gyrators. The gyrators provide protection from most undesired signals. The gyrator filters eliminate the need for expensive crystal and ceramic IF filters, which take more space on a printed circuit board.

- Eliminate passive IF and image reject filter – the IF SAW filter, IR Filter and subsequent stages are replaced with LPFs and baseband amplifiers that are amenable to monolithic integration. The LNA need not drive a 50 Ω load because no image rejection filter is required.
- Increased ADC dynamic range because of limited filtering.
- Good SSB digital modulation.
- Reduced component numbers.
- Reduced power consumption – now the filtering and gain can take place with DC, where gain is easier to achieve with low power, and filtering can be accomplished with on-chip resistors and capacitors instead of the expensive and bulky SAW filters.
- High level of integration – the zero-IF topology offers the only fully integrated receiver currently possible. This fully integrated receiver solution minimizes required board real estate, the number of required parts, receiver complexity, and cost. Most zero-IF receiver architectures also do not require image reject filters, thus reducing cost, size, and weight.
- Good multistandard ability – the placement of the filter at baseband (usually split between the analog and digital domains) permits multiple filter bandwidths to be included at no penalty in board area because the filtering is accomplished on chip. Thus, direct conversion is the key to multimode receivers for the future.

Problems of this Architecture and Possible Alternative Design Solutions

Several well known issues that have historically plagued direct conversion receivers are self-detection due to LO-RF leakage, DC offset, and AM detection.

- Local oscillator leakage. One of the most well known problems in direct-conversion receiver architecture is spurious LO leakage. This arises because the LO in direct conversion receivers is tuned exactly to the desired input signal frequency, which is the center of the LNA and antenna pass band. Owing to improper isolation, a small fraction of this LO signal leaks through the mixer and travels towards the input signal side, passes through the LNA, and reaches the antenna (Figure 6.18). Then it radiates out through the antenna. This becomes an in-band interferer for other nearby receivers tuned to the same band, and for some of them it may even be stronger than the desired signal. Regulatory bodies, such as the FCC strictly limit the magnitude of this type of spurious LO emission. Each wireless standard and the regulations of the Federal Communications Commission (FCC) impose upper bounds on the amount of in-band LO radiation – typically between 50 and 80 dBm. The issue is less severe in heterodyne and image-reject mixers because their LO frequency usually falls out of the reception band. It also leaks to other side of the entire receiver signal chain, which appears as a

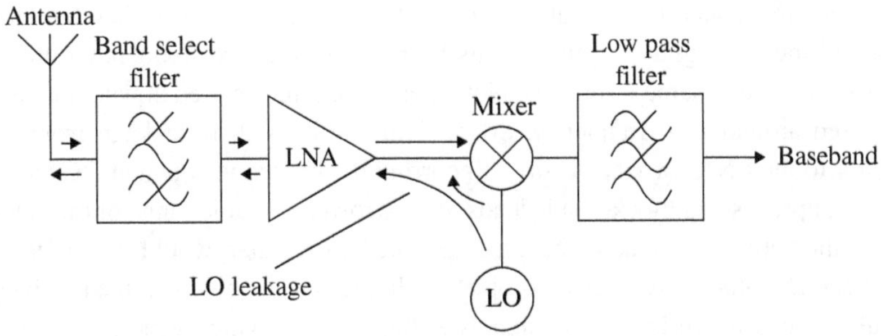

Figure 6.18 LO leakage

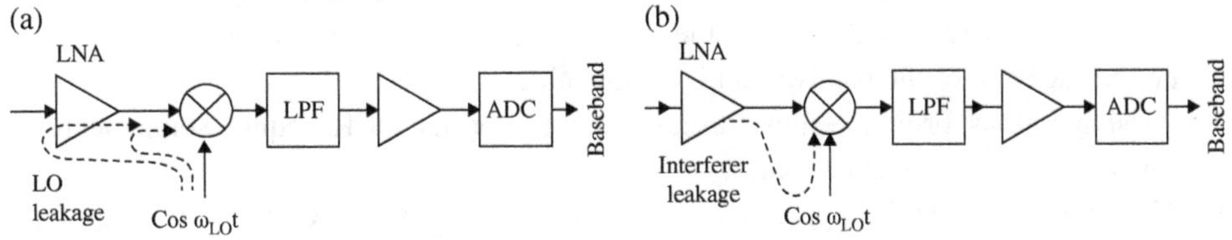

Figure 6.19 (a) Self-mixing LO. (b) Interferers mixing

DC offset. The problem of LO leakage becomes severe as more sections of RF transceivers are fabricated on the same chip.

- Design option. With differential local oscillators, the net coupling to the antenna can approach acceptably low levels.

- Self reception – because the local oscillator is tuned to the RF frequency, self-reception may also be an issue (Figures 6.19a and b).
  - Design option. Self-reception can be reduced by running the LO at twice the RF frequency and then dividing by 2 before injecting into the mixer. Because the zero-IF local oscillator is tuned to RF frequencies, the receiver LO may also interfere with other nearby receivers tuned to the same frequency. However, the RF amplifier reverse isolation prevents most LO leakage to the receiver antenna.
- DC offset problem. The basic operation of a direct-conversion receiver can be described as mixing an input signal frequency of ($f_C + f_m$), where (f_m) is the bandwidth of the modulation, with a local oscillator at f_{LO}, yielding an output at $f_{MIXOUT} = (f_C + fm - f_{LO})$ and ($f_C + fm + f_{LO}$). The second term is at a frequency twice that of the carrier frequency and can be filtered out very easily by the channel select filter. But, the first term is much more interesting, since $f_{LO} = f_C$, and substitution yields: $f_{MIXOUT} = f_{LO} + fm - f_{LO} = fm$. That means the modulation has been converted to a band from DC to the modulation bandwidth, where gain, filtering and A/D conversion are readily accomplished. The DC-offset problem occurs when some of the on-channel LO (at f_C) leaks to the mixer RF port, creating the effect that: $f_{LO} - f_{LO} = 0$ (or DC). This can corrupt wanted information that has been mixed down around zero Hz.

So, when the leaked LO signal appears at the input of the mixer, then the leaked LO signal and the LO signal are mixed, this will result in a zero frequency output or DC output as they are of same frequency. Note that in DCR the desired downconverted signal is centered around zero frequency. So, self-mixing caused by leakage from the local oscillator to the LNA (or vice versa) will corrupt the baseband signal at DC and saturate subsequent processing blocks. This leads to a narrower dynamic range of the electronics, because the active components become saturated easier than would be the case with a zero offset. DC offsets are a severe problem in homodyne receivers. If the receiver moves spatially, it receives reflected LO signals at the antenna, which generates time varying offsets. Causes of DC offset are either drift in the baseband components (e.g. op amps, filters, A/D converters), or DC from the mixer output caused by the LO mixing with itself or with the mixers acting as square law detectors for strong input signals. DC offsets from various sources lie directly in the signal band, and in the worst case they can saturate the back end of the receiver at high gain values.

- Design options. From the above discussion, we infer that DCRs require some means of offset removal or cancellation.

- AC coupling. A possible approach to remove the offset is to use AC coupling, –high-pass filtering, in the downconverted signal path. However, since the spectrum of random binary (or M-ary) data exhibits a peak at DC, such signals may be corrupted if they is filtered with a high corner frequency. One technique is to disregard a small part of the signal band close to DC and employ a high-pass filter with very sharp cutoff profile at low corner frequencies. This requires large time constants and, hence, large capacitors. It is only practical for wide-band applications (WCDMA), where the loss of a few tens of hertz bandwidth at DC does not degrade the receiver performance significantly. The system can either be AC-coupled or can incorporate some form of DC notch filtering after the mixer. But, for narrow band applications (GSM), this would cause large performance losses. A low corner frequency in the HPF may also lead to temporary loss of data in the presence of wrong initial conditions. If no data is received for a relatively long time, the output DC voltage of the HPF droops to zero. Now if data is applied, the time constant of the filter causes the first several bits to be greatly offset with respect to the detector threshold, thereby introducing errors. A possible solution to the above problems is to minimize the signal energy near DC by choosing "DC-free" modulation schemes. A simple example is the type of binary frequency shift keying (BFSK) used in pager applications.
- Offset cancellation. In time division multiple access (TDMA)-based wireless systems (like GSM), each mobile station communicates with the base station for a short period of time (a time slot) in a time frame. In the remaining time of that frame, it stays idle. This means that no RF transmission or reception occurs during that time. So, during these idle time slots, the offset voltage in the receive path can be stored on a capacitor and, later, this offset voltage is subtracted from the received signal during actual signal reception. Figure 6.20 shows a simple example, where the capacitor stores the offset voltage between two consecutive TDMA bursts while introducing a virtually zero corner frequency

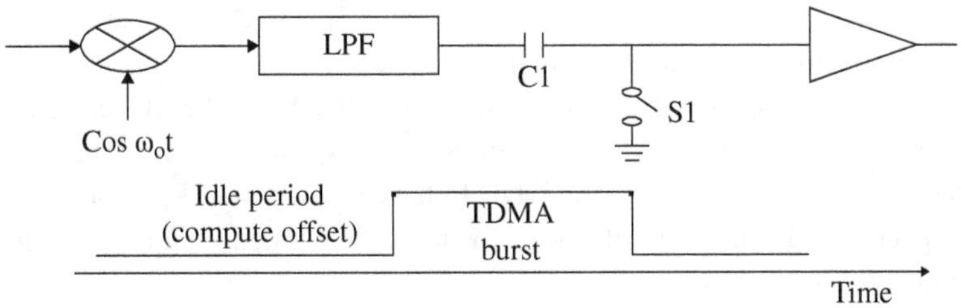

Figure 6.20 Offset cancellation

during the reception of data. But the major issues are the thermal noise, as interferers may be stored along with the offsets and this makes it difficult to cancel. This occurs because of the reflections of the LO signal from nearby objects and this signal must be included in offset cancellation. Hence, the antenna cannot be disconnected (or "shorted") during this period. Although the timing of the appearance of the actual signal (the TDMA burst or slot) is well defined in a frame, interferers can appear at any time, which make the cancelation difficult. A possible approach for alleviating this issue is to sample the offset (and the interferer) several times and take the average.

Shielding and other layout techniques are often used to reduce this effect. Another approach is to convert an off-channel (or even out-of-band) LO signal to an on-channel LO inside the chip, reducing leakage paths. Also, operating the LO at half (or twice) the necessary injection frequency is a good solution for single-band applications; a regenerative divider simplifies multiband designs. Once the DC offset due to LO-RF leakage has been reduced, a second problem arises: inherent DC offset in the baseband amplifier stages and its drift over temperature. Here, the best solution is to employ extreme care in the design of the gain stages and to make sure that enough gain – but not too much – is provided. Excessive gain in the baseband section can cause offsets that can be corrected momentarily but that may drift excessively and require additional temperature compensation. There are three possible methods by which offsets may be handled in the receiver: continuous feedback, track-and-hold, and open loop. The continuous-feedback scheme (in software or hardware) attempts to null DC error at the mixer output. That generally requires tight coupling between the baseband processor and software and makes it difficult to mate an RF IC from one vendor with a baseband controller and software from another vendor. In the "track and hold" method, the DC offset is estimated just prior to the active burst (track mode) and then stored (hold mode) during the active burst. Such schemes are generally completely integrated with the radio IC and can be made transparent to the user by locally generating all the necessary timing signals. Practical issues with the scheme include dealing with multislot data (GPRS) where the baseband gain may be changing on a slot-by-slot basis (without adequate time to recalibrate) and also ensuring that the DC estimate obtained during the track mode is accurate. Such schemes can be implemented in either digital or the analog domains. Latest generation radios using the open-loop approach have substantially lower DC offsets and can operate with lower performance A/D converters (typically 60 to 65 dB of available dynamic range), without any special software requirements.

- Need for high Q voltage controlled oscillators (VCOs). As neither image rejection filter nor channel select filtering is done prior to mixing, all adjacent channel energy is untreated. This requires the LPF and ADC to have a sharp cutoff profile and high linearity, respectively. In the view of low-Q values of integrated components, this implies tougher design challenges.
- Even order distortion. Even-order distortion, especially second-order nonlinearity, can degrade the direct-conversion receiver's performance significantly, because any signal containing amplitude modulation generates a low-frequency beat at baseband.
  - Design options. Because of the inherent cancellation of even-order products, differential LNAs and double-balanced mixers are less susceptible to distortion. However, the phenomenon is critical for balanced topologies as well due to unavoidable asymmetry between the differential signal paths. But the problem is, if the LNA is designed as a differential circuit, it requires higher power dissipation than the single-ended counterpart to achieve a comparable noise figure.
- Flicker Noise (1/f) noise. Since the downconverted spectrum is located around zero frequency, the noise of devices has a profound effect on the signal, a severe problem in MOS implementations.
  - Design options. The effect of flicker noise can be reduced by a combination of techniques. As the stages following the mixer operate at relatively low frequencies, they can incorporate very large devices (several thousand microns wide) to minimize the magnitude of the flicker noise. Moreover, periodic offset cancellation also suppresses low-frequency noise components through correlated double sampling. A bipolar transistor front end may be superior in this respect to an FET circuit, but it is also possible to use auto-zero or double-correlated sampling to suppress flicker noise in MOS op-amp-based circuits.
- I/Q mismatch. As discussed earlier, in I-Q modulation, to achieve maximum information, we should take both parts of signal. It is done by a method called quadrature downconversion. The principle of this method is that the signal is at first divided into two channels and then downconverted by an LO signal, which has a phase shift of 90° in one channel with respect to another. The vector of the resulting signal is described as: $|Signal| = \sqrt{I^2 + Q^2}$, $\arg(Signal) = \varphi = arctg\frac{Q}{I}$. The problem of the homodyne receiver, or, more concretely, of the I/Q (in-phase / quadrature) mixer, is mismatches in its branches. Assuming a mismatch of ε for the amplitude and θ for the phase, we can estimate the error, caused by these mismatches. In this way we get:

$$E_{IQ} = \left|\frac{S_{ideal} - S_{miss}}{S_{ideal}}\right| \approx \left(\frac{1}{2} + \varepsilon\right)\theta^2 \tag{6.8}$$

For typical values of $\varepsilon = 0.3$ and $\theta = 3^0$ this gives an error of $1.5 \cdot 10^{-3}$. I/Q modulation requires an exact 90° phase shift between RF and LO signal or vice versa. In either case, the error occurs in a 90° phase shift and mismatches between the amplitudes. Signals corrupt the downconverted signal constellation, thereby increasing the bit error rate. All sections of the circuit and paths contribute to gain and phase error. This will show the resulting signal constellation with finite error. This effect can be better seen by examining the downconverted signals in the time domain. Gain error simply appears as a nonunity scale factor in the amplitude. Phase imbalance, on the other hand, corrupts one channel with a fraction of the data pulses in the other channel, in essence degrading the signal-to-noise ratio if data streams are uncorrelated. But, the mismatch is much less troublesome in DCR than in image-reject architectures.

- Need of AGC and AFC. Sensitivity and rejection to some undesired signals, such as intermodulation distortion, can be difficult to achieve in DCR to enhance the performance. The active gyrator filters compress with some large undesired signals. Once the gyrator is compressed, filter rejection is reduced, thus limiting the protection. Zero-IF receivers typically require an automatic gain control (AGC) circuit to protect against large signal interference that compresses the gyrator filters. Zero-IF receiver limitations require tighter frequency centering of the LO and RF frequencies. Significant offsets in the RF or LO frequencies degrade bit error rate (BER). One solution for zero-IF designs is to add automatic frequency control (AFC). This prevents the centering problem by adjusting the frequency of the LO automatically.

Applications of this Architecture

Different modulation schemes exhibit different susceptibility to the problems in DCR. Quadrature-phase shift keying (QPSK) modulated spread spectrum schemes like CDMA and WCDMA have almost no signal energy near DC and are more immune to DC offsets. This architecture is particularly suited for the DS-SS (direct sequence spread spectrum) standard because of the wide channel bandwidth, and the removal of a small amount of energy near zero for DC offset compensation will not have much impact on the overall received energy. Conversely, Gaussian minimum shift keying (GMSK) modulated GSM signals do have a DC component in the data and are under time constraints placed by the TDMA system. For this reason, the GSM signal cannot simply be AC coupled at baseband, nor can the DC offsets be filtered easily, because either of these methods would simultaneously remove wanted and unwanted signals. That is why, zero-IF DCRs are not very useful for GSM receivers. As discussed earlier, recent work using this architecture suggests that the effects of various imperfections can be alleviated by means of circuit design techniques.

The direct-conversion receiver architecture was successfully used in pagers (AC coupling allowed) and satellite receivers.

Direct conversion-based transceiver solutions currently do not benefit from the most cost-effective CMOS technologies due to the susceptibility to 1/f noise. This is because the 1/f noise in the mixer and baseband filtering stages appears directly on top of the downconverted signal in a direct conversion radio. This effectively increases the receiver noise figure (NF), especially in narrowband applications like GSM. In practice, bipolar transistors will prove more appropriate for the LNA and mixer design, with MOS transistors allocated to the subsequent baseband stages. BiCMOS designs will be forced into more expensive and larger feature-size processes, hindering radio integration roadmaps aimed at cost reduction.

Again, careful design can minimize this problem but it can still be the reason why direct conversion will not work for every application. Its monolithic integration capabilities make the homodyne architecture an attractive alternative for wireless receivers. If the RF signal is downconverted in a single step to a low (but not to DC) frequency, then limitations at DC have less impact on the receiver performance. This approach is followed in low-IF architectures, discussed next.

6.3.3 *Low IF Receiver*

The digital low-IF receiver leverages the performance advantages of the superheterodyne approach with the economic and integration advantages of the direct conversion approach. This is accomplished by band selecting and downconverting the desired RF signal to a frequency very close to the baseband (for example 100 kHz) instead of zero, as illustrated in Figure 6.21. Next the low-IF signal is filtered with a LPF and amplified before conversion to the digital domain by the analog-to-digital converter (ADC). The final stage of downconversion to baseband and fine gain control is then performed digitally.

High-resolution, oversampling, delta-sigma converters allow the channel filtering to be implemented with DSP techniques rather than with bulky analog filters. The signal can then interface directly to a digital BBIC input or a digital-to-analog converter (DAC) can be used to output analog I and Q signals to a conventional baseband integrated circuit (BBIC).

Like the DCR, the digital low-IF receiver is able to eliminate the off-chip IF SAWs necessitated by the superheterodyne approach. While the digital low-IF approach does encounter an image frequency at the adjacent channel, an appropriate level of image rejection can still readily be achieved with a well designed quadrature downconverter and integrated I and Q signal paths. This avoids the need for external image reject filters. At the low-IF frequency, the ratio of the analog channel filter center frequency to the channel

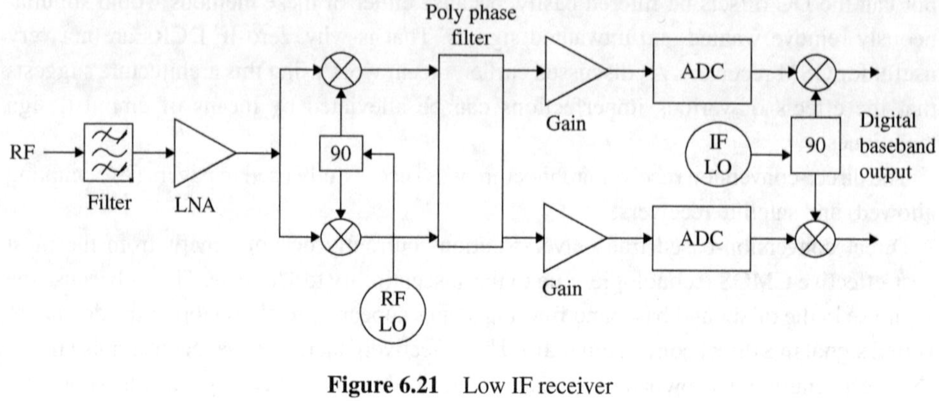

Figure 6.21 Low IF receiver

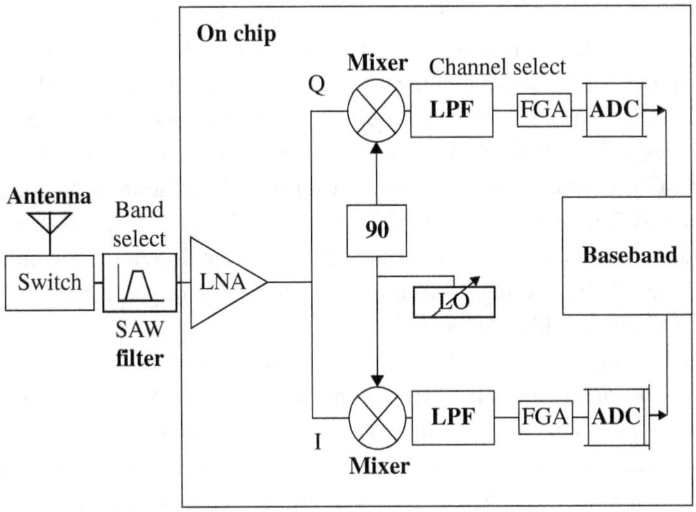

Figure 6.22 Integrated low-IF RF converter

bandwidth (moderate Q) enables the on-chip integration of this filter. Followed by amplification, the signal is then converted to the digital domain with an ADC. This ADC requires a higher level of performance than the equivalent DCR implementation because the signal is not at baseband. A digital mixer operating at 100 kHz can then be used for the final downconversion to baseband where digital channel filtering is performed (Figure 6.22).

The migration of these traditionally analog functions into the digital domain offers significant advantages. Fundamentally, digital logic is immune to operating condition variations that would corrupt sensitive analog circuits. Using digital signal processing improves design flexibility and leverages the high integration potential, scalability, and low cost structure of CMOS process technologies. While the digital low-IF receiver does add a downconversion stage (mixer and filter), it is possible to implement this functionality in an area smaller than that occupied by the analog baseband filter of the DCR architecture because the extra stage is digital. Digital low-IF receivers will also find it easy to comply with the developing DigRF BBIC interface standard for next generation transceiver applications.

The digital low-IF architecture described curtails issues associated with DC offsets. The desired signal is 100 kHz above the baseband after the first analog downconversion, so any DC offsets and low frequency noise due to second-order distortion of blockers, LO self-mixing, and 1/f noise can easily be eliminated by filtering. Once in the digital domain and after the downconversion to baseband, DC offsets are of negligible concern. The desired signal is no longer as small and vulnerable, and digital filtering is successful in removing any potential issues.

With DC offset issues avoided at the system level, digital low-IF receivers will greatly relax IP2 linearity requirements and will still meet the critical AM suppression specification with relative ease.

Manufacturers adamantly demand the most reliable, easy-to-implement, and low-cost components and ICs for each handset function. The digital low-IF receiver's immunity to DC offsets has the benefit of expanding part selection and improving manufacturing. At the front end, the common-mode balance requirements on the input SAW filters are relaxed, and the PCB design is simplified. At the radio's opposite end, the BBIC is one of the handset's largest BOM contributors. It is common for a DCR solution to be compatible only with its own BBIC in order to address the complex DC offset issues. Fortunately, digital low-IF based transceiver solutions can empower the system designer with multiple choices when considering BBIC offerings. This is because there is no requirement for BBIC support for complex DC offset calibration techniques.

In addition to flexibility, digital low-IF based transceivers may be able to capture notable sensitivity improvement from the BBIC. Many BBICs for GSM systems employ DC filtering as a default to compensate for large DC drifts that may occur when they are coupled with a DCR based design. When these same BBICs are paired with low-IF transceivers, such filtering is not needed. The handset designer is then in a position to work with the BBIC vendor to disable the unwanted filtering in software. This has the benefit of regaining the valuable signal bit energy around baseband frequencies that had been thrown away by the filtering. The handset designer can then enjoy a potential sensitivity enhancement of 0.2 to 0.5 dB for little expense!

6.3.3.1 Advantages

Here, the gain and filtering are done at a lower frequency than the conventional high-IF superheterodyne. That reduces the power and opens up the possibility of integrating the filter components on chip, thus reducing the total number of components. If the gain stage is AC-coupled, any issues relating to DC offsets should be eliminated. The main advantages are: (i) no image frequency problem; (ii) the LPF can be integrated in the IC/digital module; (iii) it eliminates passive IF and image reject filters; (iv) there is a high level of integration; (v) there are fewer components than in a superheterodyne receiver; (vi) there is a reduced DC offset problem; (vii) there is less 1/f noise compared to a zero IF receiver.

6.3.3.2 Disadvantages

(i) There is a greater baseband processing power requirement (MIPS). (ii) ADC requires a higher level of performance than the equivalent DCR implementation because the signal is not at baseband. (iii) The receiver's polyphase filter requires more components than an equivalent low-pass filter used in a DCR. We know that $\cos(\omega t) = \cos(-\omega t)$. So, the negative frequency cannot be identified. As mentioned before, we want to discriminate between positive and negative frequencies in order to realize chip selectivity. This is not possible with real signals but is possible with two-dimensional signals or complex signals. We can imagine positive and negative frequencies as being phasors rotating in the complex plan in opposite direction. The complex signals used in a receiver are called polyphase signals, and they consist of a number of real signals with different phases. A quadrature signal consists of two real signals with an $\pi/2$ phase shift. The polyphase bandpass filter ensures the rejection of the mirror frequency and provides the antialiasing necessary in the DSP, which does the final downconversion to baseband and demodulation of the signal. The wanted signal is multiplied with a single positive frequency at f_{LO}. The mirror signal will be mixed down from f_{mirror} to $-f_{IF}$ and the wanted signal at f_{IF}. With a polyphase filter it is possible to discriminate between the negative and positive frequencies and therefore, the mirror frequency will be filtered out. (iv) Image cancellation is dependent on the LO quadrature accuracy. (v) In hybrid implementations, where the image-reject function is divided into analog and digital phase-shift stages, the A/D conversion process occurs at the IF frequency. That generally requires higher power than baseband converters, and more stringent control of the sampling clock because clock jitter will degrade the conversion of an IF signal.

6.3.3.3 Applications

It is most suitable for the GSM (GMSK) receiver. It can also be used in multimode receivers.

6.3.4 Wideband IF Receivers

An alternative to the low IF design is the wideband-IF architecture shown in Figure 6.23; this receiver system takes all of the potential channels and frequency translates them from the RF to the IF using a mixer with a fixed frequency local oscillator (LO1). A simple low-pass filter is used at the IF to remove any upconverted frequency components, allowing all channels to pass to the second stage of mixers. All of the channels at the IF are then frequency translated directly to baseband using a tunable, channel-select frequency synthesizer (LO2). Alternate channel energy is then removed with a baseband filtering network where variable gain may be provided.

This approach is similar to a superheterodyne receiver architecture in that the frequency translation is accomplished in multiple steps. However, unlike a conventional superheterodyne receiver, the first local oscillator frequency translates all of the received channels, maintaining a large bandwidth signal at IF. The channel selection is then realized with the lower frequency tunable second LO. As in the case of direct conversion, channel filtering can be performed at baseband, where digitally programmable filter implementations can potentially enable more multistandard-capable receiver features.

In contrast to the previous architectures, the first local oscillator frequency is fixed. All available channels are converted to intermediate frequency, resulting in a wide bandwidth at IF. Upconverted frequency components are removed by a simple low-pass filter. Channel selection and filtering are done at IF. Due to the lower operation frequency, the requirements for the tunable LO and low-pass filter in the second downconversion stage are relaxed. Hence, a narrow channel can be selected and filtered without off-chip components.

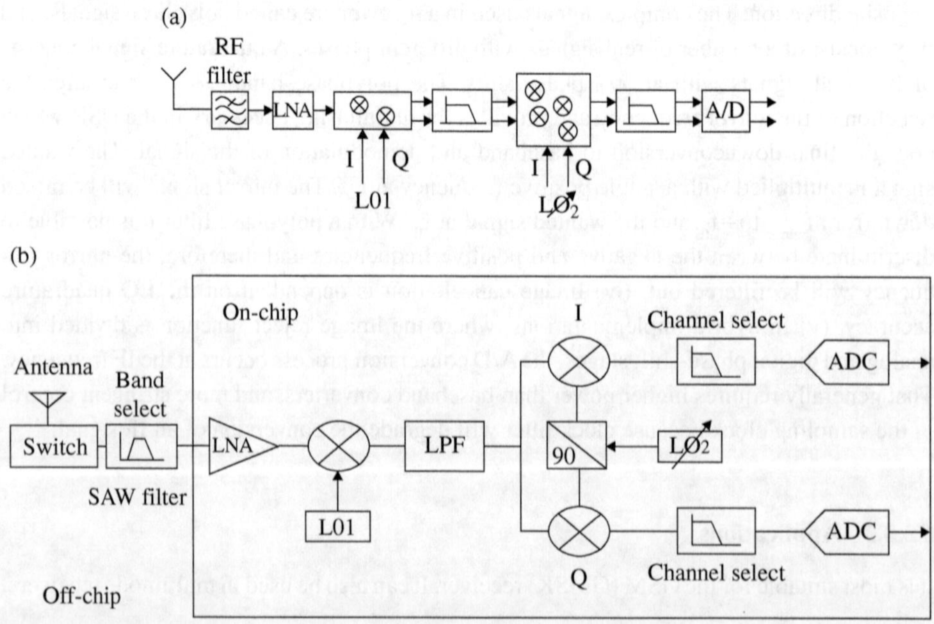

Figure 6.23 (a) Wide IF RF converter. (b) On-chip implementation of wide IF receiver

Furthermore, filtering can be performed partly in the digital domain, which adds to multi-standard operation capabilities of this architecture. This flexibility comes to the expense of higher linearity requirements of the ADC.

The wideband IF architecture offers two potential advantages over a direct conversion approach with respect to integrating the frequency synthesizer. The foremost advantage is the fact that the channel tuning is performed using the second lower frequency, or IF, local oscillator and not the first, or RF, synthesizer. Consequently, the RF local oscillator can be implemented as a fixed-frequency crystal-controlled oscillator, and can be realized by several techniques that allow the realization of low phase noise in the local oscillator output with low-Q on-chip components. One such approach is the use of wide phase-locked loop (PLL) bandwidth in the synthesizer to suppress the VCO contribution to phase noise near the carrier. Note that the VCO phase noise transfer function has a high-pass transfer function close in to the carrier and the bandwidth of suppression is related to the PLL loop bandwidth. Moreover, as channel tuning is performed by the IF local oscillator, operating at a lower frequency, a reduction in the required divider ratio of the phase-locked loop necessary to perform channel selection results. The noise generated by the reference oscillator, phase detector, and divider circuits of a PLL all contribute to the phase noise performance of a frequency synthesizer. With a lower divider ratio, the contribution to the frequency synthesizer output phase noise from the reference oscillator, phase detector and divider circuits can be significantly reduced. Moreover, a lower divider ratio implies a reduction in spurious tones generated by the PLL. An additional advantage associated with the wideband IF architecture is that there are no local oscillators operating at the same frequency as the incoming RF carrier. This eliminates the potential for the LO reradiation problem that results in *time-varying* DC offsets. Although the second local oscillator is at the same frequency as the IF desired carrier in the wideband IF system, the offset that results at baseband from self-mixing is relatively constant and is easily cancelled.

As the first local oscillator output is fixed and different from the channel frequencies, the problem of DC offset is alleviated in the wideband-IF architecture. The self-mixing in LO1 or LO2 still exists and results in constant DC offsets that can be removed either in the analog or digital domain. Isolation from the channel selection oscillator (LO2) to the antenna is much larger than in the heterodyne case. This greatly reduces problems associated with time-varying offsets. Using a fixed frequency at LO1 allows for phase noise optimization for this oscillator. Frequency conversion to IF introduces images again. These can be removed using a Weaver architecture, but mismatches between the I and Q paths limit the image suppression.

Additional components from the second conversion stage also inevitably result in larger power consumption. These problems are balanced by good monolithic integration capabilities and improved multistandard prospects due to programmable filtering in the DSP.

6.3.4.1 Advantages

(i) Allows for high level of integration. (ii) Relaxed RF PLL specification – VCO could be made on chip. (iii) Channel selection performed by IF PLL lowers the required divider ratio. (iv) Good multistandard ability. (v) Alleviated DC offset problem.

6.3.4.2 Disadvantages

(i) Increase of 1 dB compression point of the second set of the mixer. (ii) Increased ADC dynamic range requirement because there is limited filtering in comparison with the heterodyne receiver.

Applications

Feasibility has not been proven for GSM but it can be used for satellite radio receivers.

6.4 Receiver Performance Evaluation Parameters

Optimizing the design of a communications receiver is inherently a process of compromise. There are several factors that govern the performance of a radio receiver.

Selectivity and Sensitivity

The most important characteristics of a receiver are its sensitivity and selectivity. Sensitivity expresses the level of the smallest possible input signal that can still be detected correctly (i.e. within a given BER). Selectivity, on the other hand, describes the receiver's ability to detect a weak desired signal in the presence of strong adjacent channels – so-called interferers. So, sensitivity is the lowest signal power level that the receiver can sense and selectivity is the selection of the desired signal from the many (which are received by the antenna). For a good receiver, the selectivity and sensitivity should be higher than the reference level.

The "sensitivity" of a mobile wireless receiver is most commonly defined as "the minimum input signal level ($S_{minimum}$) required to produce a specified output signal at a defined signal-to-noise ratio (S/N) and it is expressed as the minimum signal-to-noise ratio times the mean noise power." This defined as:

$$S_{minimum} = (S/N)_{minimum} .(NF).k.T_o B \quad (6.9)$$

where: $(S/N)_{minimum}$ = minimum signal-to-noise ratio needed to process a signal;
B = receiver bandwidth (in Hz);
NF = noise figure (also known as noise factor);
k = Boltzmann's constant = 1.38×10^{-23} J/K
T_o = absolute temperature of the receiver input (in K) = 290 K.

Although, there is no standard definition of sensitivity level, most commonly two methods are used to express the sensitivity level of a mobile receiver.

The first one is as above in Eq. (6.9), where $S_{minimum}$ is used to express the sensitivity level of the receiver.

But, in mobile receiver design, the receiver sensitivity is most commonly expressed as an absolute receive power level in dBm. Generally, it is a negative number (and expressed in dBm), and here more negative value indicates "better" sensitivity level, which means that −109 dBm sensitivity is a "better" sensitivity level than −103 dBm.

Sensitivity level can also be expressed with a log scale, using the following expression:

$$RX_{sens}(dBm) = N_{thermal} + NF + E_b / N_0 \quad (6.10)$$

In Eq. (6.10), $N_{thermal}$ is the input thermal noise power generated due to the temperature of the receiver system, NF is the total noise figure of receiver, and E_b/N_0 is the normalized energy / bit required by the detector. In general, design decisions have an impact on all three terms in Eq. (6.10).

Now putting the thermal noise power value in the equation (6.10):

$$\begin{aligned} RX_{sens}(dBm) &= 10\log(kTB/1mW) + NF + E_b / N0 \\ kTB &= 1.381 \times 10^{-23} * 293 * (n * Rs) \text{ Watt,} \end{aligned} \quad (6.11)$$

Rs is symbol rate and, here, considering the GSM mobile system as reference, Rs = ($13*10^6$/48) = 270.833 kHz and n = number of bit per symbol = 1 for GMSK, which is the modulation type used in GSM system.

Insertion of these values in Eq. (6.11) results in:

$$
\begin{aligned}
RX_{sens}(dBm) &= 10.\log\left(1.381\times10^{-23}*293*1/10^{-3}\right)+10\log\left(n*Rs\right)+NF+Eb/N0 \\
&= 10\log\left(404.633*10^{-20}\right)+10.\log\left(n*Rs\right)+NF+E_b/N_0 \\
&= -174+10.\log.\left(n*Rs\right)+NF+E_b/N_0 = -174+10.\log\left(1*270.833*10^{3}\right) \\
&\quad +NF+E_b/N_0 \\
&= -119.67+NF+E_b/N_0
\end{aligned}
\tag{6.12}
$$

From equation (6.12), it can be concluded that the receiver sensitivity performance mainly depends on the following.

(a) RF Noise Figure (NF)

The equation which relates the noise figure and sensitivity is expressed as:

$$S=F.k.T_0B.\left(S_0/N_0\right),$$

where S = sensitivity in Watts; F = numeric system noise figure; B = receiver bandwidth in Hz, S_0/N_0 = receiver's output SNR (numeric).

From equation 6.12, it is evident that, if the RF NF is reduced, then the sensitivity of the receiver improves – the sensitivity level reduces. But the reduction of the receiver RF NF requires expensive RF circuits and components. To satisfy those criteria, the cost and size of the RF module will also increase proportionally. Today, in a multiband and multi-RAT mobile phone, the same RF front-end unit is mostly used for all supported radio access technologies (RAT) – for GSM, WCDMA and LTE. Due to cost advantages, size and power reductions, the same components are also reused across these RATs and frequency bands. For this reason the NF further increases and hence worsens the receiver's sensitivity level.

(b) Energy per Bit (E_b/N_0) Requirement

This expresses the "energy per bit to noise power spectral density ratio" in the baseband receiver for the proper demodulation and bit decoding of the received signal. This factor can be reduced in baseband by using various complex signal-processing algorithms. But, there is a minimum level that can be achieved for different modulation schemes using linear receivers for an AWGN channel.

So, from the above equations and derivations it is evident that the first factor NF is related to the analog RF module and the second factor E_b/N_0 is related to the digital baseband module. If both these factors are improved, then the receiver's sensitivity performance will improve significantly.

Image Rejection (IR)

This measures the ratio of the desired signal to the undesired image. The higher the IR the better the receiver.

Phase Noise

Phase noise describes an oscillator's short-term random frequency fluctuations. Noise sidebands appear on both sides of the carrier frequency. Typically, one sideband is considered when specifying phase noise, thus giving single sideband performance. Thus, low phase noise is crucial for oscillators in receiver systems.

Receiver Nonlinear Performance

Amplifiers usually operate as linear devices under weak signal conditions and become more nonlinear and distorting with increasing drive level. The amplifier efficiency also increases with increasing output power; thus, there is a system-level tradeoff between the power efficiency or battery life and the resulting distortion. The receiver's nonlinear performance should be good.

Processing Power to Drive Different Applications

A higher MIPS is always desirable to drive different complex processing but there should be a tradeoff between cost and power consumption.

Cost and Size

These are the most important driving factors for the design.

Complexity

The implementation of receiver architecture should be simple.

6.4.1 Receiver Architecture Comparison

The parameters for different receiver architectures are compared in Table 6.5.

6.4.2 Other Feasible Architectures

There are other architectures / subarchitectures that are currently used or being developed. Examples include the simple detector receiver (or envelope detector as shown in Figure 6.24), Hartley, or Weaver.

6.4.3 Path to Future Receivers

Future RF architectures will be able to receive any type of signal despite its bandwidth and dynamic range. Multistandard radios that are supported in SDR are capable of receiving a huge range of bandwidths combined with very different power level.

6.5 RF Transmitter

The RF transmitter module modulates the baseband digital signal to baseband analog signal, then upconverts the baseband analog signal to the RF frequency signal, then finally amplifies the signal and transmits it via the antenna (Figure 6.25). The transmitter mainly consists of

Table 6.5 Receiver architecture comparison

Parameter	Super heterodyne	DCR	Low-IF
Transceiver IC process Technology	Bipolar BiCMOS, GaAS	BiCMOS, CMOS (rarely chosen)	CMOS
Integration	Low	High	High
Off-chip IR filter	Required	Not required	Not required
IF-filtering	Requires IF SAW	On-chip LPF (may need external capacitors)	On-chip
Noise-figure	10.7 dB	same	same
Image rejection	−11 dB	−25 dB	−28 dB
Second intercept point (IP2)	N/A	43 dBm	18 dBm
Factory IP2 calibration	N/A	43 dBm	18 dBm
Third intercept point (IP3)	−19 dBm	same	same
DC offsets	Not there, easily filtered	Yes, there and inherently susceptible	Not there and easily filtered
Flicker noise (1/f)	No	Yes	No
LO-self mixing	No	Yes	No
Interferer leakage	No	Yes	No
RCVR DC offset calibration	No	Yes	No
BBIC DC offset calibration support	No	May be required	No
Option to disable DC filtering	N/A	No	Yes – potential sensitivity improvement
Die size	Large	Small	Moderate to small
Power consumption	Moderate	Small	Small
Component selection	Difficult	Moderate	Easy
PCB layout	Moderate	Difficult	Easy
Cost reduction roadmap	Difficult	Moderate	Easy
cost	High	Low	Moderate
Solution risk	Low	Moderate	Low

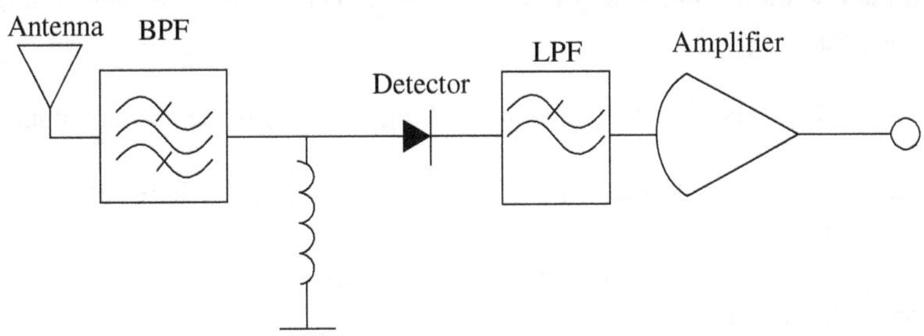

Figure 6.24 Envelope detector configuration

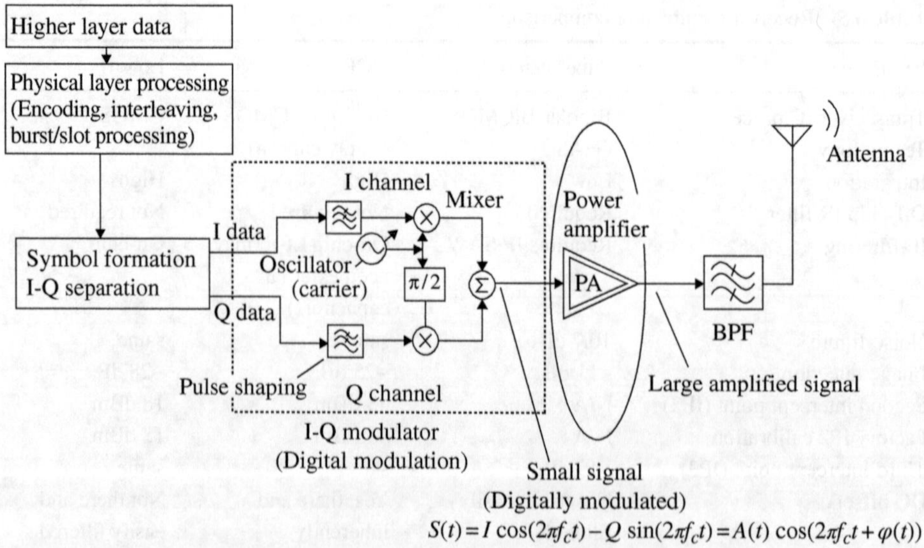

Figure 6.25 Transmitter architecture

(i) a digital modulator, which modulates the baseband digital signal to baseband analog signal; (ii) a pulse-shaping filter or root-raised cosine filter, which shapes the modulated analog signal to limit the transmission bandwidth; (iii) an RF upconverter, which converts the low frequency analog signal to the RF frequency; (iv) a power amplifier, which amplifies the signal; (v) an RF filter, which attenuates the out-of-band signal; (vi) duplexer for Tx-Rx separation, and (vii) an antenna for converting the amplified RF electrical signal to an em wave.

As shown in Figure 6.25, at every transmit time interval (TTI) the information bit reaches the physical layer for processing. After processing, the processed burst bits (slot data) are used for symbol formation according to the modulation technique (e.g. digital modulation) used for transmission. The bits in a symbol are separated into I and Q paths and are input to the I-Q modulator. Symbols are vectors and are represented in an IQ plane. The placement of symbol constellation points in the IQ plane is based on the modulation format. Next, it is pulsed shaped (filtered to make the signal band limited by removing the abrupt transitions in it). The baseband signal is then frequency upconverted to RF by using a quadrature modulator. In the quadrature modulator, the I and Q signals are modulated with the in-phase and the quadrature-phase carriers respectively. The outputs are then summed up to yield the modulated signal:

$$S(t) = I\cos(2\pi fct) + Q.\sin(2\pi fct),\text{ where fc is the carrier frequency.}$$

This can also be represented as $S(t) = A(t)\cos(2\pi f t + \varphi(t))$, where $A(t) = \sqrt{(I^2 + Q^2)}$, $\varphi(t) = \tan^{-1}(Q(t) / I(t))$.

The modulated RF signal is then amplified by a power amplifier before it is fed to antenna (via a duplexer).

6.5.1 Power-Limited and Bandwidth-Limited Digital Communication System Design

Communication system design involves tradeoffs between performance and cost. Performance parameters include transmission speed, accuracy, and reliability, whereas cost parameters include hardware complexity, computational power, channel bandwidth, and required power to transmit the signal. Generally, a communication system is designed based on:

- a bandwidth-limited system, or
- a power-limited system.

In bandwidth-limited systems, spectrally efficient modulation techniques can be used to save bandwidth at the expense of power, whereas in power-limited systems, power efficient modulation techniques can be used to save power at the expense of bandwidth. In a system designed to be both bandwidth and power limited, error correction coding (channel coding) can be used to save power or improve error performance at the expense of bandwidth. Trellis-coded modulation (TCM) schemes can also be used to improve the error performance of bandwidth-limited channels, without an increase in bandwidth.

Generally, there are two main causes of error in communication systems: noise and distortion. Here, let us consider distortion. This occurs in an ideal baseband channel, only if the bandwidth of the transmitted signal exceeds the bandwidth of the channel and also at the transmitter's power amplifier module.

In order to come up with a proper design tradeoff amongst accuracy, transmitted power and transmission speed or bandwidth requirement, we need to examine how the baseband data pulse's energy is distributed throughout the frequency band. The time and frequency band representations of single rectangular pulses of height, A and width τ, are shown in Figure 6.26.

From this, we can calculate and plot an average normalized power spectral density of a series of *n* (number) of such pulses, as shown in Figure 6.26, and its value will be:

$$G(f)=\{nA^2\tau^2 sinc^2(\pi.f.\tau)/n\tau\}=A^2\tau sinc^2(\pi.ft) \quad (6.13)$$

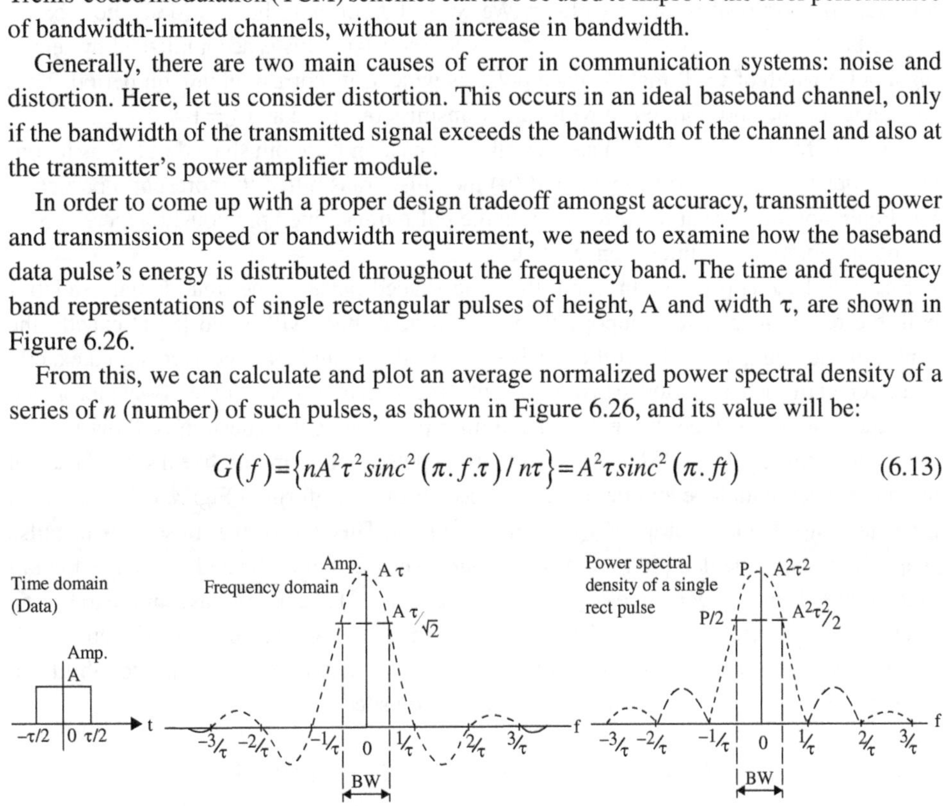

Figure 6.26 Rectangular pulse, its frequency domain representation and power spectral density

From Figure 6.27(a), it is obvious that most the power lies in the main lobe and inside the calculated BW. We can quantify the accuracy of received signal for an ideal baseband channel by stating the percentage of the transmitted signal's power that lies within the frequency band passed by the channel. From Eq. 6.13, it is evident that the accuracy (which is dependent on the transmitted power spectral density (G (f)) at a predefined level), bandwidth or speed of transmission (τ), and the amplitude of the data pulse are interdependent. If the transmission rate is higher, which requires more bandwidth to maintain the accuracy at the same level (say 95%), we need to increase the data pulse's amplitude or power level and vice versa. More BW means more average transmitted power will lie inside the frequency band, so the amplitude of the signal A (the transmitted power level) can be reduced. So, more available BW requires less transmitted power for the same level of accuracy (desired power in the selected band).

If we increase the amplitude of the pulse, this will lead to more power consumption, more interference, and more nonlinear distortion in the amplifier. So, we need to investigate how we can increase transmission speed without reducing accuracy or increasing bandwidth and amplitude of the transmitted pulse. We need to consider special shaped pulses, which require less bandwidth than rectangular pulse. We know bandwidth is inversely proportional to pulse width τ. We want the pulses to be as wide as possible to reduce bandwidth, but we don't want the pulses to overlap. This is accomplished by selecting a pulse width of $\tau = T$, making each pulse as wide as its corresponding bit period. Thus we can relate the optimum pulse width and transmission speed as $\tau_{opt} = 1/r_b$.

Our requirement should be (i) a spectrally compact and smooth shaped pulse, as it contains lower frequency components, and (ii) the pulse transmitted to represent a particular bit should not interfere at the receiver with the pulse transmitted previously – there should not be any intersymbol interference (ISI).

Keeping these two points in mind, the sync-shaped pulse in the time domain satisfies both the requirements. It is important to observe that there will be no ISI at exactly the center of each bit period. So, at the receiver we need to sample the received signal exactly in the center of each bit period to avoid ISI. But if the receiver is not completely synchronized with the transmitter, then this will cause timing jitter. Now, the question is "how we can reduce the timing jitter or ISI?" We need to use a pulse that is smooth like a sync pulse but has a narrower main lobe and flatter tails. Consider the waveform in Figure 6.27(b), which is a sync-shaped pulse multiplied by a damping factor. This is known as raised cosine pulse shape. The larger the damping factor β, the narrower the main lobe and the flatter the tail of the pulse. Thus a larger value of β means less effects of ISI and less susceptible to timing jitter. But, the greater the value of β, the greater is the bandwidth. The roll-off factor $\alpha = \beta/(r_b/2)$ allows us to express the tradeoff of additional bandwidth for less susceptibility to jitter in a manner that is independent of transmission speed.

$$P(t) = A\,sinc(\pi.t.r_b)/[cos(2.\pi.\beta.t/1-(4\beta t)^2] \qquad (6.14)$$

The tradeoffs for selecting the pulse shapes for binary PAM are shown in Table 6.6.

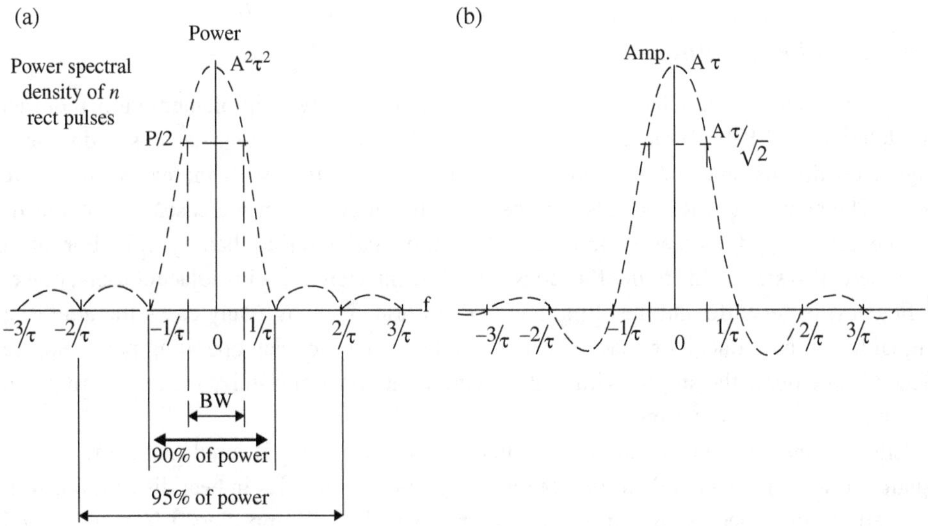

Figure 6.27 (a) Power spectral density of *n* rectangular pulses. (b) Damped sync pulse

Table 6.6 Advantages and disadvantages of using different pulse shapes

Pulse shape	Bandwidth	Advantage	Disadvantage
Rectangular $\tau = n/r_b$	$2.r_b$ (95% in-band power)	No ISI, minimum susceptibility to jitter	High bandwidth requirement
Sync $\tau = 2/r_b$	$0.5.r_b$ (100% in-band power)	Low bandwidth, no ISI only if receiver is perfectly synchronized.	Susceptible to timing jitter
Raised cosine (freq domain) $\tau = n/r_b$	$r_{b/2.}(1+\alpha)$ (100% in-band power)	No ISI, less susceptibility to jitter than sync pulse.	Requires more bandwidth than sync pulse but less than rectangular pulse

Obtaining tighter synchronization requires more complex equipment within the receiver. Thus, transmitting raised cosine pulses will reduce the receiver's complexity relative to transmitting sync pulses. A filter, composed of discrete components, is designed to produce an impulse response resembling a time-delayed version of the raised cosine pulse. Then a series of narrow pulses are input to the filter, one pulse per bit period, with a positive narrow pulse representing each "1" whereas a negative narrow pulse represents each "0." The drawback of analog method is that it requires large numbers of discrete components.

Generating raised cosine shaped pulses using digital circuitry is much easier than using analog circuits. As we have observed, design parameters like accuracy, transmitted power, BW, data rate, complexity, are all interrelated. So, designing an optimum system requires proper tradeoffs among these parameters and proper application of the system.

6.5.2 Investigation of the Tradeoffs between Modulation and Amplifier Nonlinearity

The choice of modulation has always been a function of hardware implementation, required modulation, and BW efficiency. Amplifiers usually operate as linear devices under weak signal conditions and become more nonlinear and distorting with increasing the drive level. The amplifier efficiency also increases as the output power increases, but that leads to the problem of increased nonlinear distortion and reduced battery life. For most commercial systems, this tradeoff is constrained by interference with adjacent users, power efficiency, battery life, and resulting signal distortion. Thus, in many cases the amplifier signal levels are reduced or "backed off" from the peak efficiency operating point. So, we need to investigate the amplifier-modulation combination to minimize the energy required to communicate information.

Linear transmitter power amplifiers, like Class A or Class B amplifiers, offer good-quality, low-output signal distortion but with significant penalties in heat dissipation, size, and efficiency whereas, nonlinear amplifiers like class-C amplifiers offer very good efficiency and low heat dissipation but introduce more distortion into the output signal. As class C and AB amplifiers offer good efficiency so, for better power usage purposes, this type of amplifier is generally used as a RF transmitter power amplifier.

A higher level of modulation is used to carry more information bits per symbol. But every time a modulation level is doubled, an additional 3 dB of signal energy is needed to maintain equivalent demodulator bit error rate performance.

In the case of GSM, the modulation used is GMSK, where the Gaussian filtering ensures that the modulation is a constant envelope. The disadvantage is that decision points are not always achieved, resulting in a residual demodulator bit error rate. TDMA systems have always required close control of burst shaping, the rise and fall of the power envelope either side of the slot burst. In GPRS, this process has to be implemented on multiple slots with significant variations in power from burst to burst. As OFDM shows its high sensitiveness to nonlinear effects, so it requires more linear amplification than other modulation schemes. A multicarrier modulated signal has very large peak power, so the influence of a nonlinear amplifier becomes large. An increase in peak power leads to input signal saturation, which leads to nonlinear amplitude distortion and that leads to out-of-band radiation and degradation of the BER.

The modulation schemes used for WLAN, UMTS and GSM can be broadly divided into two categories: peak-to-average ratio, and error vector magnitude.

Peak-to-Average Ratio

The ratio of the maximum peak power divided by the average power is known as peak-to-average ratio (PAPR). If the PAPR is high then the power amplifier in the transmitter is operated at a relatively lower power level so that the peaks in the signal are not distorted by amplifier moving into the saturation region. In a multicarrier transmission (like OFDMA), multiple sinusoids (carriers) are added together and the resulting signal exhibits constructive and destructive behavior. The higher the number of these sinusoids, the higher is the PAPR. In WCDMA, where many orthogonal codes are summed up, such multicode signal transmission show a large variation in the envelope, and nonlinearity can cause problem. So, here as well, the efficiency of a high power amplifier is limited due to high PAPR.

Error Vector Magnitude (EVM)

As discussed above, to achieve good power efficiency, the power amplifier should work around its compression point, which makes the output signal distorted nonlinearly. These nonlinear distortions generate in-band interference, which results in amplitude and phase deviation of the modulated vector signal. To measure the error in the symbol vectors, error vector magnitude (EVM) is used to analyze in-band distortion. It is the measure between the ideal reference target symbol vector and the transmitted measured symbol vector. EVM = error vector (E) / transmitted measured symbol vector (Po). Generally, an I-Q constellation diagram is used to define signal and error vectors. It measures the modulation quality of the signal and indicates modulation accuracy.

6.5.2.1 Constant Envelope (Nonlinear) Modulation

In this case the signal envelope is fixed. It employs only phase information to carry the user data, along with a constant carrier amplitude. This allows the use of nonlinear amplifier stages, which can be operated in class AB or C, so good power efficiency can be achieved. The most common standard employing nonlinear modulation is Global Standard for Mobile Communication (GSM), which uses Gaussian Minimum Shift Keying (GMSK) with a BT factor of 0.3 and raw data rate of 270.833 kbit/s.

6.5.2.2 Nonconstant Envelope (Linear) Modulation

In this case the signal envelope varies with time. Information is conveyed in both the phase and amplitude of the carrier so the transmitters should not distort the waveform at all. Hence amplifier stages must be operated in a linear class A fashion. QPSK and BPSK modulation types are used for the UMTS and WLAN OFDM systems. That means, WLAN and UMTS use nonconstant envelope (linear) modulation whereas GSM uses nonlinear or constant envelope modulation.

In the case of a multimode system, the modulation schemes for different modes are already defined. So, we have no scope to change the modulation scheme but we can find the most suitable transmitter architecture and power amplifier for this multimode terminal solution.

6.6 Transmitter Architecture Design

Transmitter design has not resulted in a single preferred architecture suitable for all applications, due to the differing requirements for linear and nonlinear modulation schemes.

6.6.1 Nonlinear Transmitters

The favored architecture for constant-envelope transmitters is the offset phase-locked loop. This utilizes an on-frequency VCO, which is modulated within a phase-locked loop. A block diagram of this architecture is shown in Figure 6.28.

A modulated carrier is generated at an IF of f1 using an IQ vector modulator. The modulated carrier is applied to the phase-locked loop, which modulates the VCO phase in order to track the phase of the feedback part of the phase comparator. The output signal, fout, is converted back down to f1 using a mixer with an LO at f2 such that f1 = fout − f2, for comparison in the phase detector.

This architecture is used in most of the current GSM phones, as it provides optimum power efficiency, cost, and performance by minimizing the amount of filtering required at the output. It also readily extends to future GPRS enhancements. The offset approach eliminates the problems of LO leakage, image rejection, and spurious sideband conversion associated with heterodyne architectures, reducing the filtering requirements at the PA output. The PLL has a low pass response to the vector modulator, which invariably has a high wideband noise floor. The noise is therefore rejected by the loop before the signal reaches the PA. The output is also protected from the high noise figure of the offset mixer, which is not the case in heterodyne architectures. Since the signal is of constant amplitude, it is possible to apply power control within the power amplifier stages, which follow. This allows the main transmitter to be optimized for power consumption.

6.6.2 Linear Transmitters

Linear modulation transmitters are required to preserve both the phase and amplitude of the signal. The consequence of this is that the offset phase-locked loop transmitter cannot be used as it only transmits phase information. It would be possible to apply the amplitude

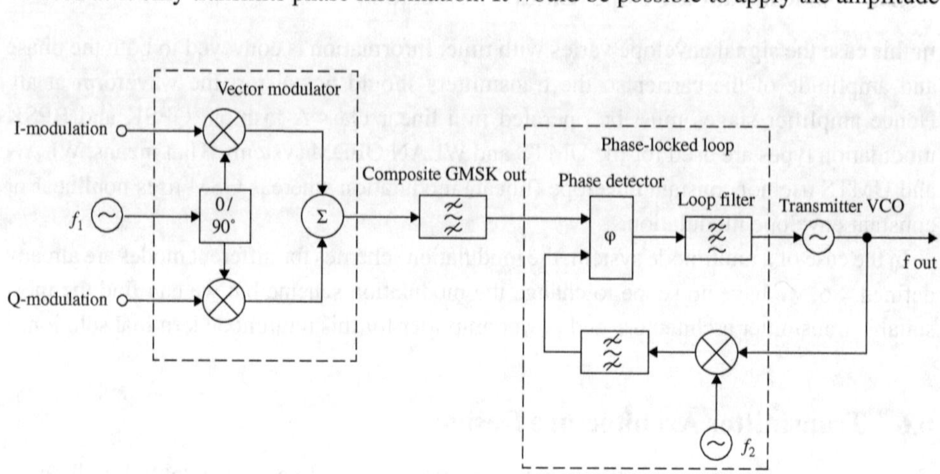

Figure 6.28 Transmitter architecture for nonlinear modulation schemes

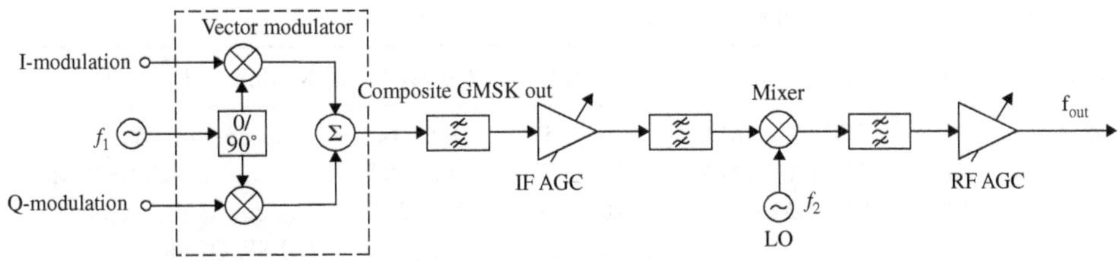

Figure 6.29 Transmitter architecture for linear modulation schemes

modulation component at the VCO output but there are technical difficulties associated with this technique, in particular AM-to-PM conversion in the power amplifier, which have yet to be solved to give a viable solution. Instead a conventional heterodyne architecture is usually employed, comprising an IF modulator and an upconversion mixer.

The power control requirements of the standards usually call for power control to be distributed through the transmitter module. This is because the required dynamic range calls for more than one variable gain stage. For a cellular system the final transmit carrier frequency can be up to 2 GHz. Variable gain amplifiers at the final transmit frequency are difficult to implement with large gain control ranges. So, it necessary to perform some of the gain control earlier in the transmitter chain.

The transmitter architecture for a linear scheme is shown in Figure 6.29. With the offset PLL architecture, the carrier modulation is achieved at an IF f1. Here, the baseband I and Q signals contain information in both phase and amplitude. The signal is band-pass filtered to remove unwanted products and wideband noise from the vector modulator output, and a variable gain stage enables some power control (subject to carrier leakage limitations). This signal is then upconverted in a mixer using a LO at frequency f2. The output is filtered for image rejection and so forth, and a further variable gain stage is used to give the total required dynamic range. The distribution of the power control needs to be carefully planned to maintain the SNR along the chain. In particular the vector modulator and upconvert mixer generate wideband noise levels, which need to be considered in the transmitter-level plan. The subsequent power amplifier is required to be linear for the entire dynamic range of the transmitter, which can lead to power inefficiency. However, some transmitters do switch the PA bias from high to low power to help this situation.

6.6.3 Common Architecture for Nonlinear and Linear Transmitters

The future of wireless communication can be considered as a plethora of heterogeneous systems operating together with current and legacy technologies. A mobile handset designed to operate on multiple RF bands is called multiband phone and when it is designed to operate for different cellular standards it is called multimode phone. One viewpoint sees the next generation of wireless communication as characterized by the seamless integration of a wide variety of systems (cellular – GSM, GPRS, EDGE, UMTS and WLAN) for

Table 6.7 Use of different modulation techniques for different systems

Mode	Duplex	Multiple access	Transmit band	Receive band	Modulation
GSM-900	FDD	TDMA/FDMA	890–915	935–960	Constant envelope (GMSK)
GSM-1800	FDD	TDMA/FDMA	1710–1785	1805–1880	Constant envelope (GMSK)
UMTS	FDD	W-CDMA	1920–1980	2110–2170	Nonconstant envelope (BPSK/QPSK)
WLAN	TDD	OFDMA/CSMA	2400–2483.5	2400–2483.5	Nonconstant envelope (QPSK)

efficient service provision. WLAN handsets offer a significant improved data rate over cellular handsets. However, they have a very limited range and access nodes can only be found in high-use areas. On the other hand, a cellular handset, which has a larger range at the expense of data rate, can solve this problem of insufficient range. In case of cellular handsets, several standards have been introduced and those are going to coexist. There are also different frequency bands introduced at different geographical locations. So, one major challenge will be to find out a robust solution to incorporate both WLAN and different cellular modes (GSM-900/1800 and UMTS) in a single piece of user terminal equipment (UE), which should have the capability to select the appropriate mode in any given coverage area, ideally with a seamless interface between different modes of operation.

For each standard, the handset must be able to transmit and receive in conformity with the ANSI/IEEE or 3GPP standards. Table 6.7 tabulates different parameters used for different modes of operation. There are number of issues that will arise when so many functionalities for different standards coexist in a small, single item of UE. Here, we will mainly focus on the transmitter design challenges.

Unfortunately, conventional GSM/GPRS transmitter architectures are designed to deliver constant-envelope half-duplex signals and have little in common with UMTS architectures, which generate envelope-varying full-duplex signals. As a result, combining these transmitters within one multimode handset can be an expensive proposition. However, by changing the transmitter's modulation system from quadrature (I/Q) to polar, the architectures can be designed to deliver constant-envelope and envelope-varying signals. Polar transmitter-based architectures have no requirement for linear radio-frequency (RF) circuitry, which means that circuits can be designed with an emphasis on optimizing efficiency rather than linearity.

This describes a new architecture for implementing multimode multiband transmitters, which provides all the necessary functionality for both linear and nonlinear modulation schemes. There are a number of different architectures that can be employed to implement a multimode transmitter. However, the tradeoffs between power consumption, linearity, baseband complexity, and implementation issues have resulted in a favored architecture for supporting both nonlinear and linear modulation. The architecture of a multimode transmitter supporting the linear and nonlinear modulation scheme is shown in Figure 6.30.

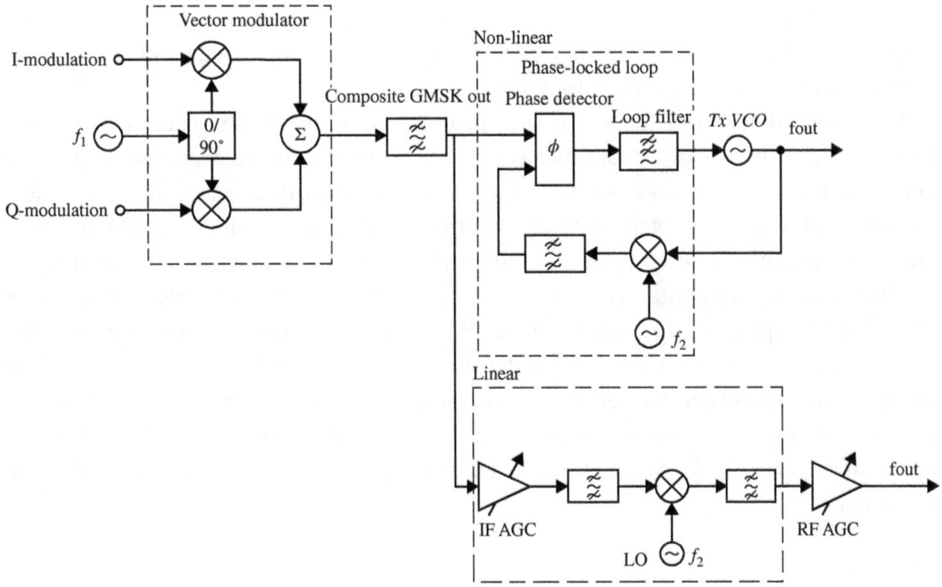

Figure 6.30 Multimode linear-nonlinear transmitter

As discussed, there are several problems associated with the power amplifier design. Its advantages and disadvantages are as follows:

- Advantages: (i) the circuit BW should not exceed the signal BW; (ii) it has a wide output power dynamic range; (iii) there is no problem if the RF signal amplitude becomes zero; (iv) design and manufacturing experience is common; (v) it generally supports any signal type.
- Disadvantages: (i) RF circuits are generally not linear, leading to compensating design complexity; (ii) Low DC to RF energy efficiency due to a back off requirement; (iii) high PA operating temperature from internal power dissipation; (iv) high broadband output noise floor; (v) difficulty maintaining modulation accuracy; (vi) the possibility of self-oscillation; (vii) gain is dependent on frequency. To address the PA efficiency problem with envelope-varying signals, polar modulation techniques were proposed.

6.6.4 Polar Transmitter

The polar transmitter transform the digital I/Q baseband signals from the Cartesian domain to the polar domain. The quadrature equivalent signal representation is: $S(t)=I(t)\cos\omega_c t+Q(t)\sin\omega_c t$, which can be represented in the polar domain as $A(t)\cos(\omega_c t+\Phi(t))$. In this case the amplitude and phase components of the signal are processed independently. The phase information extracted from the original signal (either constant envelope or non-constant envelope) is transformed into a constant envelope signal. This is achieved by phase modulating with the help of a phase-lock loop designed to output the desired transmit frequencies. The resulting signal may now be amplified by compressed amplifiers without concern of distorting amplitude information.

The extracted amplitude information is quantized into control bits. These bits are used to modulate a digital power amplifier (DPA). Each bit is a digital representation of the amplitude envelope. The control bits are used to switch amplifier elements of the DPA into on or off states. The extracted amplitude information is quantized into control bits. These are used to modulate a digital power amplifier (DPA) and switch on or off the elements of the DPA. Each bit is a digital representation of the amplitude envelope. Fewer quantization states can be implemented for decreased amplitude modulation resolution and more quantization states are implemented for greater resolution. The digitized amplitude envelope and the phase-modulated RF carrier are synchronized and recombined within the DPA to produce linear and efficient RF transmission. Existing developments of polar modulated transmitters generally fall within three major categories: polar loop, polar modulator, and direct polar.

Polar Loop

Here feedback control is used to correct the output signal into its desired form (Figure 6.31). One advantage of such a polar loop is improved PA efficiency over the best linear systems, gained from operating the power amplifier much closer to saturation. Additional benefits from this compressed PA operation include a low wideband output noise floor and also, usually, a reduction of circuit oscillation tendencies with varying output load impedance. Disadvantages include the need for a precision receiver within the transmitter, control loop bandwidths that must greatly exceed the signal bandwidth, a restricted output power dynamic range, maintaining stability of the feedback control loops across the output dynamic range, and the lack of circuit design techniques when operating with strong circuit nonlinearity.

Polar Modulator

As shown in Figure 6.32, in this type of transmitter, the output signal is amplified from the output of this polar modulator using conventional linear amplifier devices. The advantages and disadvantages of polar modulator transmitters are given in Table 6.8.

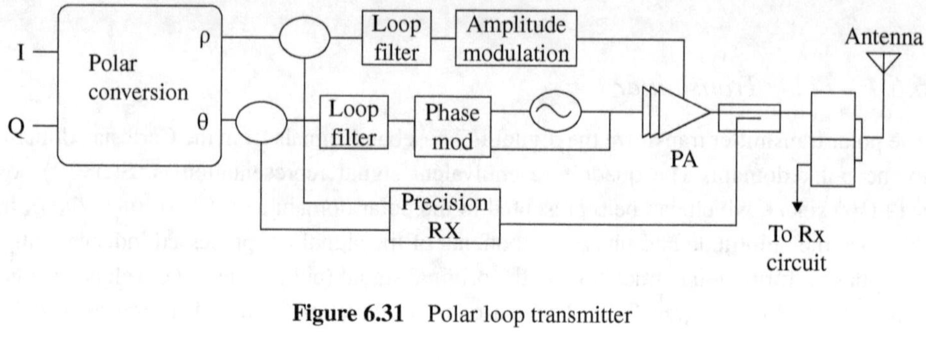

Figure 6.31 Polar loop transmitter

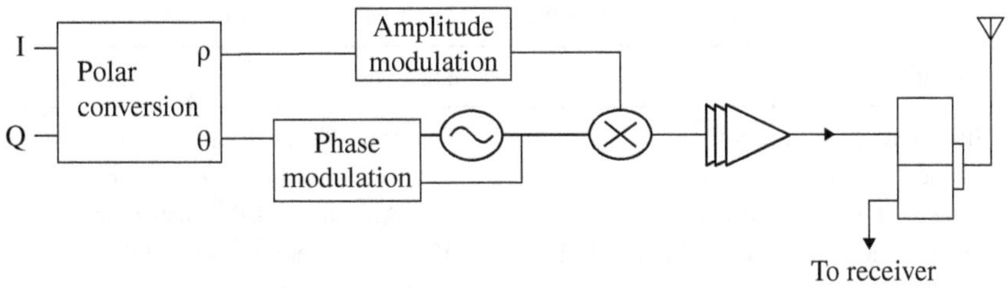

Figure 6.32 Polar modulator transmitter

Table 6.8 Advantage and disadvantages of polar modulator transmitter

Advantages	Disadvantages
The modulator noise is much lower than a quadrature modulator	Linear PA provides no efficiency benefit
Predistortion applied to the polar modulator	Time alignment of amplitude modulation and phase modulation is required
Generating output power = 0 is not an issue	Lack of manufacturing experience for polar modulator transmitter
Sigma-delta ADC can be used along with this	
Modulation accuracy is good	

Direct Polar Transmitter

Another method that removes the feedback from the PA is direct polar transmitter. This is shown in Figure 6.33. The advantages and disadvantages of direct polar transmitters are shown in Table 6.9.

All signals have constant-envelope phase components, so linear RF circuitry is not required in either architecture. Between the UMTS and GSM modes, most things are implemented digitally within the polar modulator, and therefore have little impact on how the radio design is actually implemented.

6.6.5 Power Amplifier (PA)

A power amplifier is used in the transmitter path to amplify the RF signal, which is passed to the antenna to transmit via the air. The amplified power is delivered to the load. In this case the load is an antenna circuit, which has an impedance value of Z_0 (50 Ω). As shown in Figure 6.34, a small input AC signal will cause the base current to vary above and below the DC bias point (Q point), which will then cause the collector current (output AC signal) to vary from the DC bias point set as well as the collector-emitter voltage to vary around its DC bias value. So, the amplifier amplifies the small input AC signal to a large output AC signal and it is able to do so by taking the DC power connected to the amplifier circuit (for biasing). The efficiency of a power amplifier is defined as:

$$\text{Power-aided efficiency}\left(\text{PAE}\right)=\left(\left(P_{\text{RF output}}-P_{\text{RF input}}\right)/P_{\text{dc input}}\right)*100\% \qquad (6.15)$$

Table 6.9 Advantages and disadvantages of direct polar transmitter

Advantages	Disadvantages
Its efficiency is higher than the best linear transmitter	Power amplifier (PA) characterization required within the transmitter
Linear RF circuit is not required	AM/AM and AM/PM distortion
Unconditional Power amplifier stability	Time alignment of AM and PM paths
Modulation accuracy	Lack of manufacturing experience

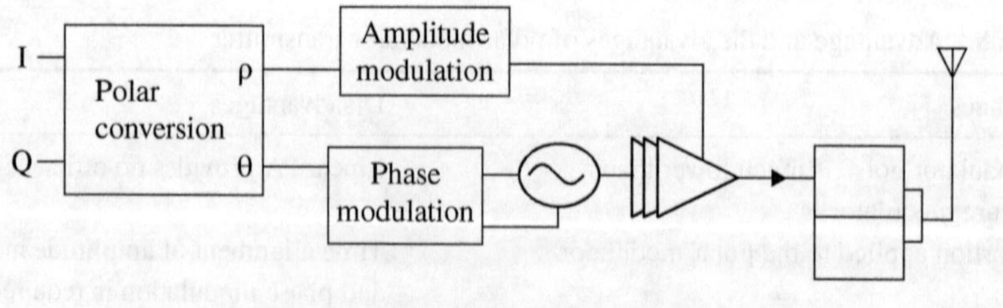

Figure 6.33 Direct polar transmitter

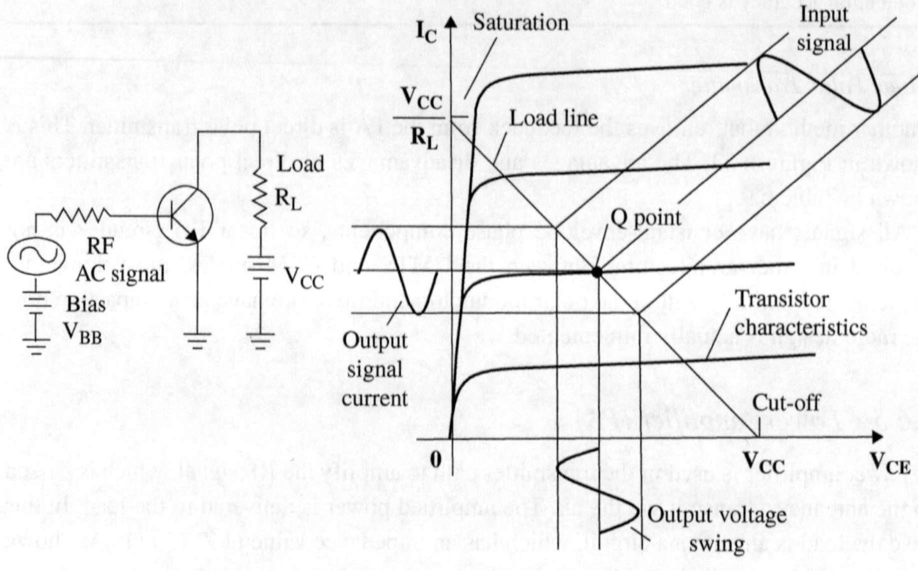

Figure 6.34 Amplifier biasing and signal variation

where, $P_{RF\,out}$ and $P_{RF\,in}$ are the output and input AC signal power to the amplifier and $P_{dc\,input}$ is the DC power consumed by the amplifier in the amplification process. Biasing a transistor amplifier is the process of setting the DC operating voltage and current (Q point) to the correct level so that any AC input signal can be amplified correctly by the transistor.

When the transistor is biased such that the Q point is near the middle of its operating range – which is approximately halfway between cutoff and saturation – it is said to be operating as a Class A amplifier. In this condition, the amplifier can amplify the input signal over its entire input cycle and saturation can be avoided. Some amplifiers are optimized to add minimum distortion to the signal, while other designs are optimized for high output power or efficiency. For example, a class C amplifier is biased in such a way that it conducts for less than 180° of the input signal. So it works mostly in the cutoff region and they will only amplify a signal large enough to get the transistors out of cutoff. They are very nonlinear and are only used in certain applications, such as RF amplification, to achieve higher efficiency. They generate a lot of distortion, so they must be followed by good filters to remove the unwanted harmonics and other distortion products. It is desirable to run the power amplifier as close to saturation as possible to maximize its power efficiency, and then employ a linearization technique to suppress the distortion introduced in this near-saturated region.

Various linearization techniques are applied to nonlinear PAs to get a good linearity and modest efficiency. These include: (i) RF amplifier PA backoff – the backoff is the distance between the saturated point and the average power level. Increasing the backoff of the power amplifier means that the signal is contained better in the linear range, and thus the effects of nonlinearities are reduced, although power efficiency is reduced as well; (ii) RF predistortion; (iii) Cartesian feedback; (iv) polar feedback; (v) feed forward; (vi) envelope elimination and restoration, etc.

6.6.5.1 Envelope Tracking

As explained, if the Q point is fixed towards the saturation region (refer to Figure 6.34) then for a small increase in the base current there will be a larger amplification of signal that means amplifier efficiency will be higher. On the other hand, if it is going to saturation then it will distort the input signal and to avoid that amplifier has to do power backoff. So, the goal is to set the Q point such that that it does not go into saturation or cutoff when an input AC signal is applied. There are various new techniques available for amplifier biasing to obtain optimum power efficiency with minimum distortion. If, instead of using fixed biasing, bias point is dynamically varied based on the input signal power level then saturation can be avoided, and at the same time, maximum efficiency can be achieved. In case of the average power tracking method, the tracking module computes the average input power level over every slot duration and, based on that, it sets the biasing point for that slot duration dynamically. In the case of the envelope tracking method (Figure 6.35) it dynamically computes the power level and then sets the biasing level accordingly so that it avoids the saturation and at the same time keeps the bias point at optimum possible point towards the saturation region to get maximum efficiency (Figure 6.36). Theoretically the envelope tracking method can provide an efficiency of 100%.

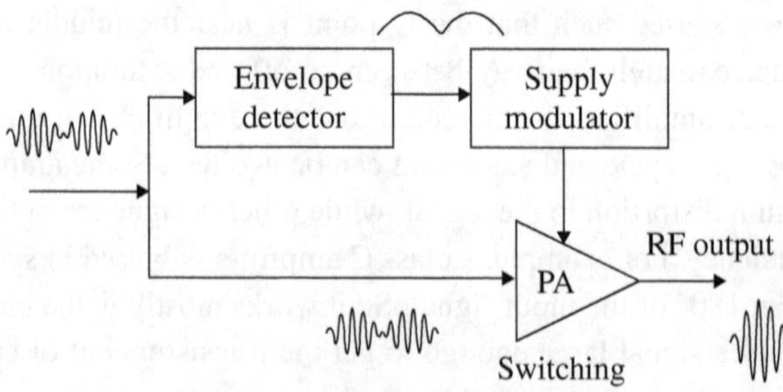

Figure 6.35 Envelope tracking method

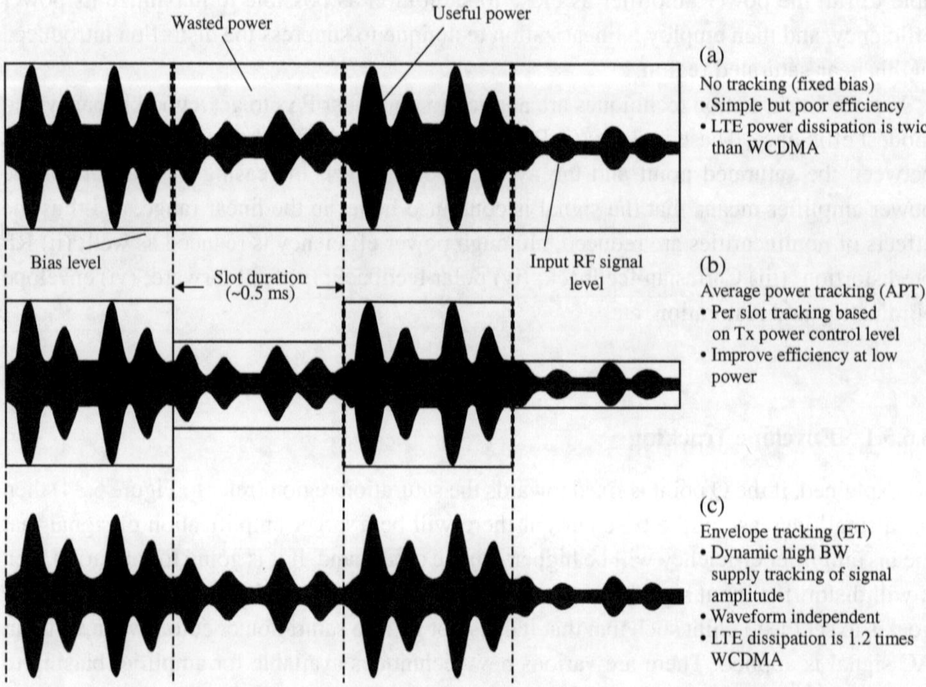

Figure 6.36 Different tracking methods and efficiency level: (a) no tracking (bias level constant line); (b) average tracking (bias level constant line over a slot); (c) envelope tracking (bias level along the envelope)

6.7 Transmitter Performance Measures

The performance of transmitter is generally measured taking these following factors into account: amplifier power efficiency, amplifier nonlinear distortion, modulated signal power efficiency and signal bandwidth efficiency. This is based on the total energy needed per transmitted bit, and bandwidth efficiency.

The average total power consumed by the amplifier is:

$$\begin{aligned} P_t &= P_{dc} + P_{in} \\ &= P_{rf}\left(1+\left(1-P_{ae}\right)P_{dc}/P_{rf}\right) \end{aligned}$$

where P_{ae} is the instantaneous power-added efficiency of the amplifier $P_{ae}(t) = (P_{rf}(t) - P_{in}(t)) / P_{dc}(t)$.

Transmitter figures of merit are dependent on the following parameters: (i) spectral efficiency; (ii) power efficiency; (iii) spurious emission; (iv) power level.

6.7.1 Design Challenges

Transmitter architectures for multiple standards include direct upconversion, translation loop, and modulation through phase-locked loop and polar loop. The trend has been towards further digitizing to reduce the analog content in the total transmitter chain. Key challenges include current drain, dynamic-range requirements, and cost. Loop-phase modulation using a sigma-delta modulator shows promise in terms of low power consumption and a simpler architectural approach. For systems like CDMA and W-CDMA, separation of AM and PM components is required. That leads to polar loop architectures, which are gaining wider use, but challenges remain in their use for wideband systems, where alignment of the AM and PM components and the effect on spectral distortion are critical. While direct modulation has the advantage of compatibility with multiple standards, the challenges of meeting noise-floor requirements remain. Multimode phones require several bulky SAW filters to attenuate the receiver band noise. Signal digitization in the transmitter could include I and Q oversampled D/A converters to ease the requirements of the reconstruction filter. There are no blockers in the transmitter, which eases the design of the converter somewhat. Sufficient dynamic range to meet spectral mask requirements still has to be considered in the transmitter chain.

The final stage in the transmitter chain is the power amplifier, which transmits close to 3 W at maximum output power in certain systems. Maintaining efficiency at such power is critical; traditionally, PAs have been designed in GaAs or InGaP.

Recent trends point toward CMOS power amplifiers that can potentially enable on-chip integration with the rest of the transmitter and lower system costs. However, challenges remain in terms of efficiency, thermal behavior and isolation.

6.8 LTE Frequency Bands

The LTE frequency band is tabulated in Table 6.10. Many of the LTE frequency bands are already in use and some are new. The FDD LTE frequency bands are paired to allow simultaneous transmission on two frequencies. As in TDD LTE, the uplink and downlink share the same frequency bands so, it is unpaired.

Table 6.10 LTE frequency bands (DL and UL) (3GPP TS 36.101)

LTE Band	Uplink Operating Band F(UL) Low – F(UL) High (MHz)	Downlink Operating Band F(DL) Low – F(DL) High (MHz)	Duplex Mode
1	1920–1980	2110–2170	FDD
2	1850–1910	1930–1990	FDD
3	1710–1785	1805–1880	FDD
4	1710–1755	2110–2155	FDD
5	824–849	869–894	FDD
6	830–840	875–885	FDD
7	2500–2570	2620–2690	FDD
8	880–915	925–960	FDD
9	1749.9–1784.9	1844.9–1879.9	FDD
10	1710–1770	2110–2170	FDD
11	1427.9–1447.9	1475.9–1495.9	FDD
12	699–716	729–746	FDD
13	777–787	746–756	FDD
14	788–798	758–768	FDD
15	Reserved	Reserved	FDD
16	Reserved	Reserved	FDD
17	704–716	734–746	FDD
18	815–830	860–875	FDD
19	830–845	875–890	FDD
20	832–862	791–821	FDD
21	1447.9–1462.9	1495.9–1510.9	FDD
22	3410–3490	3510–3590	FDD
23	2000–2020	2180–2200	FDD
24	1626.5–1660.5	1525–1559	FDD
25	1850–1915	1930–1995	FDD
26	814–849	859–894	FDD
27	807–824	852–869	FDD
28	703–748	758–803	FDD
29	Downlink Only	717–728	FDD
30	2305–2315	2350–2360	FDD
31	452.5–457.5	462.5–467.5	FDD
32	Downlink Only	1452–1496	FDD
33	1900–1920	1900–1920	TDD
34	2010–2025	2010–2025	TDD
35	1850–1910	1850–1910	TDD
36	1930–1990	1930–1990	TDD
37	1910–1930	1910–1930	TDD
38	2570–2620	2570–2620	TDD
39	1880–1920	1880–1920	TDD
40	2300–2400	2300–2400	TDD
41	2496–2690	2496–2690	TDD
42	3400–3600	3400–3600	TDD
43	3600–3800	3600–3800	TDD
44	703–803	703–803	TDD

Further Reading

Das, Sajal Kumar. (2000) *Microwave Signals and Systems Engineering*, Khanna.

Das, Sajal Kumar. (2010) *Mobile Handset Design*, John Wiley & Sons, Ltd.

Stern, H. P. E. and Mahmoud, S. A. (2004) *Communication Systems: Analysis and Design*, Pearson Education.

Varrall, G. and Belcher, R. (2003) *3G Handset and Network Design*, John Wiley & Sons, Inc.

7

Software Architecture Design

7.1 Introduction

A mobile terminal, being an embedded system, is a combination of hardware and software. It needs software to drive the hardware components. Figure 7.1(a) shows typical software components required to control any embedded device. Each software component in the stack uses a higher level of abstraction to separate the code from the hardware device. Typically, in a mobile terminal, the software part consists of several modules, like the boot loader, the initialization code, the protocol stack, device drivers, a real-time operating system (RTOS), and a general operating system (OS). Apart from those, audio/video-related software, a Bluetooth stack and some other application software (such as gaming or a calculator) are also housed in a mobile phone device, which is shown in Figure 7.1(b). As discussed in the previous chapter, inside a mobile terminal there is commonly an application processor for driving applications and a modem protocol processor for driving protocol processing. As protocol processing requires real-time processing, an RTOS is used here, whereas in application processor general purpose OS is used. Both the processors have boot software and a mechanism for communication to exchange the data.

7.2 Booting Process

Generally, in a personal computer (PC) with a CPU (e.g. Intel processor) inside, after power on the hardware setup is done by the basic input / output system (BIOS), which sets the hardware and loads the software bootloader or OS kernel. However, as mobile terminals do not have BIOS, so, they rely on a firmware to set up the hardware before the operating system

Mobile Terminal Receiver Design: LTE and LTE-Advanced, First Edition. Sajal Kumar Das.

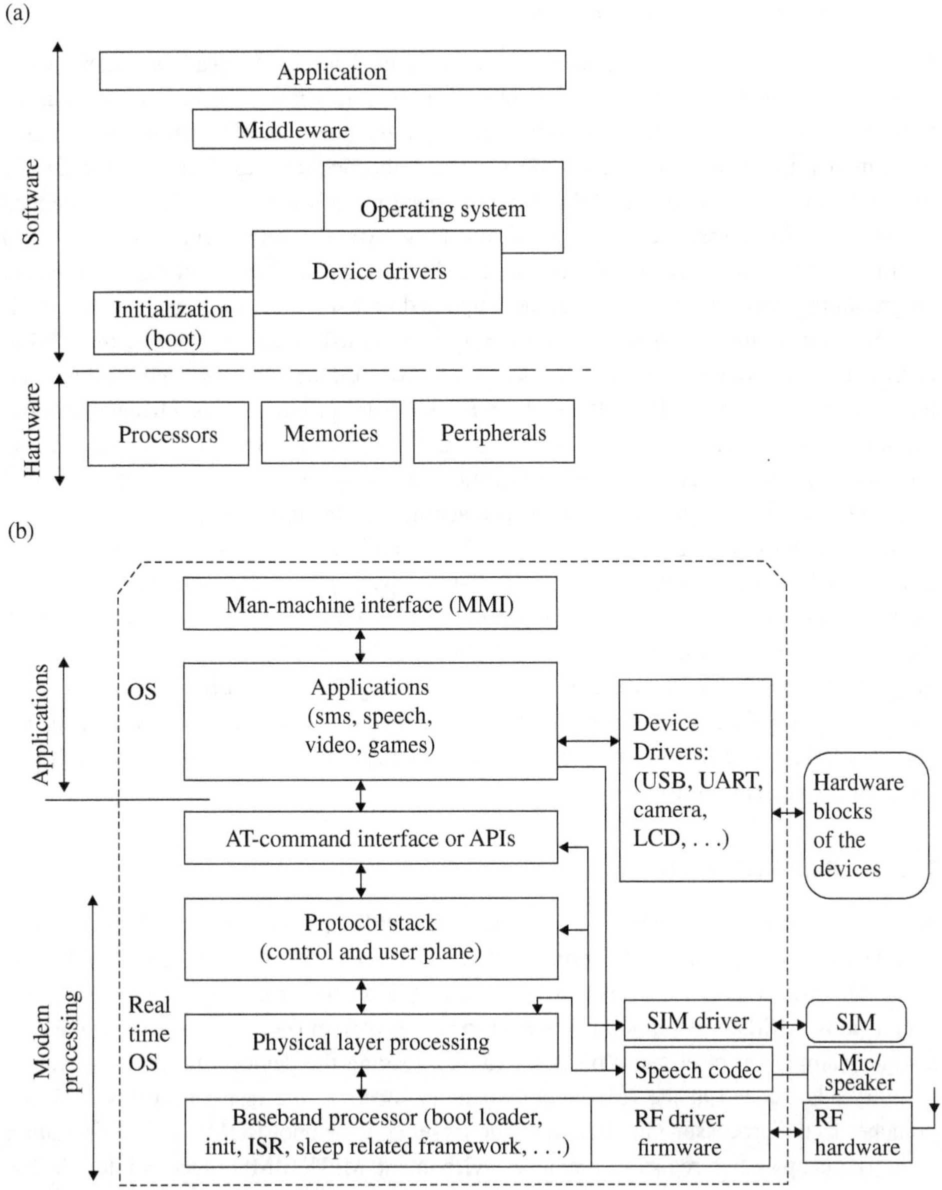

Figure 7.1 (a) Typical software components of an embedded system. (b) Software architecture inside a mobile terminal

kernel is loaded and started. The boot is often started from read-only memory (ROM). An embedded ARM processor-based system like a mobile phone can have a series of different bootloaders performing these tasks, where some of the tasks may be performed by proprietary software provided by the SoC chip vendor and some are general-purpose bootloaders.

7.2.1 *Initialization (Boot) Code*

If the development for an embedded device is completed and the software program for that device is not going to change throughout its lifetime, then the software program can be burned/programmed permanently in the memory of the device once by using conventional programmer. that is not the case for most of the embedded devices, including the mobile phone, because the software and the firmware residing in it is most likely to be changed time to time for several reasons, like firmware/software development, bug fixes, field upgrades, or end customer modifications. In such cases, the customer will require a special programming device to burn the upgraded/updated new software program, which is difficult. So, as a solution to that, a small program is provided, which will not be required to change at a later point in time, and this is used to download the main software program into the processor's memory. This allows the main software program to be changed without needing any conventional programmer. This small program is known as boot loader. It is responsible for low-level hardware initialization, and provides a stable mechanism to load and boot an operating system. This code is machine architecture specific.

The bootloader code is normally stored in ROM, so it is called as firmware and it is the first code to be ported and executed on a new platform when power is applied to the embedded hardware system. The boot is often started from a ROM embedded in the "system on a chip" (SoC) or similar persistent storage, such as NOR flash. The ROM code then reads a firmware image or directly the bootloader from a persistent storage such as flash memory. This uses the crypto capabilities and ensures that only certified software can be downloaded. An embedded ARM processor-based system can have a series of different bootloaders performing these initial tasks, as discussed below.

7.2.1.1 Boot ROM Code (Stage 1)

A hardware bootloader generally called the boot ROM is provided by the chipset vendor (preloaded into the processor's internal ROM), which is hardwired when the device is manufactured. After a power on reset, which causes the processor core to jump to the reset vector (most ARM cores support two vector locations 0×0 or 0×FFFF0000, controlled via a signal sampled at reset and a bit in CP15), by loading the device with the reset vector memory address and at the reset vector memory location, the instruction is stored that branches to the processor initialization code (reset code or boot ROM). The initialization code (i) sets up system registers, memory environment, MPU, MMU, stack pointer, the bss section; (ii) initializes the CPU clock; (iii) configures external bus; (iv) carries out low-level peripheral initialization, and so forth. These are required before jumping to perform the complex tasks. Reset must run as supervisor mode because the processor does not know the status of the register at the time of execution, so the CPU is put into supervisor mode.

Memory Remapping

One of the tasks of the initial reset code is memory remapping. At the time of power up, the processor jumps to fixed location 0×0. This is important to ensure that, at the time of power up, there is some executable code present at this location. To ensure this, some nonvolatile memory should be mapped to this address. But at a later point of time, if the vector table is stored in ROM, the execution of interrupts will be very slow as ROM access is slower than RAM (compared to RAM, ROM requires more wait states), and it cannot be modified by software code. So, for faster and more efficient execution of interrupts, it is better if interrupt handlers and the vector table is located in RAM at address 0×0. However, if RAM is mapped to address 0×0 at the power on of the processor, being a volatile memory, it won't contain any executable code. Thus, it becomes important that, at the time of startup, ROM is located at the 0×0 address and then, during normal execution, RAM is remapped to this location. Memory remapping can be achieved through hardware remapping, which is changing the address map of the bus. This can also be achieved through MMU. Generally, at the time of power on reset, apart from the reset vector, the rest of the vector table contains just the dummy handler, which is a branch instruction that causes an infinite loop (because this vector table is used very briefly and later on replaced by a vector table in the RAM after the memory remap operation mentioned above).

7.2.1.2 Software Bootloader (Stage 2)

The boot ROM code should be small in size (as it is stored in internal ROM), so if the size exceeds a specified limit a custom boot routine will need to be written, which is referred to as the second-level bootloader, or the secondary bootloader, or the device bootloader, or the software bootloader. The main task of software bootloader is to set up the execution environment (setting up the vector table – handlers for each entry in the vector table, stack, critical I/Os etc., copying initialization values for initialized variables from ROM to RAM, and resetting all other variables to zero), to load the OS and pass over the execution (see Figure 7.2). The bootloader must set up and initialize the external DDR memory (setting up the controller, refresh rate, clock, etc.) before loading an image to it. The OS image can then be loaded from flash (in case of NAND flash) to RAM. It must also perform bad block management while accessing the flash memory. The OS image may be compressed in which case it needs to be decompressed before the program counter can be modified to point to the operating system entry point address. After the system setup, the bootloader's responsibility would be to look for an OS to boot. Again, like boot ROM, if the OS is not already loaded to flash, it will load this from the boot media in flash and execute it in place in the case of NOR flash (see Chapter 5), or place it in RAM in case of NAND flash. It will also place some boot parameters in memory for the OS to read when it starts up if required. After all the necessary system setup, the bootloader will pass over the execution to OS and go out of scope.

Normally, the device bootloader is located in external nonvolatile storage such as flash memory. For systems using NAND flash, the bootrom will load this boot loader to internal RAM (as execution is not possible directly from NAND flash) and set the program counter at the load address of the SW bootloader in RAM. For systems using NOR flash, control is transferred to the external flash (NOR flash is XiP – "execute in place"). The boot ROM

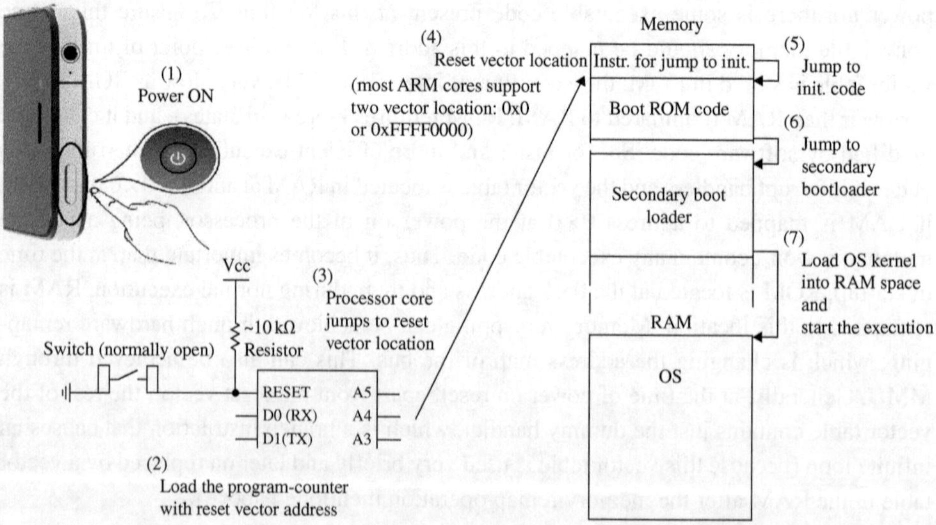

Figure 7.2 Booting process (steps 1 to 7)

will detect the boot media using a system register. This is to determine where to find the software bootloader. A particular sequence of probing for boot media is followed, as defined by the manufacturer. This includes the order of looking for the bootloader in external NOR/NAND flash, or probing for some specific characters on UART /USB for establishing connection with downloader to download the binary in flash. If no bootloader is found in any external memory, boot ROM listens for a download request on UART/USB port to start the download process. Thus, during the probing process, if the flash has already been programmed, the software bootloader will be detected to be available in flash; if not it will be downloaded to the flash by the boot ROM.

Diagnostics software provides a useful way to identify basic hardware malfunctions quickly. Debug capability is provided in the form of a module or monitor that provides software assistance for debugging code running on a hardware target.

Examples of embedded systems bootloaders are Das U-Boot and Barebox. The most well known bootloaders for Linux are GRand Unified Bootloader (GRUB) and Linux Loader (LILO).

The load process has to take into account the image format. The most basic image format is a plain binary format but a plain binary image does not contain any header or debug information. A popular image format for ARM-based systems is executable and linking format (ELF). This format was originally developed for UNIX systems and replaced the older format, called common object file format (COFF). ELF files come in three forms: relocatable, executable, and shared object.

There are generally two types of booting occurs in a system: (i) warm boot: pressing start button while system is already in the ON state (restart), and (ii) cold boot: pressing the start button while system is OFF (power ON then start).

7.2.1.3 Operating System Loading

At this stage it relinquishes control and hands over control of the platform to an operating system or application. That includes updating the vector table and modifying the program counter (*pc*) to point to the new image. For more sophisticated operating systems, such as Linux, relinquishing control requires that a standard data structure be passed to the kernel. The booting process for systems using Linux OS is explained below.

Stage 1: System Startup (Boot Monitor)
Before the Linux kernel starts, initialization of base hardware (includes setting up the CPU clock, memory, detecting or setting up the location and the size of the RAM, detecting the machine type, etc.) has to be completed and the kernel image should be loaded at the right memory address, boot parameters should be initialized. The boot monitor code initializes the memory controllers and configure the main board peripherals, set up a stack in memory, copy itself to the main memory DRAM, reset the boot memory remapping, remap and redirect the C library I/O routines and next loads and passes the control to the software bootloader. In modern systems, the boot process may include additional functionality – like, setting up additional hardware and peripherals, setting up a secure environment such as ARM TrustZone and verifying the loaded software images as a part of a secure boot.

In a multiprocessor system, when the system is powered on or reset, all CPUs fetch the next instruction from the reset vector address to their PC register. It is the first address in memory (0×00000000), where the boot code exists. Generally, CPU0 (or any one of the CPUs) continues to execute the boot monitor code as described above, whereas the other CPUs (CPU1, CPU2, and CPU3) execute a WFE (wait for event) instruction, which is actually a loop that checks the value of the SYS_FLAGS register. The other CPUs wait and start executing meaningful code only during the Linux Kernel boot process when indicated by CPU0.

The boot monitor application shipped with the board is similar to BIOS in the *pc* with limited functionalities and cannot boot a Linux kernel image. So, another bootloader like U-Boot is needed to complete the booting process.

Stage 2: Bootloader (U-Boot)
The software bootloader sets up the C environment and runs the rest of the initialization code. It configures the system's main memory, sets the Linux machine type (MACH_TYPE), loads the kernel image at the correct memory address, enters the kernel with the appropriate register values, and initializes the boot parameters to pass to the kernel. The bootloader must pass parameters to the kernel in form of tags (known as "ATAGs" these are binary encoded key value data structures defined in *linux/include/asm/setup.h*), to describe the setup it has performed, the size and shape of memory in the system, and optionally, numerous other values as described in Table 7.1.

After passing some boot parameters to it, the bootloader is ready to launch the Linux kernel image from a prespecified location. Next, it calls the kernel image by jumping

Table 7.1 Linux kernel parameter list

Tag name	Description
ATAG_NONE	Empty tag used to end list
ATAG_CORE	First tag used to start list
ATAG_MEM	Describes a physical area of memory
ATAG_VIDEOTEXT	Describes a VGA text display
ATAG_RAMDISK	Describes how the ramdisk will be used in kernel
ATAG_INITRD2	Describes where the compressed ramdisk image is placed in memory
ATAG_SERIAL	64-bit board serial number
ATAG_REVISION	32-bit board revision number
ATAG_VIDEOLFB	Initial values for vesafb-type framebuffers
ATAG_CMDLINE	Command line to pass to kernel

directly to the "start" label in the *arch/arm/boot/compressed/head.S* assembly file, which is the start header of the Linux kernel decompressor.

Finally, the bootloader installs itself at the upper end of the SDRAM area and allocates memory for use by malloc() and for the global board info data. In the low memory, the exception vector code is copied. Now, the final stack is set up.

After this stage, the kernel decompresses itself and then the processor-dependent (CPU0) kernel code executes, which initializes the CPU and memory. And finally, the processor-independent kernel code executes, which starts up the ARM Linux SMP kernel by booting up all other ARM CPU cores and initializes all the kernel components and data structures.

7.3 Operating System

The operating system performs basic tasks, such as recognizing input from the keyboard, sending output to the display screen, hardware abstraction, resource management, memory management, device management, job accounting, user interface, and driving applications. Today, every general-purpose processor and DSP should have an operating system (OS) to run the software programs. So, in a smart phone, there could be several OSs based on the requirements – one OS could be running in DSP for processing cellular modem physical layer tasks, one in an ARM processor for modem protocol processing, and one more in another ARM processor for application processing. As modem processing requires real-time processing, so the RTOS is used in DSP and ARM where receiver-signal processing and UE protocol processing is performed. A general OS is used in other processors where applications are running, which is commonly known as mobile phone OS.

A mobile phone operating system is an OS especially designed for smartphones, tablets, PDAs, or other mobile devices. Mobile operating systems combine features of a personal computer operating system with other features useful for mobile or handheld devices, which include touchscreen, cellular modem, Bluetooth, Wi-Fi, GPS mobile navigation, camera, video camera, speech recognition, voice recorder, music player, near field communication, infrared blasters, and USB. This OS must have very small code footprint, as memory size and resources are a big constraint. The power consumption should also be minimal.

The most commonly used mobile operating systems (OS) in modern smartphones include Android, iOS, Symbian, BlackBerry OS, Bada, Windows Phone, Sailfish OS, and Tizen. Some of these are discussed below.

7.3.1 Commonly Used Mobile Operating Systems

7.3.1.1 Android

Android, developed by Google Inc. and the Open Handset Alliance, is an open source operating system based on the Linux Kernel for mobile devices. It was initially developed by Android Inc., which was purchased by Google in 2005. Most Android software is free and open source, but a large amount of software on Android devices (like Google Search, Play Store, and Google Music) are proprietary and licensed. Google releases the Android code as open source under the Apache License. Android's releases are named after sweets or dessert items: Cupcake (1.5), Donut (1.6), Eclair (2.0), Frozen Yogurt ("Froyo") (2.2), Ginger Bread (2.3), Honeycomb (3.0), Ice Cream Sandwich (4.0), Jelly Bean (4.1), (4.2), (4.3), Kit Kat (4.4), Lollipop (5.0). In 2008, HTC introduced Android OS-based mobile phones (Dream); since then Android's worldwide market share increased to 85% (global smartphone market share in 2015). Presently, it has the largest installed base worldwide on smartphones. Android 2.0 (1.0, 1.5, 1.6) were used exclusively in mobile phones and tablets, whereas 3.0 was primarily targeted for tablet.

The Android system architecture consists of: (i) a modified Linux kernel; (ii) open-source libraries, which are written in C and C++; Android Runtime uses core libraries that manage most of the core functions of Java.; (iii) an application framework that manages services and libraries coded in Java for the application development; (iv) the applications that run on it. As a virtual machine, it uses Dalvin, which enables it to execute Java applications. The application code is executed inside a restricted area called a sandbox, which affects some specified operations like local file-system access. This helps to make the system more secure and stable as applications access the core operating system in a controlled and restricted way. It is a multiprocess system, in which each application (and parts of the system) runs in its own process. The booting procedure (as explained earlier) in a system using the Android OS is shown in Figure 7.3.

Today, Android has a large community of developers writing applications ("apps") that extend the functionality of devices (Android Open Source Project). Software developers write programs primarily in a customized version of Java, which requires the use of a special software development kit (SDK, freely available for download from the Internet) to create applications for an Android device, and apps can be downloaded from online stores. Android provides specific application programming interface (API) modules to the developers.

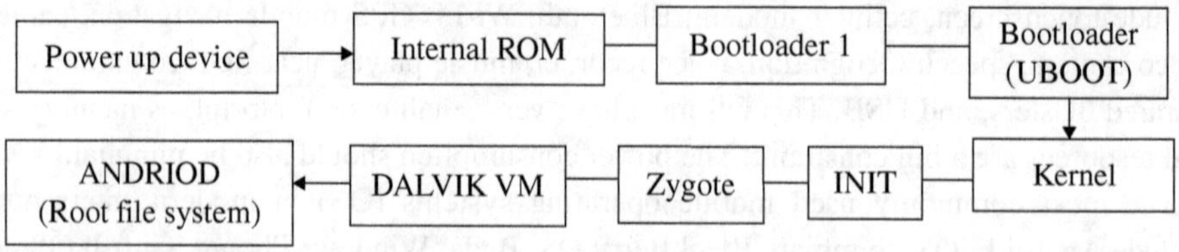

Figure 7.3 Andriod OS booting process

Android provides several advantages to the developers: (i) the application framework can be reused and replaced by the required components; (ii) reliable and enhanced data storage (using SQLite framework); (iii) support for 2D and 3D graphics (OpenGL ES 1.0); (iv) support for common media file formats (MPEG, MPEG3, MPEG4, H.286, AAC, AMR, JPG, PNG, GIF and more), so it is easy to create media common applications; (v) support for GSM, EDGE, 3G, LTE, Wi-Fi network, GPS, a navigational compass, Touch-Unlock, and accelerometer applications; (vi) the open-source Web-Kit engine-based Web browser. Apart from those, the Android development environment includes a device emulator, a debugger, a performance profiling tool, and an Eclipse IDE plug-in, which makes development easy. On the other hand, Android provides the advantage of getting low-cost smart phones to the consumers.

7.3.1.2 iOS

iOS is a mobile operating system, developed and distributed by Apple Inc. It was originally released in 2007 for the iPhone and iPod Touch, and later it was extended to support other Apple devices such as the iPad and Apple TV. It is derived from OS X (used on Apple computers), which is built on an open-source Darwin core OS, related to the UNIX OS. Unlike Google's Android and Microsoft's Windows CE, Apple does not license iOS for installation on non-Apple hardware. iOS has the second largest installed base worldwide on smartphones behind Android and, as of September 2014, its global market share was 11%.

iOS uses a similar sandboxing model to Android. The iOS system architecture is identical to the MacOSX architecture and consists of the following components:

- a core OS layer – this is the kernel of the operating system;
- core services layer – this is for fundamental system services; it is subdivided into different frameworks and based on C and Objective C;
- a media layer – this is for high-level frameworks, which are responsible for using graphic, audio, and video technologies;
- the Coca Touch layer – this is an Objective-C based framework and provides a number of functionalities that are necessary for the development of an iOS application like user interface management.

The security implementation includes a daemon called the security server, which implements several security protocols. The iOS security APIs are located in the core services layer and are based on the services in this layer. Applications on the iPhone call the security services APIs directly rather than going through the Cocoa Touch or media layers. The networking applications can also access secure networking functions through the CF network API, located in the core services layer.

7.3.1.3 Windows Phone

Windows Phone is a closed-source and proprietary mobile phone OS developed by Microsoft in 2010. It has the third largest installed base on smartphones, having a market share of 4.2%.

It includes Microsoft services such as OneDrive and Office, Xbox Music, Xbox Video, Xbox Live games, and Bing, and integrates with many other non-Microsoft services such as Facebook and Google accounts. Windows Phone's released versions are: Windows Phone 7, 7.5, 7.8, 8 (GDR1, GDR2 and GDR3), and 8.1 (GDR1 and GDR2). For sandboxing, Windows Phone uses the same model like Android and iOS.

7.3.1.4 BlackBerry OS

BlackBerry 10 (based on the QNX OS) is a closed-source and proprietary OS developed by BlackBerry.

Bada OS is developed by Samsung Electronics for smart phone and tablets. MeeGo OS was from the Linux Foundation, a nonprofit organization, and it is open source and GPL. WebOS is from LG Electronics, although some parts of it are open source. It is running on the Linux kernel, and was initially developed by Palm, which launched with the Palm Pre. Later it was acquired by HP and in 2011 two phones (the Veer and the Pre 3) and a tablet (the TouchPad) running webOS were introduced.

Figure 7.4 shows the market share of different mobile OSs in the first quarter of May 2015.

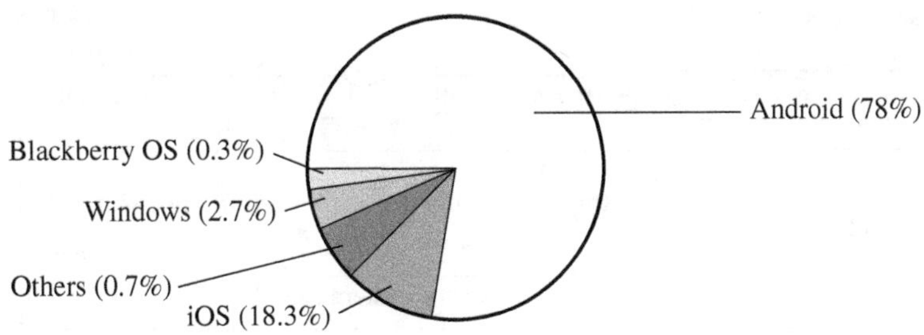

Figure 7.4 Market share of different mobile OSs in Q1, May, 2015

7.3.2 *Real-Time Operating System*

A real-time operating system (RTOS) performs its functions and responds to external events (such as interrupts) within a specified period of time (~20 μs). RTOS is usually more efficient, predictable, easier to maintain, and less buggy. It should implement task priority levels, so that important tasks may be executed at a higher priority. Real-time systems should allow a task's priority to be changed during run time. ThreadX, RTLinux, and Nucleus are commonly used RTOS in mobile phones.

7.3.3 *OS Operation*

As discussed earlier, a bootstrap program must locate and load the OS into memory. The OS then starts executing the first process such as "init" and waits for some event to occur. Typically, in an OS, there are two separate modes of operations – user mode (when the system is executing on behalf of the user application) and kernel mode (also known as supervisor or system mode or privileged mode), and in CPU hardware a bit is provided to set the mode bit (0 for kernel and 1 for user). Typical components of OS kernel are interrupt handlers to service interrupt requests, a scheduler to share processor time (CPU execution time) among multiple processes, a memory management system to manage process address spaces, and system services such as networking and interprocess communication. Applications running on the system communicate with the kernel via system calls. At the boot time the hardware starts at the kernel mode, and then the operating system is loaded and starts the user application in user mode. The occurrence of an event is generally signaled by an interrupt from either hardware or a software process. The hardware sends a signal to the CPU by way of a system bus or interrupt line, whereas the software triggers an interrupt (known as a trap) by executing a special operation system call. Whenever a trap or hardware interruption occurs the hardware switches from user to kernel mode and jumps to the interrupt reset vector, stopping the normal flow of execution. The vector location contains the starting address of the interrupt service routine (ISR) and interrupt handlers are written accordingly. Once the ISR is completed, it returns to the normal flow of that process and sets the mode as user mode (see Figure 7.5).

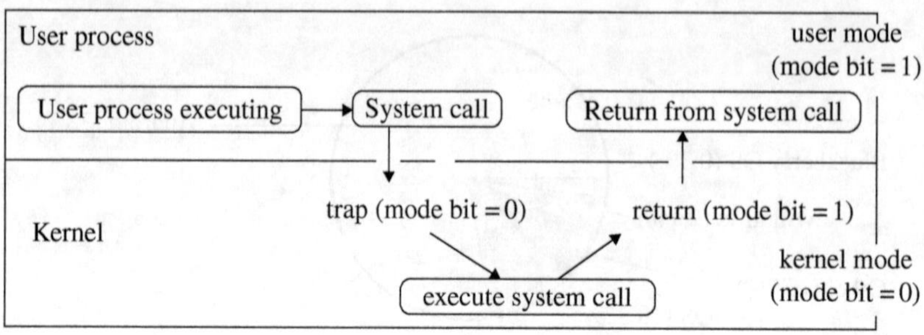

Figure 7.5 User mode to kernel mode transition

The scheduler is an algorithm that decides which task should be executed next. The device driver framework provides a consistent interface between different hardware peripherals. The memory handling involves setting up the system and task stacks (based on whether the task is static or dynamic). A static task is defined at build time and is included in the operating system image. For this task, the stack can be set up during operating system initialization. A dynamic task is loaded and executed once the OS is loaded and executing. A preemptive operating system requires a periodic interruption, which is normally produced by a counter / timer device on the target hardware. As part of the initialization stage, an operating system sets the periodic interrupt frequency.

7.3.4 Selection of an Operating System

The prices and features of an OS vary. Most companies charge for purchasing their "development environment," which allows us to develop code that will run on their OS. Some companies also charge for each product we build that includes their operating system software. This is usually called "target fee" or "runtime license fee." Some companies offer "runtime licenses" free of cost. It is always better to settle this issue before development because a user might have invested a lot of time and effort writing software to work with a particular OS only to find, when about to launch the product in the market, that the "runtime license" fee has substantially increased and is no longer affordable. We also need to decide whether the OS source code is required or not and, if so, whether it is freely available or not.

Some OS features should be checked before selecting any OS and vendor. These include: preemptive task scheduling, time-slice scheduling, round-ribbon scheduling, parallel processing, intertask messages, memory management, interrupt management, timer management, and OS code size.

Check for the feature cost and whether it includes a C/C++ compiler, an assembler, source code, a runtime license, a development license, and so forth.

Check the availability of the libraries – C runtime libraries, DSP math libraries, an image processing library, an X Windows library, Ethernet or communication libraries, and so forth.

Check the availability of debugging features – basic debugging, performance timer, debugger costs, and so forth.

7.4 Device Driver Software

A device driver is a software program that controls a specific type of device, which is connected to mobile device processor. Here, the device hardware details are abstracted and, in place of that a software interface is provided to enable OS and other computer programs to access the device's hardware functions through bus or other communication subsystems connected to the hardware – see Figure 7.1(a). A device driver acts like a translator between the device and the programs that use the device. Each device has its own set of specialized commands, which is known to its driver. In contrast, most programs access devices using generic commands like *init, open, read, write*, and *close*. The driver therefore accepts generic commands from a program and then translates them into specialized commands for that device. Now, instead of putting code in each application, we write to control each device and then share the code amongst different applications to share the same hardware amongst various applications. In many cases, device drivers work within the kernel layer of the operating system. So, instead of accessing a device directly, an operating system loads the device drivers and calls the specific functions in the driver software in order to execute specific tasks on the device. When a device connected to mobile processor is activated for use, the device driver specific to it is installed, and a device object is created on the host software, which is designed to control the device through operating system generic calls.

7.5 Speech and Multimedia Application Software

Today's mobile phone supports voice, audio, and video playback features. Ideally, each of these applications generates huge number of bits from the information source, which needs to be reduced and controlled at the source level using different source coding techniques. Uncompressed multimedia data (for graphics, audio and video) requires considerable storage space and transmission bandwidth. Despite the rapid progress in mass-storage density, processor speeds, and digital communication system performance, demand for data storage capacity and data-transmission bandwidth continues to outstrip the capabilities of available technologies. The recent growth of data-intensive multimedia-based applications needs more efficient ways to encode signals and images for storage and communication technology.

Modem software helps to transfer data from one mobile (or network) to another distant mobile or server. So, the data generated at the speech encoder side is transmitted and, on the receiver side, the received bits are passed to the speech decoder to reproduce the voice. When we talk in front of a microphone, the voice is turned into a digital bit stream using waveform encoding (source coding). The vocoder's main job is to reduce the data rate. In wire line this is achieved in time domain by using time domain compression techniques, but in digital cellular handsets a speech synthesis codec is used in the frequency domain. On the transmitter side, the source encoding is done by describing each sample in terms of frequency coefficients, and compression is achieved by exploiting similarities between samples. On the receiver side, the decoder uses frequency coefficients to rebuild or synthesize the harmonic structure of the original voice sample. For the source encoding and decoding functions, the mobile contains the source codec unit. This can be implemented in the software (which generally runs on DSP) or in hardware logic.

7.5.1 Speech Codec

A speech codec is a special type of audio codec designed especially for human voice encoding and decoding. Here, by analyzing vocal tract sounds, instead of sending the sound waves, a recipe is sent to the receiver end for rebuilding the sound. The speech codec is able to achieve a much higher compression ratio, which results in a smaller amount of digital data transmission. Speech quality as produced by a codec is a function of transmission bit rate, complexity, delay, and bandwidth. Speech coding differs from other forms of audio coding, as speech is a much simpler signal than most other audio signals. The speech signal is limited to a bandwidth of 300 to 3400 KHz (whereas audio signals are limited to a bandwidth of 0 to 20 000 Hz – the audible range) and there is lot of statistical information available about the properties of speech. The speech signal varies quite infrequently, resulting in a high degree of correlation between consecutive samples. This short-term correlation is due to the nature of the vocal tract. Long-term correlation also exists due to the periodic nature of the speech. This statistical redundancy can be exploited by introducing prediction schemes, which quantize the prediction error instead of the speech signal itself. On the other hand, the shortcomings in the capability of humans to receive sounds leads to the fact that lot of information in the speech signal is perceptually irrelevant. That means that human ear can't differentiate between the changes of magnitude below a certain level and can't distinguish frequencies below 16 Hz or above 20 KHz. This can be exploited by designing optimum quantization schemes, where only a finite number of levels are necessary.

Speech coding methods can be classified as:

- waveform coding;
- source coding; and
- hybrid coding.

Source codecs try to produce a digital signal by modeling the source of the codec, whereas waveform codecs do not use any knowledge of the source of the signal but instead try to produce a digital signal whose waveform is as identical as possible to the original analog signal. Pulse code modulation (PCM) is the simplest and purest waveform codec. Hybrid codecs attempt to fill the gap between waveform and source codecs. In the early 1960s, when telephones were using the first digital signal transmission technique, PCM was used to generate a 64 kbps digital bit stream from an analog voice signal. In a PCM codec, a voice signal is sampled at a rate of 8 KHz, with each sampled voltage level converted to 8 bits, so the total number of bits generated per second is 8 * 8000=64 kbits. Narrowband speech is typically sampled 8000 times per second, and then each speech sample must be quantized. If linear quantization is used then about 12 bits per sample are needed, giving a bit rate of about 12 * 8 kbits/s=96 kbits/s. However, this can be reduced easily by using nonlinear quantization. For coding speech it was found that with nonlinear quantization 8 bits per sample was sufficient for speech quality, which is almost indistinguishable from the original. This gives a bit rate of 64 kbits/s and two such nonlinear PCM codecs were standardized in the 1960s. In America the μ law coding standard, and in Europe the slightly different A law compression is used. Because of their simplicity, excellent quality and low delay, both these codecs are still widely used today. Sun Microsystems Inc. has released code to implement the G711 A-law and μ-law codes into the public domain and this was modified by Borge Lindberg.

The most common speech coding scheme is code excited linear prediction (CELP) coding, which is used, for example, in the GSM standard. In CELP, the modelling is divided in two stages: a linear predictive stage that models the spectral envelope, and a codebook-based model of the residual of the linear predictive model.

7.5.1.1 Speech Codecs

Full-Rate Codec (FR)

The full-rate speech codec in GSM is described as regular pulse excitation with long-term prediction (LPC-RPE codec – GSM 06.10 RPE-LTP). It is a full-rate speech codec (FR) and operates at 13 kbits/s. The encoder has three major parts: (i) linear prediction analysis (short-term prediction); (ii) long-term prediction; and (iii) excitation analysis.

The encoder processes 20 ms blocks of speech and each speech block contains 260 bits (188+36+36=260), as depicted in Figure 7.6(a). So the rate is 260 bits/20 ms=13 000 bits/s=13 kbits/s.

Generally, the input speech is split up into frames of length 20 ms, and, for each frame, a set of 8 short-term predictor coefficients is computed. Each frame is then further split into 4×5 ms subframes (4 * 5=20 ms) and for each subframe the encoder finds a delay and a gain for the codec's long-term predictor. The linear predictor part of the codec uses 36 bits and linear prediction uses a transfer function of the order of 8. The long-term predictor estimates pitch and gain four times at 5 ms intervals. Each estimate provides a lag coefficient and gain coefficient of 7 bits and 2 bits, respectively. Together these four estimates require 4 * (7+2) bits=36 bits. The gain factor in the predicted speech sample ensures that the synthesized speech has the same energy level as the original speech signal.

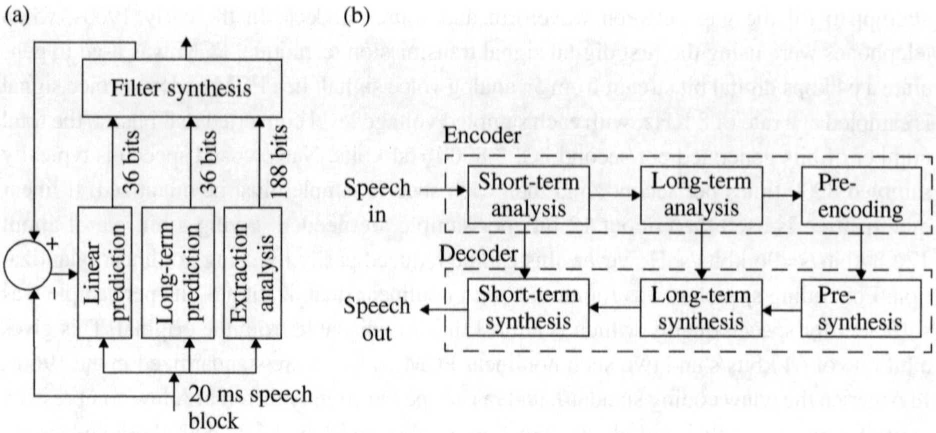

Figure 7.6 (a) The GSM full-rate LPC-RPE codec. (b) Block diagram of GSM speech encoder and decoder

The remaining 188 bits are derived from the regular pulse excitation analysis. After both short- and long-term filtering, the residual signal, which is the difference between the predicted signal and the actual signal, is quantized for each 5 ms subframe.

At the decoder, the reconstructed excitation signal is fed through the long-term and then the short-term synthesis filters to give the reconstructed speech as shown in the Figure 7.6(b). A postfilter is used to improve the perceptual quality of this reconstructed speech.

On the network side the situation is slightly more complicated as speech signals are usually coded using an eight-bit A-law pulse-code modulation (PCM) format in order to be compatible with the PSTN or ISDN. So, before the speech signal is passed to the speech coder on the network side, it must first undergo an eight-bit A-law PCM to 13-bit uniform PCM conversion.

Half-Rate (HR) Codec

GSM has also defined the half-rate version of the GSM codec. This is a vector self-excited linear predictor (VSELP) codec at a bit rate of 5.6 kbit/s. It is a close relative of the CELP codec family. The difference is that VSELP uses more than one separate excitation codebook. These are separately scaled by their respective excitation gain factors. The GSM half-rate vocoder operates in one of four different modes (0, 1, 2, 3) based on the grade of the voice detected in the speech. The speech spectral envelope is encoded by using 28 bits per 20 ms frame for vector quantization of the LPC coefficient and the four synthesis modes corresponds to different excitation modes.

Enhanced Full-Rate (EFR) Speech Codec

The enhanced full-rate speech codec is defined by the European Telecommunications Standards Institute (ETSI). It has a bit rate of a bit rate of 12.2 kb/sec and uses the algebraic code excited linear prediction (ACELP) algorithm, which is an analysis-by-synthesis algorithm.

7.5.1.2 AMR Codec (AFS/AHS)

The adaptive multirate (AMR) codec is the speech codec standard for GSM phase 2+, which adaptively changes the source rate based on the quality of the wireless channel. The AMR speech codec was proposed by ETSI in June 1996 to improve speech quality in mobile phones and to compensate for the GSM slow power control. AMR is based on EFR speech codec; it incorporates multiple submodes for use in full-rate or half-rate mode. These are determined by the channel quality. The two options for AMR logical speech channels are adaptive full-rate speech (AFS) and adaptive half-rate speech (AHS). In order to provide best speech quality, variable portioning between speech and channel coding bit rates is selected based on the variation of the channel conditions. According to the channel quality, the receiver can request (or command) the transmitter AMR to adjust the speech coding rate to allow for a higher or lower channel coding rate in response. So, if the channel quality deteriorates, progressively lower codec rates are requested, otherwise, if channel conditions improve, higher codec rates are requested. The codec rate requests and commands are transmitted as often as every 40 ms, using in-band signaling. The AMR rate requests / commands indicate the channel quality and they are transmitted more often than RxQUAL and RxLEV.

The AMR codec (Narrow Band) uses a set of eight codec rates (4.75, 5.15, 5.90, 6.70, 7.40, 7.95, 10.2 and 12.2 kbit/s) for speech encoding. For every 20 ms input speech frame, the codec rate can be switched to a different codec rate. In GSM, only a subset of the possible codec rates is used during a connection. This subset is referred to as the active codec set (ACS) and contains at least one and at most four of the possible AMR codec rates. The network decides on an active codec set of up to four code modes in AFS and AHS. This active code set is initially signaled to the MS during call setup via layer 3, signaling channel assignment / immediate assignment / channel mode modify / handover message (see GSM spec. 4.08). It is possible to change the ACS during a connection.

Codec Mode Information

The codec mode information sent on the downlink is: (i) codec mode indications (CMI) – CMI is used to indicate to the peer AMR about the codec rate to be used for decoding the received speech frame; (ii) codec mode commands (CMC) – code mode command instructs the AMR on the MS side about the codec mode to be applied.

The codec mode information sent on the uplink is: (i) codec mode indications (CMI) – as mentioned above; (ii) codec mode requests (CMR) – these inform the other end (the BTS end) about the preferred codec mode. That means, based on the channel quality, MS requests for the preferred rate to the network.

The codec mode indications and codec mode command / requests are sent alternatively, on consecutive speech frames. Codec mode changes only every second speech frame – for example, the signaling of CMI and CMR messages (CMI and CMC) is alternated in the uplink (downlink) resulting in a 40 ms signaling interval for each type of message. Codec mode information is transmitted in-band in the speech traffic channel. The details of the in-band coding can be found in standard 5.03, section 3.10.7. The codec rate to be applied for encoding each input speech frame needs to be provided to the codec at every 20 ms. The codec rate to be used for decoding every frame has to be provided. In GSM, the codec rate information is transmitted inband every 20 ms with the encoded speech frames. The robust AMR traffic synchronized control channel (RATSCCH) mechanism is used to modify the AMR configuration on the radio interface without interruption of the speech transmission. During regular speech transmission (in the middle of a speech burst) RATSCCH replaces (steals) one TCH/AFS or two TCH/AHS speech frames (see 3GPP TS 25.009). In all nonspeech cases the RATSCCH should be handled like speech.

Channel Quality Measure and Link Adaptation

The receiver side performs link-quality measurements, which are known as quality indicators. The details concerning the reference performance are in standard 5.05. The CMC/CMR messages are generated based on some channel quality metric such as the estimated received

C/I, and then the estimated C/I is compared against codec mode switching thresholds. A quality indicator is defined as a normalized C/I ratio based on actual C/I estimates from the equalizer or raw BER estimates. The equalizer either estimates C/I (SNR) or this has to be derived from the raw bit errors from the channel decoder. If the equalizer estimates C/I (SNR) then it should communicate this information to the Vo-coder for postprocessing and to determine the code mode requests by comparing with the threshold levels. The MS and BTS should continuously update their quality indicator on a frame by frame basis.

The quality indicator is fed directly into the UL mode control unit in case of uplink adaptation. This unit compares the quality indicator with certain thresholds and generates a codec mode command indicating the codec mode to be used on the uplink. The codec mode command is then transmitted in-band to the mobile side, where the incoming speech signals are encoded using the corresponding codec mode.

For downlink (DL) adaptation, the DL mode request generator within the mobile compares the DL quality indicator with certain thresholds and generates a codec mode request indicating the preferred codec mode for the downlink. The codec mode request is transmitted in-band to the network side where it is fed into the DL mode control unit. This unit generally grants the requested mode or sometimes ignores it. The resulting codec mode is then applied for encoding of the incoming speech signal in downlink direction.

Both for uplink and downlink, the presently applied codec mode is transmitted in-band as a codec mode indication together with the coded speech data. At the decoder, the codec mode indication is decoded and applied for decoding of the received speech data.

WideBand (AMR-WB)

WideBand (AMR-WB) is a speech-coding standard developed after the AMR, using the same technology as ACELP. This provides excellent speech quality due to a wider speech bandwidth of 50–7000 Hz compared to narrowband speech codecs, which in general are optimized for POTS wire-line quality of 300–3400 Hz. AMR-WB is codified as G.722.2, an ITU-T standard speech codec. The wide version of AMR supports codec rates of 6.6 kbps, 8.85 kbps, 12.65 kbps, 15.25 kbps, 15.85 kbps, 18.25 kbps, 19.85 kbps, 23.05 kbps, and 23.85 kbps.

7.5.2 Voice Support in LTE

LTE is designed for data-only services but it can support voice service using IMS (IP Multimedia System). As LTE and IMS systems will not be deployed overnight, operators are choosing some alternative options to support this. There could be several cases:

- LTE for data-only services – VoIP applications like Skype and Google talk can be used here, but these third-party applications cannot control the quality of service in all load conditions.
- LTE for data-only services with 2G-3G network for voice – when LTE network does not support IMS, then, being in the LTE network, one cannot make a normal voice call.

So, 3GPP provides a circuit switch fall back (CSFB) procedure through which an LTE device can fall back to legacy 3G or 2G networks depending on coverage and capacity in relevant networks (3GPP TS 23.272). This is being considered as an interim solution to ease the transition for VoLTE.

- Data and voice both services on LTE – these are mainly to provide voice services over LTE and include Voice over LTE (VoLTE) and Voice Services over LTE via Generic Access Network (VoLGA).

7.5.3 Audio Codec

The audio codec uses a time-domain-to-frequency-domain transform to expose redundancy in the input signal. Contents are converted using various different compression algorithms, such as those from Microsoft, Advanced Streaming Format (ASF), Real Audio (rm), or MPEG-1 Audio Layer 3 Protocol (MP3). Here, some commonly used file formats are : WAV – Waveform Audio, MIDI – Music Instrument Digital Interface, AAC – Advanced Audio Coding, ASF – Advanced Streaming Format, MP3- MPEG-1 Audio Layer 3 Protocol. Of these only the MP3 file format will be discussed in detail here.

7.5.3.1 MP3

The MP3 is a special format used to compress digital audio, keeping the audio quality as high as possible. This is a lossy compression technique but the loss can be hardly noticed because the compression method tries to control it. By using mathematical algorithms, it will only lose those parts of sound that are hard to hear even in the original form. This way the audio can be compressed up to 12 times, which is really significant. MP3 encoding tools analyze the incoming source signal, break it down into mathematical patterns, and compare these patterns with psychoacoustic models stored in the encoder itself. The encoder can then discard most of the data that doesn't match the stored models, keeping that which matches. This configuration is based on a "tolerance" level. The lower the data storage allotment, the more data will be discarded, and leads to poorer the audio sound quality.

MP3-encoded files are composed of a series of very short *frames*, one after another, much like a filmstrip. Each frame of data is preceded by a *header*. The header contains extra information about the compressed data frame. In some encodings, consecutive frames may hold information for each other. For example, if one frame has leftover storage space, whereas the next frame is running short of free space, then they may team up for optimal results. At the beginning or end of an MP3 file, extra information about the file itself, such as the name of the artist, the track title, the name of the album from which the track came, the recording year, genre, and personal comments may be stored. This is called "ID3" data.

The frame header is constituted by the very first four bytes (32 bits) in a frame. The first 11 bits of a frame header are always set and they are called "frame sync." The exact meaning

Table 7.2 MP3 header format

Sign	Length (bits)	Position (bits)	Description
A	11	(31 to 21)	Frame sync (all bits set)
B	2	(20, 19)	MPEG Audio version ID: 00 – MPEG Version 2.5, 01 – reserved, 10 – MPEG Version 2 (ISO/IEC 13818-3), 11 – MPEG Version 1 (ISO/IEC 11172-3)
C	2	(18, 17)	Layer description: 00 – reserved, 01 – Layer III, 10 – Layer II, 11 – Layer I
D	1	16	Protection bit: 0 – Protected by CRC (16bit crc follows header), 1 - Not protected
E	4	(15, 12)	Bit-rate index: 8 to 448 Kbps
F	2	(11, 10)	Sampling rate frequency index: 8000 to 44 100 (values are in Hz)
G	1	9	Padding bit: 0 – frame is not padded, 1 – frame is padded with one extra slot. Padding is used to fit the bit rates exactly.
H	1	8	Private bit. It may be freely used for specific needs of an application, for example if it has to trigger some application specific events.
I	2	(7,6)	Channel Mode: 00 – stereo, 01 – joint stereo (Stereo), 10 – dual channel (Stereo), 11 – single channel (Mono)
J	2	(5,4)	Mode extension (only if joint stereo)

of each bit in the header ID is defined in the standard ISO/IEC 11172-3 and these are mentioned in Table 7.2. Frames may have a 16-bits long CRC check just after the frame header. Next to that, the audio data is stored. We may calculate the length of the frame and use it, if we need to read other headers too or just want to calculate the CRC of the frame, to compare it with the one we read from the file.

7.5.4 Images

Image compression can be lossy or lossless (lossless compression involves compressing data, which, when decompressed, will be an exact replica of the original data). Different techniques, like run-length encoding, entropy coding, and deflation are used for lossless and chroma subsampling; transform coding, and fractal compression are used for lossy compression. A common characteristic of most images is that the neighboring pixels are correlated and therefore these contain redundant information. So, we need to find out the less correlated representation of the image using the spatial redundancy (correlation between neighboring pixel values), spectral redundancy (correlation between different color planes or spectral bands), and temporal redundancy (correlation between adjacent frames in a sequence of images) methods. For still-image compression, the Joint Photographic

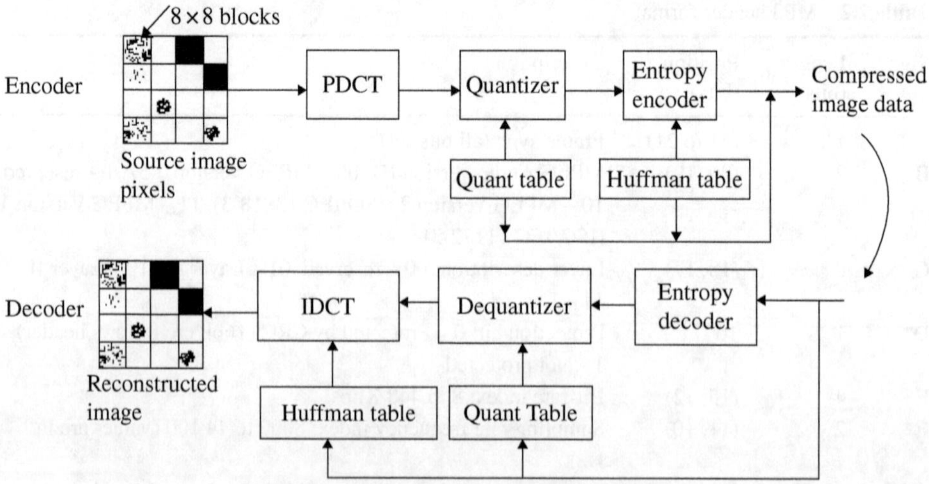

Figure 7.7 JPEG encoder and decoder blocks

Experts Group (JPEG) standard has been established by the International Organization for Standardization (ISO) and the International Electro-Technical Commission (IEC) in 1992. Generally, in JPEG, the encoders and decoders are discrete cosine transform (DCT) based. DCT can be computed with a fast Fourier transform-like algorithm in *O(n log n)* operations. The JPEG standard specifies three modes namely sequential, progressive, and hierarchical for lossy encoding, and one mode of lossless encoding. The "baseline JPEG coder" uses sequential encoding. In Figure 7.7, the key processing steps in such an encoder and decoder are shown for grayscale images and for color images compression can be approximately regarded as compression of multiple grayscale images. The DCT-based encoder can be thought of as essentially compression of a stream of 8×8 blocks of image samples. Each 8×8 block makes its way through each processing step, and yields output in compressed form into the data stream. Because adjacent image pixels are highly correlated, the "forward" DCT (FDCT) processing step lays the foundation for achieving data compression by concentrating most of the signal in the lower spatial frequencies. After output from the FDCT, each of the 64 DCT coefficients is uniformly quantized in conjunction with a carefully designed 64-element quantization table. A quantizer simply reduces the number of bits needed to store the transformed coefficients by reducing the precision of those values. An entropy encoder further compresses the quantized values losslessly to give better overall compression. At the decoder, the quantized values are multiplied by the corresponding QT elements to recover the original, unquantized values. After quantization, all of the quantized coefficients are ordered into the "zig-zag" sequence as shown in Figure 7.7. This ordering helps to facilitate entropy encoding by placing low-frequency nonzero coefficients before high-frequency coefficients. The DC coefficient, which contains a significant fraction of the total image energy, is differentially encoded. The JPEG proposal specifies both Huffman coding and arithmetic coding.

The performance of these coders generally degrades at low bit rates mainly because of the underlying block-based DCT scheme. More recently, the wavelet transform has emerged as a cutting-edge technology, within the field of image compression.

7.5.5 Video

Video compression is a combination of image compression and motion compensation. Video is basically a three-dimensional array of color pixels, where two dimensions represent spatial (horizontal and vertical) directions of the moving pictures and the third dimension represents the time domain. A data frame is a set of all pixels that correspond to a single time moment. Basically, a frame is same as a still picture. Like JPEG images, spatial encoding is performed by taking advantage of the fact that the human eye is unable to distinguish small differences in color, as easily as it can changes in brightness, so very similar areas of color can be "averaged out." With temporal compression only the changes from one frame to the next are encoded as, often, a large number of the pixels will be the same on a series of frames. The steps that are commonly followed for encoding are signal analysis, quantization, and variable length encoding. There are four methods for compression: discrete cosine transform (DCT), vector quantization (VQ), fractal compression, and discrete wavelet transform (DWT).

In 1993 the motion picture expert group (MPEG) was founded; this originally focused on producing noninteractive video compression but was later extended as MPEG-4 and MPEG-5.

- MPEG-1 – CD-ROM storage compression standard. Designed for bit rate up to 1.5 mbit/s.
- MPEG-2 – DVB and DVD compression standard. Designed for bit rate between 1.5 and 15 mbit/s.
- MPEG-3 – MPEG-2 Layer-3 audio streaming standard.
- MPEG-4 – Audio and video streaming and complex media manipulation.
- MPEG-5 – Multimedia hypermedia standard.

The MPEG-4 standard was created to be the next major standard in the world of multimedia. Unlike MPEG-1 and MPEG-2, where more focus was given to better compression efficiency, in the case of MPEG-4 the emphasis was on new functionality. The new MPEG-4 standard facilitates the growing interaction and the convergence of the previously separate worlds of telecommunications, computing and mass media.

MPEG-4 runs on the MP4 file format. It is the next generation beyond MP3. Like MP3, MPEG-4 will become the accepted standard because it extends the success of MP3 in several important ways:

- MPEG-4 enables video, even at bit rates as low as 9.6Kb/s.
- MPEG-4 enables digital rights management to protect the precious intellectual property of the content provider.

- The MPEG-4 solution provides mobile users access to full-motion news and financial stories, sports highlights, short entertainment clips and music videos, weather and traffic reports, home or work security cameras and corporate communications, from any location.

Some other commonly used video standards are:

- H.261 – an ITU standard designed for two-way communication over ISDN lines (video conferencing). It supports data rates that are multiples of 64 kbit/s. The algorithm is based on DCT and can be implemented in hardware or software and uses intraframe and interframe compression. H.261 supports CIF and QCIF resolutions.
- H.263 – based on H.261 with enhancements that improve video quality over modems. It supports CIF, QCIF, SQCIF, 4CIF and 16CIF resolutions.

7.6 UE Protocol Stack Software

LTE UE Stack is a key software element of LTE mobile terminal modems. There are different layers in protocol stack as shown in Figure 7.8 and each layer interacts with the corresponding peer layer residing on the network side, spread over different network entities (as described in Chapter 3).

The user plane in LTE UE consists of upper layers, NAS, PDCP, RLC, MAC, PHY and RF, whereas the control plane in LTE UE consists of upper layers, NAS, RRC, PHY and RF:

- Physical layer. This carries all information from the MAC transport channels over the air interface. It performs CRC attachment, coding, scrambling / descrambling, modulation / demodulation, HARQ, and MIMO, and it carries out the link adaptation (AMC), power control, cell search and other measurements for the RRC layer.
- Medium access layer (MAC). This carries out mapping between logical channels and transport channels, multiplexing of MAC SDUs from one or different logical channels onto transport blocks (TBs) to be delivered to the physical layer on transport channels, demultiplexing of MAC SDUs from one or different logical channels from transport blocks (TBs) delivered from the physical layer on transport channels, handling control elements, scheduling information reporting, error correction through HARQ, sending BSRs, priority handling between UEs by means of dynamic scheduling, priority handling between logical channels of one UE, DRX, and so forth.
- Radio link control (RLC). This operates in three modes: transparent mode (TM), unacknowledged mode (UM), and acknowledged mode (AM). The RLC layer is responsible for the transfer of upper layer PDUs, the buffer status report (in uplink), error correction through ARQ (for AM data transfer), concatenation, segmentation, and reassembly of RLC SDUs (for UM and AM data transfer). The RLC is also responsible for resegmentation of RLC data PDUs (for AM data transfer), reordering of RLC data PDUs (for UM and AM data transfer), duplicate detection (for UM and AM data

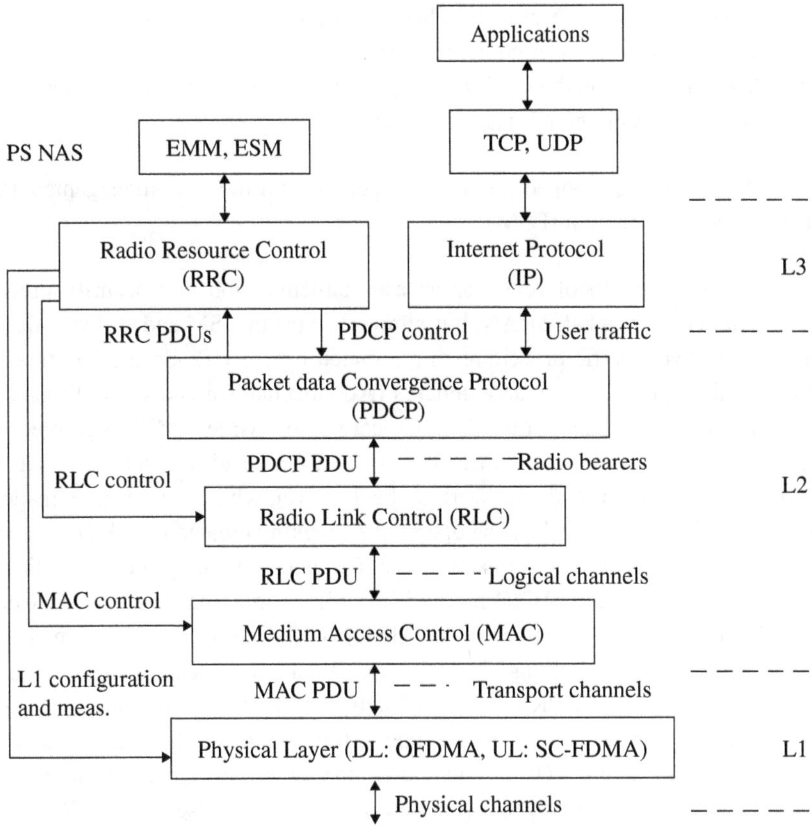

Figure 7.8 UE protocol software architecture (LTE)

transfer), RLC SDU discard (for UM and AM data transfer), RLC re-establishment, and protocol error detection (for AM data transfer).

- Radio resource control (RRC). This resides in the control plane and carries out configuration management, broadcasting of system information related to the nonaccess stratum (NAS) and the access stratum (AS), paging, establishment, maintenance, and the release of an RRC connection between the UE and E-UTRAN security functions including key management, establishment, configuration, maintenance, and release of point-to-point radio bearers.
- Packet data convergence control (PDCP). This is responsible for header compression and decompression of IP data, transfer of data, maintenance of PDCP sequence numbers, in-sequence delivery of upper layer PDUs at re-establishment of lower layers, duplicate elimination of lower layer SDUs at re-establishment of lower layers for radio bearers mapped on RLC AM, ciphering and deciphering of user-plane data and control-plane data, integrity protection and integrity verification of control plane data, timer based discarding, duplicate discarding, and so forth. PDCP is also used for signaling radio bearers (SRBs) and data radio bearers (DRBs) mapped on DCCH and DTCH type logical channels.

- Nonaccess stratum (NAS) protocols. These protocols form the highest stratum of the control plane between the user equipment (UE) and MME. The NAS protocols support the mobility of the UE and the session management procedures to establish and maintain IP connectivity between the UE and a PDN gateway.

The NAS procedures are grouped into two categories: EPS mobility management (EMM) and EPS session management (ESM).

EMM. The EMM protocol refers to access, authentication and security procedures related to mobility over an E-UTRAN. It is similar to MM in GSM and GMM in the GPRS network. The EMM-specific procedures are initiated by the UE and these define attach/detach mechanisms, the tracking area update (TAU) mechanism and so on. When an LTE device is in an active state (while communicating, or while EMM registered/ECM-connected) its location is known by the LTE network at cell level. When the UE is in an idle state its location is known by the network at the TA level, where a TA can be made up of several cells or eNBs as defined by the operator. A tracking area code (TAC) is the unique code that each operator assigns to each of the TAs and a tracking area identifier (TAI) consists of a PLMN ID and a TAC. Where, a PLMN ID, a combination of a mobile country code (MCC) and a mobile network code (MNC), is the unique code assigned to each operator in the world. The tracking area is the equivalent to the location area (LA) in GSM and the routing area (RA) in GPRS. In the LTE network, an MME entity must have updated location information about UEs in an idle state to find out in which TA a particular UE is located, so that it can page the UE for any incoming call or message accordingly. For this, the UE notifies the LTE network about its current location by sending a TAU message every time it moves between TAs. In EPS, a UE initiates a tracking area update when it detects that it has entered a new tracking area. EPS also introduces the concept of a tracking area list (TAL), which allows the provision of lists of tracking areas in UE. So, UE does not need to initiate a TAU if it enters a tracking area that is included in its TAL.

ESM. The ESM protocol supports the establishment and handling of user data in the NAS. The IP connectivity between UE and a packet data network (PDN) are defined by the PDN connection and the EPS bearer. A PDN connection consists of a default EPS bearer and possibly additional "dedicated bearers." EPS supports multiple simultaneous PDN connections, like a PDN connection to the Internet (with just a default EPS bearer), and a PDN connection to the operator's IMS (with additional dedicated bearers, if required by the service). Within a PDN connection, all EPS bearers share the same UE IP address and an APN.

Further Reading

Das, Sajal Kumar. (2010) *Mobile Handset Design*, John Wiley & Sons, Ltd.
Phillips, B. (2013) *Android Programming*, Big Nerd Ranch.

8

Battery and Power Management Unit Design

8.1 Introduction to the Power Management Unit

Mobile phones are equipped with a wide range of sensing, computational, storage, display, multimedia and communication components, each with different voltage and current requirements. Power management broadly refers to the management of power-related activates in a mobile device, which include generation, storing, distribution, conservation, and control of regulated voltages required to operate the host mobile system. The power management unit in mobile phone can be divided into two sections: (i) the power distribution and switching unit, and (ii) the charging unit:

- Power distribution section. This section is used for the distribution of voltage and current to the other components of the smartphone. This section is generally part of an analog baseband (ABB) unit. It takes power from the battery (which is commonly 3.6 V) and converts it (stepping up / down) to various voltages such as 4.8 V, 2.8 V, 1.8 V and 1.6 V, and distributing it to other components. Switching regulators, linear regulators, switched capacitor voltage converters, and voltage references are typical elements of this unit. In some implementations, a PM integrated circuit (PMIC) can be used to implement this section. To maintain high efficiency the power supply and distribution design should be integrated with the system design. In recent years the trend is towards lower supply voltages because of the processes used in manufacturing integrated circuits. In some designs, to eliminate one extra supply, the analog and digital circuitries are connected to the same supply, which challenges the low-noise requirement due to the noisy digital supply. In some designs, power management adopts unipolar (single) supplies, which requires level shifting and AC coupling.

Mobile Terminal Receiver Design: LTE and LTE-Advanced, First Edition. Sajal Kumar Das.

- Charging section. This is responsible for charging the battery of the mobile phone. It is composed of a charging integrated circuit, which takes power from an external source and charges the battery of the smartphone. Commonly it uses 6.4 V, from an external battery charger and regulates it to 5.8 V while giving it to battery. This is described in section 8.2 of this chapter.

8.2 Battery Charging Circuit

The normal battery charger takes an AC supply (220 V or 110 V) and converts it to DC output voltage (around 5 V). The output voltage is regulated by filtering the ripples. So, the circuit mainly consists of transformer, rectifier, filter, and regulator as shown in Figure 8.1.

Generally, mobile-phone batteries are charged through a proprietary charging technique. To minimize power dissipation and consequent thermal problems in the mobile phone, the charging supply is current limited and specified according to the battery's chemistry and charge-recovery requirements. A generalized battery charging and monitoring circuit inside a mobile phone is shown in Figure 8.2. The voltage developed across the RSENSE resistor is used to maintain a constant current. The voltage is monitored and controlled by the microcontroller. The temperature sensor is used to monitor battery temperature. The battery is charged with a constant current until it is fully charged.

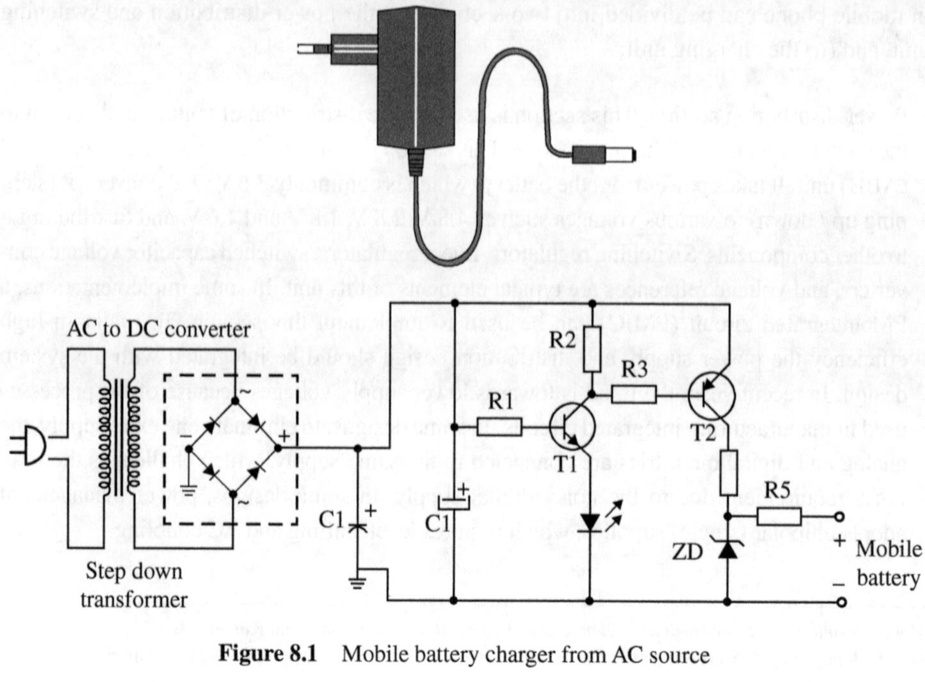

Figure 8.1 Mobile battery charger from AC source

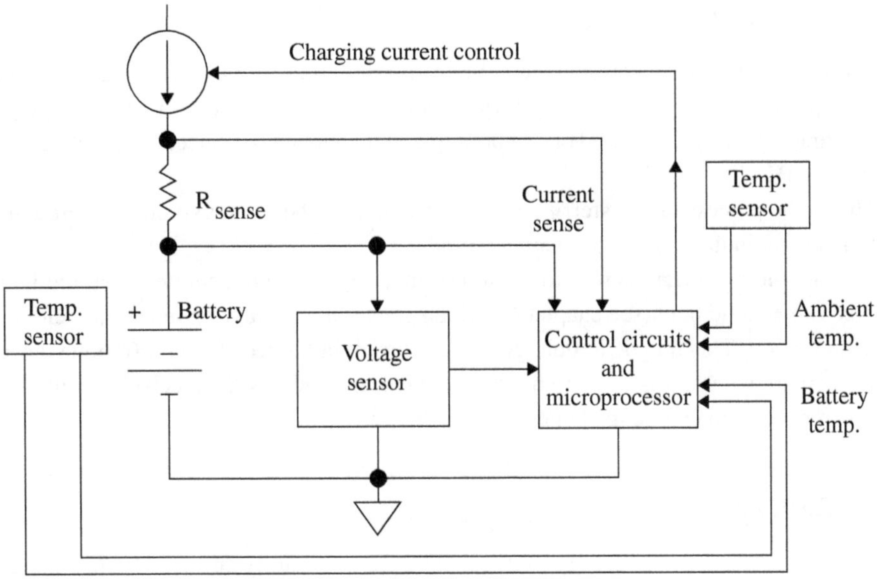

Figure 8.2 Generalized battery charging circuit

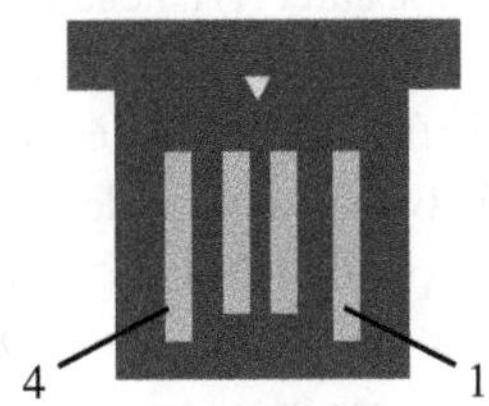

Figure 8.3 USB connector

8.2.1 Battery Charging from a USB Port

As discussed in Chapter 5, the rectangular Type A USB plug has four connector pins and a shield. As shown in Figure 8.3, the rightmost contact (number 1) carries 5 V and the leftmost contact (number 4) is ground. The two shorter middle pins are used for data transfer and have no function in the USB charger.

The USB port is unidirectional and power only flows out from the host port. The most common USB chargers are designed for single-cell Li-ion. With 5 V and 500 mA of available current, the USB bus can charge a small single-cell Li-ion pack. The USB host (computer) device supplies the current and, to prevent overload, some hosts include current-limiting circuits. Some USB chargers could be plugged into the AC mains or the cigarette lighter of a car and deliver higher peak currents than 500 mAh.

8.2.2 Wireless Charging

Wireless charging uses electromagnetic fields to safely transfer power from a transmitting source (power source) to a receiving device (mobile phone) for the purposes of charging (or recharging) a battery. It is based on the principle of magnetic resonance, or Inductive Power Transfer (IPT).

This is the process of transferring an electrical current between two objects through the use of coils to induce an electromagnetic field.

The alternating current is sent to the transmitter coil by the transmitter circuit to induce an electromagnetic field (see chapter 6), which creates a time varying magnetic field in the transmitter coil. This magnetic field generates current in the receiver coil (mobile device). This alternating current is converted into direct current (DC) by the receiver circuit, which can then be used to charge the battery.

8.3 Battery

The battery is the source of energy for the mobile phone circuitry and subsystems. Batteries are devices that convert the chemical energy stored in their active materials into electrical energy by means of an electrochemical oxidation-reduction reaction. A battery consists of one or more electrochemical cells. All electrochemical cells consist of two electrodes separated by some distance. The electrodes are: (i) a negative terminal, or anode – this permits electrons to flow out of it; (ii) a positive terminal, or cathode – this receives the electrons. The space between the electrodes is filled with an electrolyte that provides the medium for the transfer of charge, as ions, between the anode and cathode. Cells are electrically connected in an appropriate series / parallel arrangement to provide the required capacity, operating voltage, and current levels.

Electrochemical cells and batteries are categorized into two classes: (i) primary, which are nonrechargeable; (ii) secondary which are rechargeable – for these cells, the electrochemical reactions are electrically reversible.

8.3.1 Battery Working Principles

The chemical reaction at the anode releases electrons. This means that, during the discharge process, two or more ions from the electrolyte combine with the anode to form a compound and release one or more electrons. At the same time, the cathode undergoes a reduction reaction in that cathode material, ions, and free electrons combine to form compounds. As shown in Figure 8.4, when the electrical path provided by the electrolyte and an external electrical circuit connects the cathode and anode, the two simultaneous reactions mentioned earlier proceed. The electrons freed at the anode travel through the external electrical connection and react chemically at the cathode to make the cell function. In this way, the cell continues to discharge until either or both of the electrodes run out of reagents for their respective reactions. In the case of a primary cell this brings the end of its useful life as it cannot be recharged further. For the secondary cell, it means that it is required to be recharged. For secondary cells the recharge process is the reverse of the discharge process,

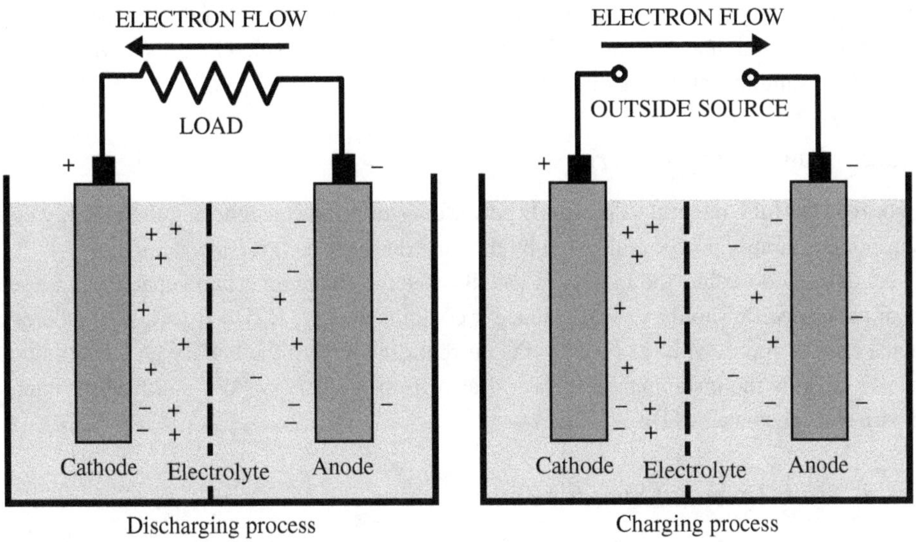

Figure 8.4 Secondary cell discharging and charging process

as discussed above. In the recharge process, an external source of direct electrical current (like an AC source) supplies electrons to the anode of the secondary cell and removes them from the cathode, forcing the chemical reactions into reverse until the cell is recharged.

- Discharge. When the battery is connected to an external load, electrons flow from the anode, which is oxidized, through the external load or circuit to the cathode, where the electrons are received and the cathode material is reduced. The electric circuit is completed in the electrolyte by the flow of anions or negative ions to the anode and cations or positive ions to the cathode. This process is called battery discharge. Self-discharge is the rate at which a battery discharges with no load due to chemical changes of the electrodes. Batteries gradually self-discharge even if not connected and delivering current through the external load. This rate increases with temperature and state of charge.
- Rate of discharge (or C rate). This refers to the amount of current that a battery can drive or sustain for 1 hour, while supplying a specified voltage range. C-rate = C / 1 hour, where C is the battery capacity.
- Battery capacity (Q or C) – it is simply expressed as the product of current (I) and time (t). $Q = I * t$. It is published by the manufacturer as a nominal rating for a given set of discharge conditions, like discharge rate (C rate), temperature, and minimal cell voltage. The battery could be discharged without damaging it until a "lowest" voltage level is reached; that lowest level is known as the minimum cell voltage.

The typical unit of battery capacity is mAh (milliamp hours) or Ah (ampere hours). This rating implies the discharge rate in milliamps (or amperes) that the battery can sustain for a period of one hour. The size and cost of the battery varies with respect to the amp-hour capacity of the pack. Most of the mobile phone battery packs have a rating of 3.6 V, 650 mAh. Whereas, the total energy stored in a battery pack is expressed in watt hours.

Watt hours = battery voltage * amp hours. The battery energy density refers to the energy in watt hours per unit mass of battery. The battery energy is described in units of Watt hours / mass, for example watt hours / kilogram.

8.3.2 Power versus Energy

Energy is the time integral of power. If power consumption is a constant, then energy consumption is simply power multiplied by the time during which it is consumed.

As described earlier, the battery is usually stated to have a certain capacity in ampere hours at a specific supply voltage. Hence it contains energy: $E\,[J] = P \cdot t = U \cdot I \cdot t$, where, U is the battery supply voltage [V], I is the current drawn from the battery [A], t is the time of use [s], P is the instantaneous power drawn from the battery [W], and E is the energy drawn from the battery [J].

8.3.3 Talk Time and Standby Time

The performance of a mobile battery is measured in terms of talk time (in dedicated mode) and standby time (in idle mode). Most commonly, there are three aspects associated with a mobile phone battery performance:

- Standby time. This indicates the total idle mode time – the time duration for which the mobile phone battery will last when it is operating in the idle mode.
- Talk time. The total time that the battery will power the mobile phone when it is in a voice call.
- Battery life. indicates how long the battery can be used – how long it can continue to change and recharge.

The talk time and standby time performance not only depend on the type of the battery used, but also on the sleep handling, clock, and some other system design parameters.

8.3.4 Types of Rechargeable Batteries and Performance Parameters

Batteries convert chemical energy directly to electrical energy. Battery performance parameters include voltage, amp hour capacity and C rate. Various factors are considered for rechargeable battery selection. These include multiple cell configurations (serial / parallel), battery capacity and voltage, cost, weight and volume, charging and discharging characteristics and time, and complexity. Generally, the important characteristics of a rechargeable battery include cell voltage, capacity, energy density, cost, memory effect, self-discharge rate, operating temperate range, and environmental concerns. There are tradeoffs to be made in selecting the battery and designing the appropriate charging circuits and, unfortunately, these considerations not only interact but often conflict.

Some mobile phones can take primary (nonrechargeable) batteries, while others can also work on "ordinary" batteries, for example, the Motorola c520 works with four AA batteries but rechargeable (secondary) batteries are most commonly used in mobile phones and other portable electronic equipment. Alkaline MnO_2, Li/MnO_2, and zinc air are examples of nonrechargeable batteries.

Table 8.1 Rechargeable battery technologies

Battery types	Sealed lead-acid	Nickel cadmium	Nickel metal hydride	Lithium metal	Lithium ion
Average cell voltage (V)	2	1.20	1.25	3.1	3.7
Energy density (Wh/kg)	35	45	55	140	100
Energy density (Wh/l)	85	150	180	300	225
Cost ($/Wh)	0.25–0.5	0.75–1.5	1.5–3.0	1.4–3.0	2.5–3.5
Memory effect?	No	Yes	No	No	No
Self-discharge (% month)	5–10	25	20–25	1–2	8
Discharge rate	<5 C	>10 C	<3 C	<2 C	<2 C
Charge/discharge cycles	500	1000	800	1000	1000
Temperature range (°C)	0 to +50	−10 to +50	−10 to +50	−30 to +55	−10 to +50
Environment concerns	Yes	Yes	No	No	No

The most popular types of rechargeable batteries in use today are the sealed lead acid (SLA), nickel cadmium (NiCd), nickel metal hydride (NiMH), methanol fuel cell, and lithium-ion (Li-Ion). Table 8.1 shows the merits and demerits of different rechargeable battery technologies.

- NiCd (nickel cadmium). This is the oldest type of battery. Generally, these are used in cordless phones and the old-generation mobile phones. Advantages: these are the cheapest variety of batteries; thus they are highly affordable and bring down the overall cost of the mobile handset. Disadvantages: as shown in Table 8.1, this type battery is very prone to the "memory effect." This is sometimes referred to as voltage depression. If it is not fully discharged before recharging, after a few cycles, the battery "learns" this low water mark, and acts as if it is discharged at this point. It must be discharged and recharged fully on every recharge cycle. The chemicals in nickel cadmium are not environmentally friendly, and the disposal of cadmium-rich waste is an increasing problem.
- NiMH (nickel metal hydride). This is a better battery type than the NiCd and is much less prone to "memory effect." Advantages: these are cheaper than Li-Ion batteries, so are affordable and bring down the overall cost of the mobile phone. They are less prone to the "memory effect" issue, and have a higher capacity in relation to their size and weight. Disadvantages: their longevity is less compared with NiCd cells. After a few hundred charge cycles, the crystals inside NiMH cells become coarser and, although they are able to provide the power for long standby times, when the extra current to sustain a call is needed, the voltage available drops rapidly, and suddenly shows low battery warnings.
- Li-Ion (lithium-ion). These are considered the most advanced and widespread cell phone batteries. Advantages: these are lighter and slimmer than the NiMH and NiCd batteries and are not subject to the "memory effect." Usually, they offer a longer standby time and talk time. Disadvantages: they are expensive. As Li-Ion offers a high capacity-to-size (weight) ratio and has a low self-discharge characteristic, it is very popular in many mobile phone devices. A Li-Ion battery of a mobile phone is shown in Figure 8.5. In Li-Ion batteries, lithium ions move from the negative electrode to the positive electrode during discharge

Figure 8.5 Li-Ion battery of a mobile phone

and back when charging. Generally, the mobile phone Li-Ion battery has three pins, and is labeled as: "+," "T," "–" for positive, thermistor, and negative respectively. The third pin ("T") is usually for an internal temperature sensor, to ensure safety during charging.

- Li-polymer (lithium-polymer). These are very similar to lithium-ion, except that they can be molded into more varied shapes, and so they can be squeezed into smaller phone casings. These are even thinner and lighter batteries.

Sealed-lead-acid and NiCd batteries have environmental concerns regarding the proper disposal as these contain hazardous metal content, whereas NiMH and Li-Ion batteries do not contain significant amounts of pollutant.

To find out the type of battery the mobile phone has, switch off the phone and remove the battery (normally on the back side of the phone). The battery type might be written on the label of the battery.

8.4 Mobile Terminal Energy Consumption

Smartphone has various wireless connectivity units including cellular modem, an application processor, a graphics processor, an audio codec, various types of memory, and an interface with the human user as shown in Figure 8.6.

Energy can be saved by identifying the main power-consuming components in a smartphone and then studying how the components' power consumption depends on relevant parameters. One method to achieve this is to define a power consumption model. One simple model that can be used for analysis is shown in Figure 8.7.

If we consider the mobile device to be a black box connected to the supply voltage as shown in Figure 8.7, then the average current consumed during different operational

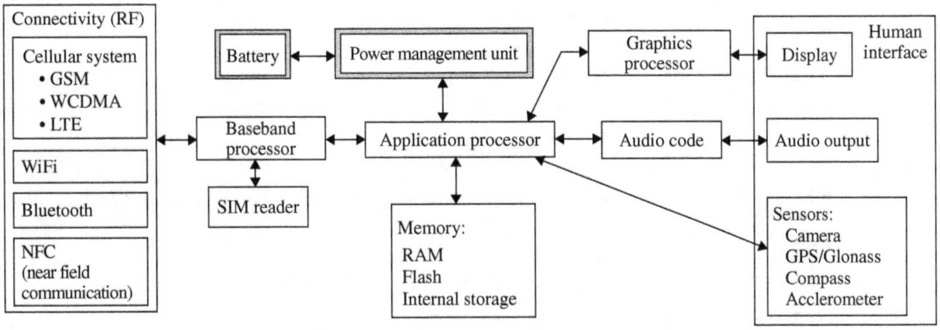

Figure 8.6 Typical components of smart phone

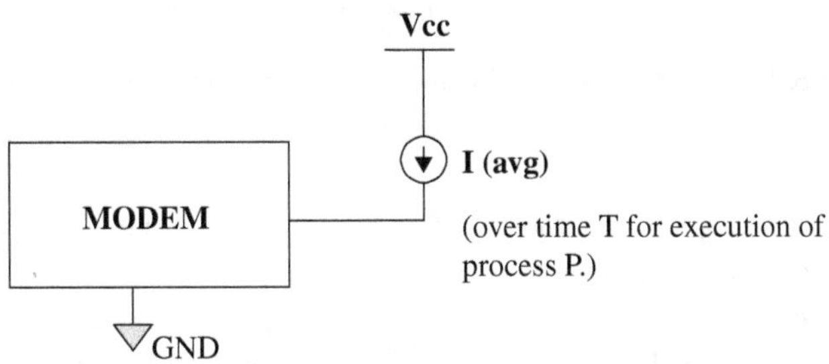

Figure 8.7 Mobile terminal as a black box model

modes like sleep, data reception and transmission, and running applications can be measured or estimated. During these different operations the average current and voltage can be measured and knowing the average operating voltage, average operating current and the duration of these intervals for any given use case, the total energy consumed for that use can be easily computed. Energy = Power * Time = (Voltage * Current) * Time.

8.4.1 System-Level Analysis of Power Consumption

The power consumption at various units of a typical LTE smart phone is shown in Figure 8.8. It shows that the majority of the power is consumed by the CPU/GPU for running different applications, followed by RF transmitter unit.

The slow development in battery capacities cannot catch up with the speed of evolution in Internet technologies and mobile devices. So, researchers recently started to work toward making every layer of the network more energy efficient. The power consumption optimization in a mobile terminal can be possible on multiple layers:

- **Application**. In the application layer, power can be saved if the smartphone's performance is dynamically adjusted according to the current needs of the running applications.

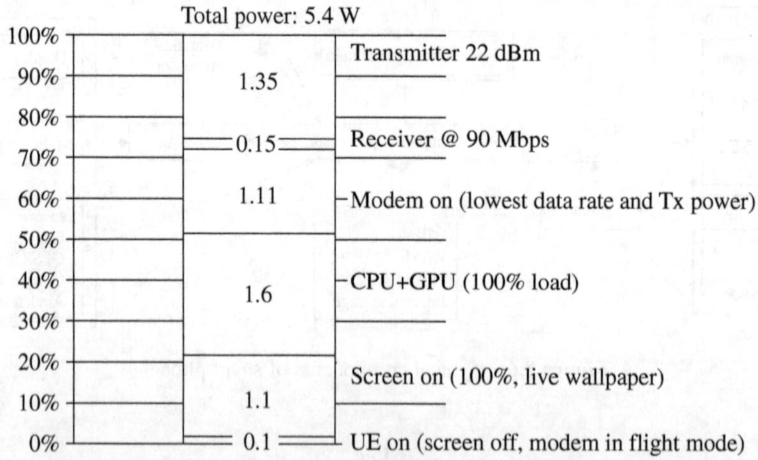

Figure 8.8 Power consumption by different components in LTE phone

It should adjust the display brightness, CPU speed, and so forth based on the needs of the application that is currently running.

- **Transport**. The data from multiple applications can be combined into one transfer. On the transport layer, it may be possible to combine data packets from multiple applications / protocols on the upper layer into one transfer. In this way the smartphone only needs to carry out one data transfer even though the data has multiple end points.
- **Network**. This layer basically consists of the Internet protocol and should ensure low overheads to reduce power consumption. To optimize the smartphone's energy consumption, this layer should adopt efficient header information and acknowledgements, and fast routing in the core network.
- **Physical and data link**. This layer should adjust network-controlled parameters such as scheduling and transmit power control; it should use low power components, and apply power management. This could, for example, imply the use of an energy-aware scheduler and transmit power control for low power consumption. On the physical layer the smartphone's instantaneous power consumption is directly related to the hardware components comprising the smartphone.

8.5 Low-Power Smartphone Design

Today, a widening gap is created between the battery technology and semiconductor or wireless data rate driving technology because of the disparity between the demands of power hungry wireless applications and the batteries to power them. Researchers are working to improve their methods, and also the power density, safety, cycle durability, recharge time, cost, flexibility and other characteristics of these batteries. Many new smartphones are already equipped with high-capacity batteries (Samsung's Galaxy Note II, claims a capacity of 3100 mAh) but due to the introduction of MIMO and Multicore SoCs the power consumption in a mobile phone is many folded.

8.6 Low-Power Design Techniques

Any power consumed by the system without any useful result is considered to be waste of power. In a power-voracious system, the power consumption will be very high, so the design should be such that the power consumption should be reduced or at least optimized. For a mobile system, the battery weight and size plays an important role. A reduction in power consumption provides several benefits: less heat will be generated (which reduces the problems associated with the high temperature), it prolongs the battery life, and device reliability increases. Efficient power management can reduce power consumption. Power management techniques can be applied at various levels in order to achieve the common objective of power reduction.

8.6.1 System-Level Power Optimization

The power consumption at the system level plays an important role in overall power consumption. Some system-related factors that govern the power supply current are operating frequency, supply voltage, and switching rate.

8.6.1.1 Clock Speed

The power consumption of a device increases with the operating clock frequency. As the clock speed increases, the current increases proportionally but the time required for executing the same operation decreases proportionally. If the clock speed is doubled, twice the current is required but half of the time is required for the same operation. Two separate cases are analyzed below (see Figure 8.9):

- Case A: at clock speed C, the processor takes T1 time to complete the process.
- Case B: at clock speed 2C the processor takes T2 time to complete the process and remaining time (T1–T2) is in idle state.

In case B, as the clock speed is greater, it takes more current I2 and time taken is less T2. But after the completion of the process, the processor is idle for (T1–T2) time. In the case A, the device takes I1 current for time interval T1. Now, if we compare the overall power consumption of these two cases – then case B looks advantageous over case A only if the idle state power consumption is very less, unless case A is better. For this reason, if the application does not require the entire MIPS capability of the device, it is more power efficient to slow down the system clock to minimize the idle time.

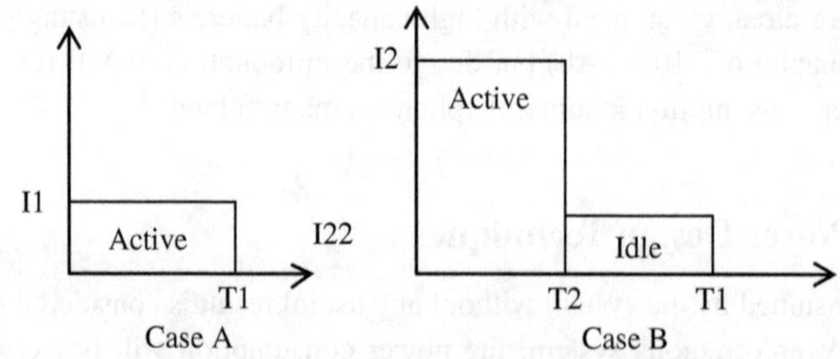

Figure 8.9 Different clock-speed scenarios

8.6.1.2 Adaptive Clocking and Voltage Schemes

As the performance requirements of a system typically vary over time as the task it is performing changes, it is wasteful to run the system at maximum performance when this is not required. Adapting the clocking frequency or supply voltage of the system to reduce performance and power during those periods can result in substantial power savings.

8.6.1.3 Switching Rate

The CMOS is the basic building block of today's integrated circuits. The CMOS current is directly proportional to the switching rate. So, we have to reduce the switching rate to the minimum possible in the all hardware design and software programming.

8.6.1.4 Supply Voltage (V_{dd})

Supply voltage also increases the power consumption. The higher the operating voltage, the greater the power consumption.

8.6.1.5 Power-Down Mode

Power-down modes can be used to turn off the device, when it is not doing any task. Several types of power-down modes are possible:

- Disable the processor clock to the CPU but allow the on-chip peripherals to remain active.
- Disable the clock to the CPU, and also disable some on-chip peripherals like the timer, standard serial ports, the TDM serial ports but keep the buffered serial ports active.
- Disable clock to the CPU, disable on chip peripherals, and PLL.

Sleep mode is an extension of selective power-down strategy. The activity of the entire system is monitored rather than that of individual modules. If the system has been idle for some predetermined time, then the entire system is shut down and this mode is called sleep mode. During sleep mode the systems inputs are monitored for activity, which will then trigger the system to wake up and resume processing. As there are some overheads in time and power associated with entering and leaving sleep mode, some tradeoffs have to be made in setting the length of the power down time.

8.6.1.6 Hardware / Software Portioning

Portioning the system's functions into hardware and software functional blocks also plays a vital role in the overall power consumption. The functionality should be portioned between hardware and software depending on the speed, MIPs availability, complexity, interfacing and power consumption.

8.6.2 Algorithmic Level

The choice of proper signal processing and control related algorithms are the most highly leveraged decision in meeting the power constraints in a mobile phone system. It has most dramatic impact. Selecting the correct algorithm reduces the power consumption and increases the operation speed significantly. The total power consumed by the device varies, based on the program activity. The ability for an algorithm to be parallelized is critical with respect to speed of execution, and basic computation complexity must be optimized in order to reduce the power.

8.6.2.1 Minimizing the Number of Operations

A good way to save power is to avoid wasteful activity. At the algorithm level the size and complexity of a given algorithm determines the activity. As each operation consumes power, the best way is to reduce the number of operations.

8.6.2.2 Minimizing the Memory Access

An algorithm or program code that uses less memory access is always amenable to low-power implementation because low memory access means low usage of the bus, which in turn reduces the power consumption. In particular, DSP algorithms that operate on blocks of input data often fetch the same data from memory multiple times during execution. A clever programmer can reuse previously fetched data to reduce the number of memory accesses required by an algorithm. If the programmer uses previously fetched data then this process reduces the memory bandwidth requirement at the expense of slightly larger code size. Fetching data from external memory also increases the power consumption.

8.6.2.3 Pipelining

We know that if we reduce the supply power level then the power consumption will be reduced very much. But one of the major abstractions for reducing the voltage level is that it reduces the speed of operation. At the algorithm level, transformations can be used to increase speed and allow lower voltages. One approach is to improve throughput by exploiting concurrency via pipelining. This enables the hardware to be operated at lower clock frequencies, and thereby lower voltages. But the disadvantage of using pipelining is that it increases the latency of the circuit and we may require more additional hardware due to having more operations in parallel.

8.6.3 Technology

By using different technology the power requirement for a device and that for a system can be reduced. We know today that, due to their small size and much smaller power consumption, CMOS logic devices are very popular in digital design. The only problem that CMOS logic suffers is its high propagation delay. There are several ways in which we can further reduce the power consumption of a CMOS device.

8.6.3.1 Threshold Voltage Reduction

As in a CMOS circuit the energy dissipation per transition is proportional to V_{dd}^2, so the reduction of the supply voltage will reduce power consumption drastically but in CMOS logic the delay increases as V_{dd} approaches the threshold voltage of the device. Even though the exact analysis of the delay is quite complex, if the nonlinear characteristics of a CMOS gate are taken into account, a simple first order equation can express the delay:

$$T_d = (C_L * V_{dd}) / (\mu Cox) * (W/L) * (V_{dd} - V_{th})^2$$

So, the reduction of V_{dd}, will reduce the power consumption but increase the delay. If the threshold voltage is reduced then the reduction of V_{dd} will not affect delay. This is one approach to reduce the threshold voltage of the device. As a significant power improvement can be gained through the use of low-threshold CMOS devices, the question of how low the thresholds can be reduced must be addressed. The limit is set by the requirement to retain adequate noise margins and increase subthreshold currents. The optimum threshold voltage must compromise between improvement of current drive at low supply voltage operation and control of subthreshold leakage.

8.6.3.2 Technology Scaling

Technology shrinkage causes the capacitance of nets to decrease. This reduction in capacitance results in less power consumption.

8.6.3.3 Minimize the Sources of Power Dissipation in CMOS Circuits

CMOS transistors are the basic building blocks of almost all today's digital devices. Most of the power consumed by the system is wasted by the CMOS circuits. There are three major sources of power dissipation in digital CMOS circuits: static power dissipation, dynamic power dissipation, and short-circuit power dissipation.

Static power consumption: when all the input levels are held in some logic level and the circuit state is not changing, there is some leakage current between the GND and V_{dd} due to reverse bias. So, there is a power loss and it is called static power loss:

$$P_S = i_S(\exp^{ev/kt} - 1) * V_{dd}$$

Short-Circuit Power Consumption

The short-circuit component of the dissipated power is due to the finite rise and fall time results in the direct current path between the supply and the ground, which exists for a short period of time during switching, especially when the nMOS threshold is less than the input voltage and pMOS threshold is higher than the input threshold. If such a situation arises there will be a conduction path between the V_{dd} and ground, so both nMOS and pMOS devices will be on.

$$Psc = Isc * V_{dd}$$

Dynamic Power Consumption

This is due to the current flow when the transistors of the devices are switching from one logic state to other. This is a result of current required to charge the capacitance (wiring cap, junction cap, input cap of the fan out gates).

$$P\ dynamic = Cpd * V_{dd}^2 * \text{input signal frequency}$$

8.6.4 Circuit/Logic

In the circuit-level optimization of power consumption several factors play a significant role, like transistor sizing, energy recovery, placing, and routing.

8.6.4.1 Place and Route Optimization

At the layout level, the place and route should be optimized so that signals that have high switching activity should be assigned short wires and signals with lower switching activities can be allowed longer wires.

8.6.4.2 Transistor Sizing

Independent of the choice of logic family or topology, optimized transistor sizing will play an important role in reducing the power consumption. It is important to equalize all the delay paths. However, the W/L ratio should be raised for all the devices.

8.6.4.3 Energy Recovery or Adiabatic Circuits

Adiabatic circuits are also known as energy recovery circuits. They resonate the load capacitance with the inductor in order to recover some of the energy used to change the capacitor's voltage. This is not widely popular as it introduces substantial delay.

8.6.5 Architecture

Architectural-level design has a dramatic impact on the overall power budget design. The general approach to compute the power dissipation of a logic circuit at the architectural level is to estimate the power of each module that makes up the circuit. At this level, the circuit is described in terms of functional modules, of different levels of complexity, and how they are interconnected.

8.6.6 Power Consumption in Microprocessors

Several factors can be considered to reduce the power consumption of a microprocessor, like an efficient memory-management unit, proper memory architecture and internal memory selection, clock speed, proper instruction set design, and efficient power management by using different power down modes.

8.6.7 Power Consumption in Memory

Selection of proper memory type and memory architecture decides a considerable amount of the total power consumption of the device. The overall memory size should also be less in order to achieve low power consumption. That indicates the code should be highly optimized.

8.6.7.1 Techniques to Optimize Power using Intelligent Refresh Mechanism for DRAM

A DRAM cell comprises an access transistor and a capacitor. Data is stored in the capacitor as an electrical charge. But, as an electrical charge leaks over time, so, DRAM must be refreshed periodically to preserve the stored data. This periodic operation negatively affects performance and power. As DRAM architecture evolved from "asynchronous" DRAM to SDRAM, refresh parameters and options were also modified accordingly. For the DRAM cell, the refresh operation functionality is accomplished by a read or write operation. This means that if the cell has been recently read (or written to) then it does not need to be refreshed again. The refreshment cycle can be varied based on different operating scenarios.

8.6.7.2 OS-Controlled Refresh

Sometimes it may happen that some portion of the memory is not used. As it is not necessary to refresh unused memory, a considerable amount of power can be saved by intelligently controlling which pages get refreshed.

Power Reduction Technique for Analog Blocks
In a wireless system, the RF processing part consumes most of the energy. So, a global approach to low-power methodology is: digitize the signal at the earliest stage; simplify the remaining analog hardware requirements, and leverage low power architectural techniques in the digital section. From a system perspective, location of the A/D conversion constitutes one of the most important design choices influencing overall power dissipation. As digital designs are particularly amenable to low power techniques, it is desirable to "go digital" at the earliest possible point in the receive chain.

Analog blocks are generally used only for filtering, amplification, demodulation, sampling, and A/D conversion. All other functions are implemented digitally at baseband.

8.6.7.3 VGA Dynamic Range Reduction

Spreading the signal wider than the coherence bandwidth of the channel reduces the likelihood that the entire signal is lost in a local deep fade. That relaxes the dynamic range requirements for the VGA.

8.6.7.4 Sampling Demodulation

Demodulation and sampling can be combined into a single step. A carefully controlled passive sampling switch replaces the standard mixer-sampler configuration. No PLL is required, yielding significant power savings at the expense of increased phase noise.

8.6.7.5 A/D Resolution Reduction

As the signal is quantized prior to correlation, precise quantization is not required. Software simulation is used to determine the appropriate A/D converter resolution.

8.6.7.6 RF Processing and Mixed Signal

RF processing is one of the most power-consuming components of the transceiver. The high-frequency active components such as mixers and synthesizers take a disproportionate fraction of the power budget in the transceivers. Some novel architecture that reduces the impact of these components or eliminates them all together is hence quite attractive.

Suppose a fourth-order maximum ratio combining diversity is implemented in the receiver. The required Eb/No value decreases from 45 dB to approximately 15 dB. In this case, the transmit power requirement drops to 4.5 mW, the power drawn from the battery decreases to 9 mW, and battery life increases to 22 000 h. This provides a powerful example of how one can trade receiver complexity for improved energy efficiency. Although the receiver will certainly consume additional power to implement fourth-order diversity, the net system energy savings will be substantial.

This trend of performing more and more computations on less and less space in less and less time comes with enormous physical challenges. One of these challenges is power dissipation, which needs to be reduced at every stage. High power consumption means high power costs and short battery lifetime for mobile applications. As discussed, various techniques can be used to reduce the power consumption of SoC ASIC designs, including dynamic frequency control, dynamic power management and the ability to idle embedded processors. An SoC ASIC external reference clock and internal clock generator can be used to provide dynamic frequency control. The designers of a low-power system should always keep few things in mind: (i) Run the clock as slowly as possible without affecting the functionality and speed requirements. Introduce an adaptive clocking mechanism to vary the clock speed as and when required. (ii) Introduce dynamic voltage and frequency scaling (DVFS) mechanism to cater to different scenarios based on different operating modes. (iii) Put some modules into sleep mode whenever possible. (iv) Use “master clear” (MCLR) to wake parts from sleep instead of the watchdog timer (WDT) if possible. (v) Do not let any inputs float. (vi) Do not drive any unnecessary loads. Minimize capacitive or inductive loads on switching I/O pins, or resistive loads on other driven pins. If a pin is not to be used you can leave it disconnected and drive it low or high, or put a pull up or down resistor on it as an input. (vii) Turn off all timers when not in use. For instance, TMR0 can be incremented from the instruction clock or an external pin. When not in use, assign it to the pin (if it is toggling at a lower rate than the instruction clock). Do not use prescalers when unnecessary. In other words, minimize the amount of logic-changing states. (viii) Turn off any other peripherals when not in use. (ix) Use dedicated hardware whenever possible. (x) Minimize the CMOS-switching activities and the number of gate counts.

The choice of Transmit Power Amplifier (PA), display size and technology and so forth will also have a major effect. Finally it is important to have a power management unit (PMU) that makes sure to power OFF unused components and scale the performance of the components, which are currently ON.

8.6.7.7 LTE UE Power-Saving Modes

A mobile terminal is supposed to monitor the control signals (Physical Downlink Control Channel (PDCCH)) continuously to be able to send and receive actual data. But continuously monitoring PDCCH becomes a waste of radio resources and battery power,

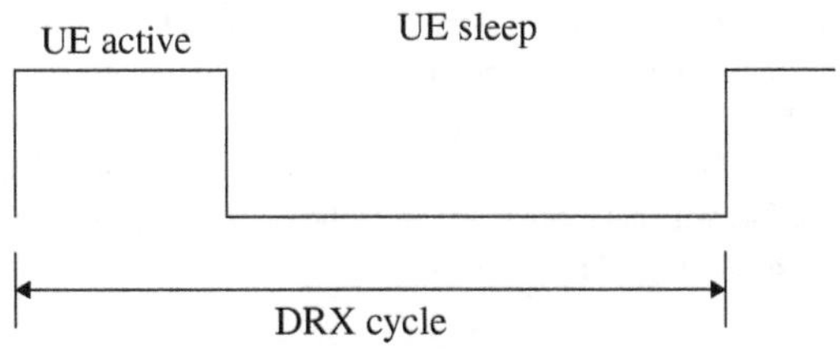

Figure 8.10 Representation of DRX cycle

particularly in the case when no uplink (UL) or downlink (DL) transmission is scheduled for long periods. Discontinuous reception (DRX) and discontinuous transmission (DTX) are possible solutions to avoid this situation, which means that the UE stays asleep and periodically wakes up only after a particular interval to monitor PDCCH for any data transfer. Idle-mode DRX has a longer cycle time than active mode. In active mode, there is a dynamic transition between long DRX and short DRX, and the durations for long and short DRX are configured by the RRC layer. Long DRX has a longer "off" period. The LTE standard has a number of power-saving modes incorporated into it that allow the UE to enter into power-saving mode (DRX cycles are specified in 3GPP TS 36.321):

- Idle mode. In this scenario, the UE does not have any active data sessions but is camped onto the network and performs regular synchronization / location-update operations. In this state, the control processor puts the UE into sleep mode and brings it out of sleep to listen to broadcast channels or transmit location update information. During the power-save mode the UE can be almost entirely shut down except for a small low-power timer block, which is configured to wake the system at the appropriate times.
- Active mode. The UE is fully active with all or most blocks powered up. A typical use case scenario would be video call, video streaming, or TCP/IP data transfer. In this mode both the ARM and DSP subsystems are powered on, supporting uplink and downlink data transfers as well as the associated signaling. Figure 8.10 shows the UE periodic activities in active and sleep modes.

Support for voice results in small packet transmission and reception (small, infrequent data transmission), which in turn allows the UE to perform power-saving operations during idle times. The ARM control processor will manage the overall power-saving scheme as it has knowledge of the scheduling of the voice packets and will thus, in turn, move the DSP in and out of power save accordingly.

Further Reading

Bircher, W. L., and John, L. K. (2007) *Complete System Power Estimation: A Trickle-Down Approach based on Performance Events*. Proceedings of the IEEE International Symposium on Performance Analysis of Systems and Software, San Jose, CA, USA, April 25–27, 2007, IEEE Computer Society, pp. 158–168.

Das, Sajal Kumar. (2010) *Mobile Handset Design*, John Wiley & Sons, Ltd.

Hewlett-Packard (2004) *Lithium-Ion Battery Technology: Getting the Most from Smart Batteries*, Hewlett-Packard Development Company, LP.

Kravets, R., and Krishnan, P. (1998) *Energy Consumption Techniques for Mobile Communication*. Proceedings of the 4th Annual ACM/IEEE International Conference on Mobile Computing and Networking, pp. 157–168.

Linden, D. (2002) *Handbook of Batteries*, McGraw-Hill.

9

4G and Beyond

9.1 Introduction to LTE-Advanced

As discussed in Chapter 3, 3GPP LTE Rel-8 version (3.9G) does not satisfy all the ITU-R 4G requirements, so LTE-Advanced (LTE-A) was introduced. LTE-A is an evolved version of LTE with a major enhancement of the LTE standard to fulfill the requirements for IMT-Advanced, and is capable of providing peak data rates of 1 Gbit/s, as shown in Table 9.1. LTE-A can be considered as 4G. It was initially specified as part of the 3GPP Release 10 specifications, with a functional freeze targeted for March 2011. Since then, LTE/LTE-Advanced technology has been enhanced continuously either by the addition of new technology components or by improving existing ones.

9.2 LTE-Advanced Features

The main goal of LTE-Advanced is to provide IMT-Advanced functionality while retaining backward compatibility. LTE-Advanced comprises a number of enhancements or additions, and some of these are discussed below.

9.2.1 Carrier Aggregation

A carrier is a high-frequency RF signal with a defined center frequency and specified bandwidth. Now, if instead of one carrier, two carriers are assigned for any transmission, the data rate can be simply doubled. So, two or more component carriers (CC) can be used to support

Mobile Terminal Receiver Design: LTE and LTE-Advanced, First Edition. Sajal Kumar Das.

Table 9.1 LTE, LTE-Advanced, and IMT-Advanced performance requirements for downlink (DL) and uplink (UL)

Parameters	Transmission path	Antenna configuration	LTE (Rel. 8)	LTE-Advanced	IMT-Advanced
Peak data rate	DL	8×8	300 mbps	1 Gbps	1 Gbps
	UL	4×4	75 mbps	500 mbps	–
Peak spectrum efficiency (bps/Hz)	DL	8×8	15	30	15
	UL	4×4	3.75	15	6.75
Capacity (bps/Hz/cell)	DL	2×2	1.69	2.4	–
		4×2	1.87	2.6	2.2
		4×4	2.67	3.7	–
	UL	1×2	0.74	1.2	–
		2×4	–	2.0	1.4
Cell-edge user throughput (bps/Hz/cell/user)	DL	2×2	0.05	0.07	–
		4×2	0.06	0.09	0.06
		4×4	0.08	0.12	–
	UL	1×2	0.024	0.04	–
		2×4	–	0.07	0.03

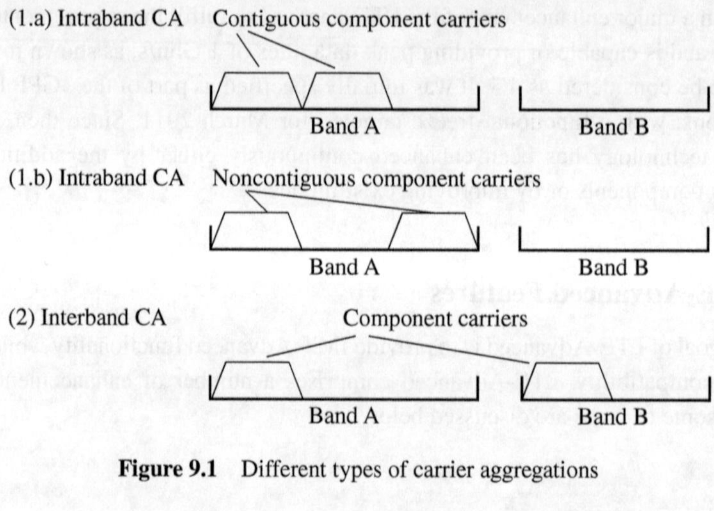

Figure 9.1 Different types of carrier aggregations

wider transmission bandwidths. This technique of using (or aggregating) two or more component carriers together for a transmission is known as carrier aggregation (CA) (see Figure 9.1). Using one carrier, we can achieve up to a certain maximum data rate, despite employing higher order modulation, coding, diversity (MIMO), and other advanced techniques. So, to overcome that limit, we introduce the concept of multicarrier transmission, which simply multiplies the data rate by the number of carriers used for a transmission. The basic idea of the multicarrier feature is to achieve a higher data rate, better resource

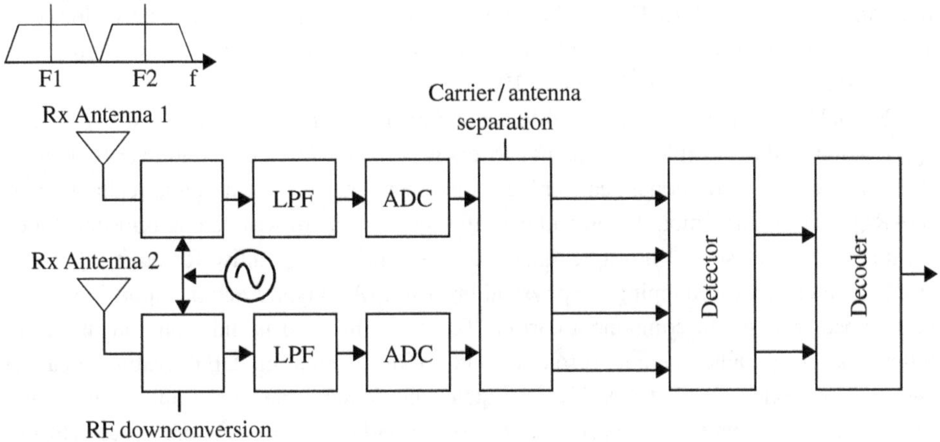

Figure 9.2 High-level block diagram of mobile receiver for carrier aggregation support

utilization, and spectrum efficiency by means of joint resource allocation and load balancing across the carriers. Several types of CA are possible, mentioned below:

- Intraband CA. Here, component carriers belong to a single frequency band – the aggregated carriers are from the same frequency band. This is of two types:
  - Contiguous or adjacent. Here, the aggregated carriers are adjacent to each other. In this case, a single RF transceiver can be used for the reception of aggregated carriers. So, the same RF module (which is used for single carrier reception) can be used with a wider analog low-pass filter and higher sampling rate ADC.
  - Noncontiguous. The aggregated carriers reside in the same frequency band but not adjacent to each other. Here, generally separate RF transceivers could be required for the reception of each carrier.
- Interband. Aggregated carriers reside in two different frequency bands. Separate RF transceivers are required for each carrier's reception (see Figure 9.2). Here, as the carriers are widely separated, so separate channel estimation and equalization are also needed. This has a greater impact on silicon real-estate size, cost and power consumption.

CA in the 4G System

For the LTE system, a maximum carrier bandwidth of 20MHz was specified in 3GPP Releases 8 and 9. Using this 20MHz bandwidth with 4×4 MIMO (e.g. four transmit antennas and four receive antennas), a peak data rate of 299.6 mbit/s can be achieved in the downlink, but this was not sufficient for LTE-A requirements. So, in 3GPP Release 10, the carrier aggregation technique was introduced in LTE to aggregate multiple cells together. This was called LTE carrier aggregation, Multicell, or Multicarrier LTE. Release-10 carrier aggregation supports peak data rates of 1 Gbps on a downlink using five carriers (20 * 5 = 100 MHz total bandwidth). The maximum size of resource blocks for a single component carrier is specified as 100 RBs (in earlier releases), although it was proposed to increase this to 110 in Rel-10.

UE could be scheduled over multiple component carriers simultaneously. The data aggregation of the multiple component carriers is performed at the medium access control (MAC) layer. Each component carrier has its independent hybrid automatic repeat request (HARQ) process, and modulation and coding schemes. There will be one transport block (in the absence of spatial multiplexing) and one hybrid-ARQ entity for each scheduled component carrier for mapping the physical layer to MAC layer. Each transport block will be mapped to a single component carrier. The downlink and uplink component carrier linkage is configurable via radio resource control (RRC) signaling. UE operating in carrier aggregation mode in RRC_CONNECTED state, should have one pair of uplink and downlink primary component carriers (PCC) that correspond to the primary serving cell (PCell) and maybe one or multiple pairs of uplink and downlink secondary component carriers (SCCs) from the same eNB that correspond to the secondary serving cells (SCells). Generally, the eNB can trigger a PCell change for UE due to mobility, load balancing, and so on. See http://www.3gpp.org/dynareport/36807.htm (accessed May 17, 2016), which summarizes a study of CA, enhanced multiple antenna transmission, and CPE.

Although CA brings considerable technical challenges owing to the cost and complexity that will be added to the UE, the CA technique has been introduced in HSPA and in EGRS systems due to its support for a higher data rate.

CA in the 3G System

To achieve a better data rate in the downlink, 3GPP Release-8 introduced a DC-HSDPA operation for two adjacent carrier cells operating in the same frequency band using contiguous intraband CA, as discussed earlier. From the higher layer perspective, each component carrier appears as a separate cell with its physical cell identifier. A cell is characterized by a combination of scrambling code and carrier frequency. Now, in that sense, two carriers along with a scrambling code can form dual cells. Dual cell (DC) HSDPA is the natural evolution of HSPA by means of carrier aggregation in the downlink. Here, the two cells belong to the same node-B and are on different carriers:

- Anchor carrier. A UE anchor carrier has all the physical channels including DPCH/F-DPCH, E-HICH, E-AGCH, and E-RGCH. This is also, known as a primary carrier (primary serving cell – PSC).
- Supplementary carrier. During dual-carrier operation in CELL DCH, the UE's supplementary carrier is the downlink carrier, which is not the UE's anchor carrier. This is also known as a secondary carrier (secondary serving cell – SSC).

Using this technique, the peak data rate is doubled from 21 mbps to 42 mbps, even without use of MIMO. Release 9 introduces DC-HSDPA in combination with MIMO on both carriers. This will allow a theoretical speed of up to 84 mbit/s. Often UMTS licenses

Box 9.1 CA in 3G evolution path

Release 5 : 5 MHz, no MIMO: 14 mbps
Release 7: Year 2007, HSPA+, 5 MHz (single carrier) 2×2 MIMO: 28 mbps
Release 8: Year 2008, DC-HSDPA, 10 MHz (dual carrier), no MIMO: 42 mbps
Release 9: Year 2009, DB and DC-HSDPA, DC-HSUPA, 10 MHz (dual carrier), 2×2 MIMO: 84 mbps
Release 10: Year 2010, 4 Carrier HSDPA, 20 MHz, 2×2 MIMO: 168 mbps
Release 11: Year 2012, 8 carrier HSDPA, 40 MHz with 2×2 MIMO or 20 MHz with 4×4 MIMO: 336 mbps

are issued in a paired spectrum of either 10 MHz or 15 MHz blocks – two or three carriers, for uplink and downlink. So, DC-HSPDA implementation becomes easy using two adjacent carriers, for operators and UE vendors. But, in many cases, operators have several frequency bands, so they want to use carrier aggregation with different frequency bands. That is why Release 9 allows paired cells operation on two different frequency bands. This is known as DB-DC HSDPA and this is interband CA as discussed above. DC HSDPA and DB-DC HSDPA are considered to be the same, the only difference being whether intraband or interband CA is used. Support for this optional feature is signaled to the network via UE capability signaling. The 3G evolution path is shown in Box 9.1

A frequency band is a specific range of frequencies with defined upper and lower frequency limits in the radio frequency (RF) spectrum, which is divided into several component carriers. Various bands are deployed in UMTS, like Band I (W-CDMA 2100 band, in a downlink from 2110 to 2170 MHz) in Europe, India, Africa, Israel, Australia, New Zealand and Brazil), Band II (W-CDMA 1900 band in North America and South America), Band VIII (W-CDMA 900 band in Europe, Asia, Australia and Thailand), and so forth.

CA in 2G Systems

A similar concept, the multicarrier downlink, has also been proposed by ST-Ericsson and by Ericsson, and is now a part of 3GPP GERAN Release 12.

9.2.2 Enhanced Uplink Multiple Access

The enhanced uplink multiple access scheme adopts clustered SC-FDMA. It is also known as discrete Fourier transform spread OFDM (DFT-S-OFDM). It is similar to SC-FDMA but it allows noncontiguous (clustered) groups of subcarriers to be allocated for transmission by a single UE, thus enabling uplink frequency-selective scheduling and better link performance. It increases uplink spectral efficiency while maintaining backward compatibility with LTE. Clustered SC-FDMA shows better PAPR performance than OFDM, so it is chosen to avoid a significant increase in PAPR. It will help satisfy the requirement for increased uplink spectral efficiency while maintaining backward compatibility with LTE.

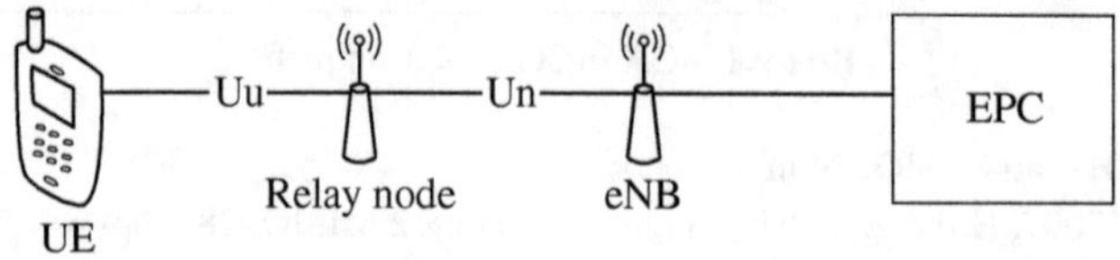

Figure 9.3 Relay node

9.2.3 *Enhanced Multiple Antenna Transmission*

In order to simplify the baseline in UE Release-8 LTE, the downlink supports a maximum of four spatial layers of transmission (4×4, assuming four UE receivers) and the uplink a maximum of one per UE (1×2, assuming an eNB diversity receiver). To improve single-user peak data rates and to improve spectrum efficiency, Release-10 specifies up to eight layers in the downlink, which, with the requisite eight receivers in the UE, allows the possibility in the downlink of 8×8 spatial multiplexing. The UE will be specified to support up to four transmitters allowing the possibility of up to 4×4 transmission in the uplink when combined with four eNB receivers. To support that, there will be changes to the UE-specific demodulation reference signal (DMRS) patterns, channel state information reference signals (CSI-RS), and associated UE feedback, and downlink control signaling.

9.2.4 *Relaying*

Relaying is introduced in 3GPP Release-10 in order to increase the coverage and throughput. Generally, a repeater just rebroadcasts the signal whereas a relay will receive, demodulate and decode the data, apply any error correction to it, and then retransmit the signal, so signal quality is enhanced in the latter case. As shown in Figure 9.3, the LTE relay is a fixed relay, which means infrastructure without a wired backhaul connection that relays messages between the eNB and UEs.

These will help (i) to provide coverage in new areas and poor coverage areas; (ii) temporary network deployment; (iii) to improve cell edge throughput; (iv) group mobility. Relays brings the advantages like (i) cost reduction – cost of a relay is less than cost of an eNB; (ii) power-consumption reduction as the required transmitter power in relay is lower than eNB.

9.2.5 *Device to Device*

Device-to-device (D2D) communication enables direct communication between nearby devices without routing the data paths through a network infrastructure. It is currently being specified by 3GPP in LTE Rel-12 and is recognized as one of the key technology components of the evolving 4G and 5G architecture. The cellular system infrastructure controls and assists the operation of D2D links, which will coexist with cellular communications within the same shared cellular spectrum. In 3GPP Release 12, the first set of features,

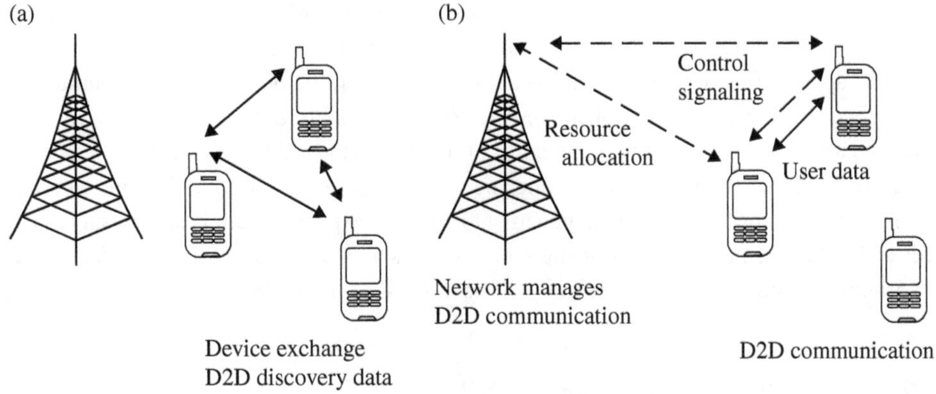

Figure 9.4 (a) D2D discovery and (b) D2D communication

namely LTE device-to-device (D2D) proximity services and the group call system enablers (GCSE) for LTE were introduced and off-network communication and mission-critical push to talk (MCPTT) are part of Release 13. Proximity services (ProSe) allow devices in close proximity to detect and communicate directly with each other, which reduces the network load, increases capacity in a given bandwidth and allows communication in areas where there is no network coverage. The ProSe has two main components: (i) D2D discovery – this helps a device to discover the presence of other D2D-capable devices in its vicinity using the LTE radio interface and to ascertain certain information about these devices wherever permitted (see Figure 9.4a); (ii) D2D communication – D2D-capable devices will use the LTE radio interface to communicate directly with each other, without routing the data through the LTE network. The network just facilitates this by controlling the radio resource allocation and security of the connections (see Figure 9.4b).

There is a split between application and LTE layers, where the application layer will mainly control the group management, floor control decisions, and legacy interoperability, and LTE will provide mobility and service continuity plus air-interface efficiency.

D2D offers several benefits to users in various applications:

- Data rates. It helps to achieve a high data rate among devices even though there is poor network coverage because of close proximity and potentially favorable propagation conditions.
- Latency reduction. When devices communicate over a direct link the end-to-end latency is reduced.
- Reliable communications. Local communication will provide high-reliability communication.
- Use of the licensed spectrum. Due to the use of licensed frequencies the interference level will be lower, thereby allowing more reliable communications.
- Power saving. Lower transmission power levels will be required because of the shorter transmission distances involved.

There are several applications feasible using D2D technology, like proximity-based services where devices detect their proximity and subsequently trigger different services such as social applications, advertisements, local exchange of information, and smart communication between vehicles. Other applications include public safety support, where devices provide at least local connectivity even in cases of damage to the radio infrastructure.

However, D2D communication implies new challenges for devices design, interference management, security, mobility management and other aspects. In addition, the success of this technology largely depends on the scenarios in which users in proximity to each other communicate and the applications that will be developed in the coming years.

9.2.6 Coordinated Multipoint (CoMP)

Coordinated MultiPoint (CoMP) is a new intercell cooperation technology introduced in 3GPP Rel-11 to enhance throughputs of UEs at cell edges. It encompasses all the required system designs to achieve tight coordination for transmission and reception. A CoMP-supported UE can communicate with more than one cell located at different points. These groups of cells act like a virtual MIMO system and "CoMP cooperating cells" are responsible for directly or indirectly transmitting data to UE and data-transmitting cells are known as "CoMP transmission points." Downlink CoMP features the following: (i) coordinated scheduling (cells cooperate with each other to allocate different frequency resources); (ii) coordinated beamforming (this allocates different spatial resources, or beam patterns, to UEs at the cell edge by using smart antenna technology), and (iii) joint processing/joint transmission (multiple cells can transmit the same data concurrently by using the same radio resources). For the uplink, the scheduling is generally coordinated from the different cell sites to improve the link performance.

eNBs command UEs about which cells' CSIs are to be measured and how by sending a CSI reference signal (CSI-RS) configuration message and, upon receiving this, UEs measure CSI – which includes the channel quality indicator (CQI) the precoding matrix indicator (PMI), and the rank indicator (RI) – and report to their serving cells. For fast cooperation among eNBs, the interconnections among the different nodes, very high-speed dedicated links (for example optical fiber, wired backbone connection) are used.

9.2.7 Heterogeneous Networks and Enhanced ICIC

To overcome the capacity challenges in a limited spectrum, operators follow the conventional route of splitting macro cells into several relatively smaller macro cells. But, this does not help to cut the CAPEX and OPEX; instead, introducing low-power cells (such as microcells, picocells, HeNBs, and relay nodes) is much more attractive because of smaller cell site footprints, ease of deployment, and lower equipment and operating costs. The LTE-A provides efficient support for a mixture of macrocells and low-power eNBs. In some cases, the macro eNBs are deployed along with low-power nodes with different

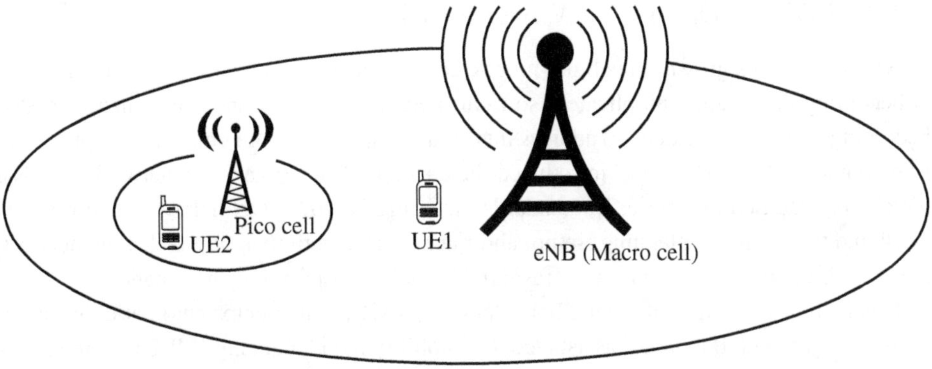

Figure 9.5 Heterogeneous network (Het-Net)

carrier frequencies, resulting in no interference issues. But, as the spectrum available for a cellular system is scarce and expensive so, in many cases, eNB and low-power nodes share the same carrier, leading to challenges in interference mitigation. Such a deployment scenario is known as an heterogeneous network (Het-Net, see Figure 9.5) and here interference mitigation techniques like intercell interference coordination (ICIC) mechanisms are very important. In 3GPP Release 8, ICIC was introduced and in 3GPP Release 10 eICIC (enhanced) was introduced as part of LTE-Advanced.

The LTE signal is present in time as well as in the frequency domain, so interference can occur in both domains. In the frequency domain interference could be reduced by allocating resource blocks from multiple neighboring cells in such a way that these never overlap. In the time domain, one way to reduce the interference is that the serving cell can stop transmitting, or reduce power, at a certain subframe so that other cells can transmit the signal during that period. These subframes with very low signal power are called "almost blank subframes" (ABSs).

In ICIC, part of the subchannel's power is reduced in the frequency domain, which allows it to receive only close to the eNB. In some schemes, no two neighbor eNB use the same set of resource blocks at a given time for cell edge users. In another scheme, all the neighbor eNBs use different power schemes across the spectrum while resource block assignment can be according to the scheme above.

eICIC mitigates interference on traffic and control channels and it uses power, frequency, and also the time domain to mitigate intrafrequency interference in heterogeneous networks. It uses the concept of ABS. If the UE is supporting the eCIC feature then that needs to be signaled between the UE and the network. So, in summary, ICIC reduces intercell interference by allocating different frequency resources (RBs or subcarriers) to UEs at cell edges, and eICIC does the same task in the time domain by allocating different time resources (subframes) through cooperation between a macro cell and small cells in a HetNet (see 3GPP TR 36.872 for E-UTRA small cell enhancement physical layer aspects). An X2 interface is used to share the information between the eNBs.

9.2.8 LTE Self-Optimizing Networks (SON)

Due to tremendous growth in cellular mobile communications and data usage, the network is becoming dense and complicated, so radio network planning and maintenance is also becoming more complicated. To address this, and to support smooth unplanned deployment of pico nodes, 3GPP Rel-8 has introduced the concept of self-organizing networks (SON), which are specified in a series of standards including 36.902. SON helps to govern a network and this includes planning, setup, and maintenance activities. It enables the network to set itself up and then manages the resources to achieve optimum performance.

This includes (i) automatic neighbor relations (ANR), which helps automatic discovery of new neighbor eNBs via UE assistance; (ii) mobility load balancing (MLB), which helps in tuning the handover thresholds between macro and pico cells based on cell loading to balance the load between macro and pico cells; (iii) mobility robustness optimization (MRO), which helps to monitor failed handovers to fine-tune mobility parameters such as handover hysteresis and trigger time.

It (i) helps to reduce the OPEX by reducing the level of human intervention in network design and operation; (ii) helps to reduce CAPEX by optimizing the use of available resources; and (iii) reduces human errors.

9.3 LTE-A UE Modem Processing

Figure 9.6 shows different blocks involved in transmission and reception with an LTE-A UE transmitter and an eNB receiver. The transmit bit-rate processing block includes a CRC attachment, code block segmentation, and code block CRC attachment, channel coding,

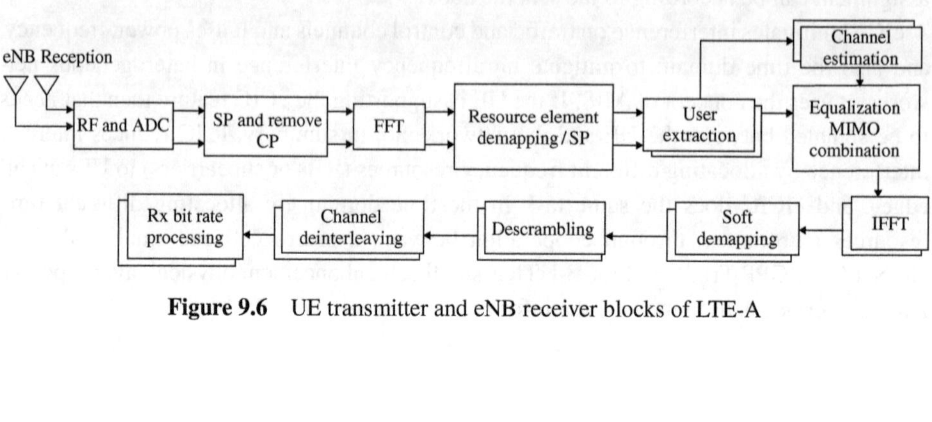

Figure 9.6 UE transmitter and eNB receiver blocks of LTE-A

rate matching, and code block concatenation. The scrambling block scrambles the bits with the UE-specific scrambling code prior to modulation to reduce the adjacent cell interference. The modulation mapper maps binary bits into complex value (I, Q) symbols for modulation schemes like, QPSK, 16-QAM, and 64-QAM. The layer mapping block maps the complex modulation symbols for each of the code words to one, two, three, or four PHY layers (based on spatial multiplexing or transmit diversity used). A DFT block performs DFT and converts the signal from the time domain to the frequency domain. The precoding block maps the complex valued modulation symbols from the layers to multiple antennas. Then pilot symbols are generated and inserted into complex value modulation symbols on each antenna port. RE mapping maps into different elements in an RB. Next N-point IFFTs are performed to convert the signal from the frequency domain to the time domain after resource element mapping. Then a cyclic prefix (CP) is inserted into every symbol and performed parallel to serial conversion (SP). The digital signal is then converted to an analog signal, amplified, and transmitted using the appropriate carrier frequency.

On the receiver side, the eNB receives the analog RF signal, downconverts and samples through an ADC. It performs serial-to-parallel conversion and removes CPs, then performs the N-point FFTs to convert the signal from the time domain to frequency. The reference and data signals are separated. The reference signal is used for channel estimation. Here, every user's signal will be extracted from different subcarriers according to their PRB configurations. The channel estimation is carried out using received reference symbols and preknown reference symbols, and the channel matrix H is estimated. Based on the estimated H, the equalization is performed on the whole slot's data. The pilot symbols are removed from the modulated symbol frame. Then the complex values modulated symbol frame is demapped into blocks. Next, M-point FFTs are performed to convert the data from the frequency domain to the time domain. The received SC-FDMA symbols are converted to soft bits according to the modulation scheme used for that channel. The inverse stage of scrambling uses a deinterleaver for rank-indication bits, HARQ bits, and PUSCH/CQI bits. Then receiver bit-rate processing is performed, which includes code-block deconcatenation, rate dematching, turbo decoding, code-block CRC removal, code-block desegmentation, and transport block CRC checking and removal.

9.4 LTE-A UE Implementation

As discussed in earlier chapters, the LTE-A UE modem architecture is more-or-less similar to the LTE UE. The LTE-Advanced modem consists of receive and transmit signal processing chains. The RF block is improved to support more bands (apart from supporting LTE bands, it will also support the 450–470 MHz band, 698–862 MHz, 790–862 MHz, 2.3–2.4 GHz, 3.4–4.2 GHz and 4.4–4.99 GHz), and supports interband CA. The signal processing is divided into layers as defined in the 3GPP specification, with layer 1 providing all of the low-level signal conditioning concerned with the successful transmission and reception of the signal. Typical functions in layer 1 include: forward error correction, interleaving and bit-stream manipulation, constellation modulation, MIMO

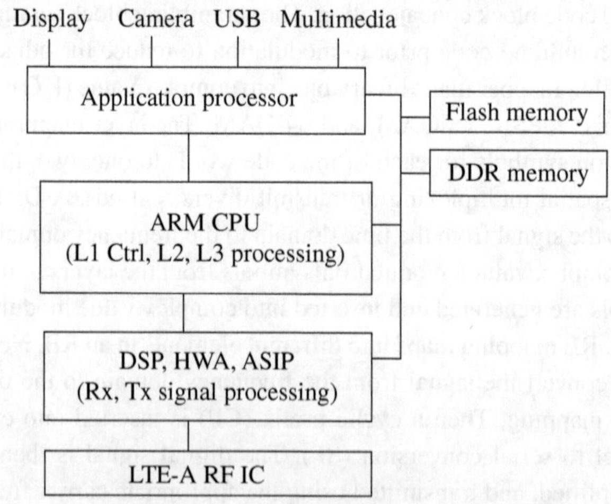

Figure 9.7 UE architecture of LTE-A

encoding, OFDM signal modulation, and RFIC signal conditioning. Generally the layer-1 signal-processing functions are implemented in DSP or ASIP or HW accelerators, and the layer-1 control and protocol layers are implemented using ARM CPU (see Figure 9.7).

9.5 Future Generations (5G)

A 5G cellular system is in the early development stages and currently it comprises research papers and pilot projects. Although the standards bodies have not yet defined the parameters needed for a 5G system, typical parameters for a 5G standard may include: (i) network capacity: 10 000 times the capacity of the current network; (ii) peak data rate ~ 10 Gbps; (iii) cell-edge data rate ~ 100 mbps; (iv) latency < 1 ms; (v) support for 100+ billion connections. That means it mainly focuses on better levels of connectivity and coverage.

The 5G network is not going to be a monolithic network technology; rather, it will be a combination of different technologies: 2G, 3G, LTE, LTE-A, Wi-Fi, M2M, and so forth, and will be designed to support a variety of applications such as IoT, connected wearables, augmented reality, and immersive gaming. It will bring new architectures like cloud RAN and virtual RAN to facilitate a more centralized network establishment, and will handle a plethora of connected devices with different traffic types.

The early blueprint of 5G pilot networks mostly comprises beamforming technology, small cell-base stations and millimeter wave. There are several key areas under research:

- *Millimeter-wave technologies.* Today, maximum carrier frequencies of around 2 GHz and bandwidths of 10–20 MHz are in common use. Having carrier frequency higher than this offers benefits like wider channel bandwidth, more spectrum availability, and reduced antenna length, which means smaller antenna array size, and so forth. Theoretically millimeter waves occupy the frequency spectrum from 30 GHz to 300 GHz (wavelength λ 10 mm to 1 mm). For 5G, carrier frequencies of above 50 GHz are being considered. Although this offers several advantages, it will bring some real challenges in terms of the circuit design, the technology, and also the way the system is used as these frequencies are attenuated very fast by obstacles. It also poses new challenges for handset development.

- *Massive MIMO.* MIMO is used in LTE but the numbers of antenna elements are limited due to the size factor. Now, use of higher carrier frequencies (millimeter wave) opens up the possibility of using many antenna elements on a single array. Massive MIMO is the solution to access thousands of arrays to be configured at the base station, which helps for accurate beam control and greater spectral efficiency. Spatial multiplexing and interference mitigation are also used here to increase system capacity.
- *Cognitive radio technology.* A cognitive radio can sense its environment, location, and intended use, and then alter its power, frequency, modulation, and other parameters so as to dynamically reuse the available spectrum. In 5G the use of cognitive radio techniques allows the infrastructure to decide automatically about the type of channel to be offered, differentiate between mobile and fixed objects, and adapt to conditions at a given time. Technologies like adaptive radio and software-defined ratio (SDR) are used in cognitive radio networks. The move towards the reconfigurability concept was initiated as an evolution of software-defined radio http://www.wirelessinnovation.org/ (accessed May 17, 2016). Its aim is to provide essential mechanisms to terminals and networks, so as to enable them to adapt dynamically, transparently, and securely to the most appropriate RAT depending on the current situation. It helps to improve or maximize the utilization of radio frequency spectrum.
- *Dense networks.* As discussed earlier, a smaller cell size provides several advantages like reuse of spectrum, capacity enhancement, and lower transmit power (which offers energy efficient communication). So, 5G ensures that small cells are deployed in the macro network. Small cells (femtocells, picocells, and microcells) are low-powered radio access nodes that operate in the licensed and unlicensed spectrums.
- *LTE-U.* The existing spectrum is not enough to carry an increasing amount of data, and one possible solution is to use the unlicensed spectrum alongside the licensed bands. 3GPP has termed this LTE license-assisted access (LTE-LAA) or more generally, LTE Unlicensed (LTE U). It enables access to the unlicensed spectrum, especially in the 5 GHz ISM band (already used by Wi-Fi devices).
- *PHY/MAC possibilities.* The physical and MAC layers will have interesting new possibilities, like new waveform formats such as generalized frequency division multiplexing (GFDM), filter bank multicarrier and universal filtered multicarrier, new multiple-access schemes (SCMA, NOMA, PDMA, MUSA and IDMA), and techniques like light-fidelity (Li-Fi), which provides transmission of data through illumination by sending very high-speed data through an LED light bulb.

The wireless industry is broadly targeting 2020 for the widespread deployment of 5G networks.

9.6 Internet of Things (IoT)

Extending the present Internet and providing connection, communication, and internetworking between devices and physical objects, or "Things," is a growing trend that is often referred to as the Internet of Things (IoT), which is the next wave in the era of communication and computing. The IoT (wearables, smart home appliances, connected cars) is expected to grow exponentially, and it will need a network that can accommodate billions of connected devices. 5G is being developed to provide this capacity, also known as "connected world."

Internet of Things (also known as machine-to-machine or M2M) applications are growing rapidly over today's cellular networks. There were 8.7 billion connected Things at the end of 2012, which grew to 10.8 billion in 2015. A number of companies believe that the ratio of connected Things to people will rise sharply and there could be as many as 50 billion connected Things in the world. To cater to that market need, Vodafone proposed a work plan for new study item on cellular IoT (IoT devices connected through cellular network), and currently specification work is under way for this. Also there is another proposal for low cost enhanced M2M single receiver and limited data rate rate (< 1.4 MHz) using LTE specification (see 3GPP TR 36.888 for more information).

The primary requirements of IoT devices are lower cost (< \$5), high coverage (20 dB below sensitivity), better capacity and the most important one is prolonged battery life (10 years or more for small primary batteries). This requires a major breakthrough in transceiver signal processing and semiconductor technology, and also an extensive exploration of system architecture to offer high integration, low power consumption and acceptable cost. Any power that is consumed by the system without any useful result is considered as waste of power. A reduction in power consumption provides several benefits. Less heat is generated (which reduces the problems associated with the high temperature), battery life is extended, and device reliability increases.

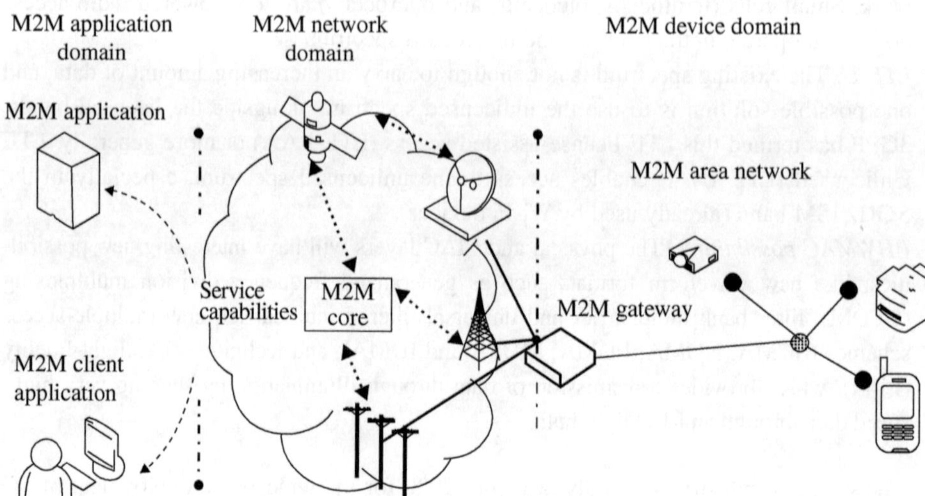

Figure 9.8 M2M (IoT) architecture (ETSI)

Figure 9.8, shows an IoT or M2M device that runs applications using M2M capabilities and network domain functions. It has sensor or actuator mounted in it. An IoT device is either connected straight to an access network or interfaced to M2M Gateways via an M2M area network. An M2M area network provides connectivity between M2M devices and M2M Gateways. M2M Gateways are equipment using M2M capabilities to ensure that M2M devices are interworking and interconnect to the network and application domain.

Future potential IoT applications are: (i) smart city (smart lighting, smart parking, traffic congestion control, waste management, air pollution control, etc.); (ii) smart electric metering; (iii) home and public building automation; (iv) factory automation; (v) monitoring structural health (buildings, bridges, historical places, etc.); (vi) object and animal tracking; (vii) medical and health monitoring; (viii) remote control.

However, IoT faces several challenges, such as signaling, security, addressing, presence detection, power consumption, bandwidth, and other issues, which are currently the subjects of research.

Further Reading

Ericsson Research Blog (n.d.), http://www.ericsson.com/research-blog/category/5g/ (accessed May 15, 2016).

Rodriguez, J. (2015) *Fundamentals of 5G Mobile Networks*, Wiley, 978-1-118-86752-5.

3GPP Reference documents for LTE-Advanced include LTE-A technical reports, study items, and requirements:

Physical Layer Aspects. TR 36.814 (Stage 2 Development), http://www.3gpp.org/dynareport/36814.htm (accessed May 17, 2016).

Requirements for LTE-Advanced. 3GPP Technical Report (TR) 36.913, http://www.3gpp.org/dynareport/36913.htm (accessed May 17, 2016).

Study Item Final Status Report. RP-100080, ftp://ftp.3gpp.org/tsg_ran/TSG_RAN/TSGR_47/Docs/RP-100080.zip (accessed May 16, 2016).

Study Phase Technical Report. TR 36.912 (Stage 1 Summary), http://www.3gpp.org/dynareport/36912.htm (accessed May 17, 2016).

Study Phase Technical Report on E-UTRA UE Radio Transmission and Reception. TR 36.807.

Index

www.ingramcontent.com/pod-product-compliance
Lightning Source LLC
LaVergne TN
LVHW062206200926
848484LV00006B/1329
* 9 7 8 1 1 1 9 1 0 7 3 0 9 *